ECLIPSES OF THE SUN

THE END OF THE TOTAL ECLIPSE, SEPTEMBER 10, 1923
From the painting by Howard Russell Butler, N. A. Shows the edge of the reappearing sun.

ECLIPSES OF THE SUN

BY

S. A. MITCHELL

PROFESSOR OF ASTRONOMY AT
THE UNIVERSITY OF VIRGINIA
AND DIRECTOR OF THE LEANDER
MCCORMICK OBSERVATORY

FOURTH EDITION

REVISED AND ENLARGED

NEW YORK M·CM·XXXV

COLUMBIA UNIVERSITY PRESS

This edition comprises all the material contained in the third edition, as well as a new chapter and certain additions to the text. Chapters in which no out-of-date scientific statements occur have been reprinted without change from the third edition.

First edition, 1923
Second edition, revised and enlarged, 1924
Third edition, revised and enlarged, 1932
Fourth edition, revised and enlarged, 1935

PRINTED IN THE UNITED STATES OF AMERICA
BY THE PLIMPTON PRESS · NORWOOD · MASS

TO MY FRIEND

EDWARD DEAN ADAMS

THIS BOOK IS AFFECTIONATELY DEDICATED

PREFACE

THIS book owes its inception to the keen scientific interest of Mr. Edward D. Adams. The magnificent painting of the corona by Howard Russell Butler, which now finds an honored place in the American Museum of Natural History, New York, was originally planned that it might provide a frontispiece illustration.

Fortunately for the author, he had the hearty coöperation of many of his friends in the preparation of the book. Professor E. E. Barnard, an intimate friend for twenty-five years, furnished all of the photographs illustrating the eclipse expedition in 1901 to Sumatra. The last work undertaken by him, a few days before his lamented death, was to dictate the captions to accompany each of the photographs. Another life-long friend, Professor Edwin B. Frost, director of the Yerkes Observatory, was most generous in his assistance. As editor of the *Astrophysical Journal*, he was instrumental in providing the blocks for many of the illustrations. Professor George B. Pegram, chairman of the Adams Fellowship Committee of Columbia University, showed continued interest during the progress of the work. Dr. W. W. Campbell, director of the Lick Observatory, kindly furnished information in advance of publication regarding the Einstein problem and provided photographs of the 1922 eclipse and with great generosity supplied many additional illustrations from the incomparable Lick series of photographs of the solar corona, while Professor H. Deslandres kindly sent for reproduction many photographs taken with the spectro-heliograph at Meudon.

The author wishes to express to these scientists his heartfelt appreciation for their assistance so freely given.

The second edition was prepared while the author was in residence for a brief stay at the Mount Wilson Observatory.

In the eight years which separated the second and third editions, scientists had a most thrilling time in their investigations of the chemical atom. It has been said with truth that discoveries have followed each other in such rapid succession that before the printer's ink was dry, a new and, at times, startling theory was superseded by another theory still newer.

As the problems of solar physics have their fundamental basis in the structure of the atom, an attempt has been made to incorporate in this book the recent important researches dealing with atomic physics.

The fourth edition, prepared while in residence again at Mount Wilson Observatory, adds to the third edition a chapter giving the results obtained from the 1932 and 1934 eclipses. Minor changes have been made in Chapter IV dealing with future eclipses.

S. A. Mitchell

Leander McCormick Observatory
May 13, 1935

CONTENTS

INTRODUCTION . . . . xiii–xviii

I. EARLY HISTORICAL ECLIPSES 1–26
Chinese Eclipses — Eclipses of Babylon — Signs of the Zodiac — Egyptian Astronomy

II. BIBLICAL AND CLASSICAL ECLIPSES 27–36
Dates of the Bible — Eclipses of Classical Literature — Eclipse of Thales — Xerxes

III. THE PREDICTION OF ECLIPSES 37–52
Metonic Cycle — Aristotle — Hipparchus — Saros — Theory of Eclipses — Ecliptic Limits

IV. THE VERIFICATION OF ECLIPSES 53–71
Long-Range Predictions — One Hundred Years of Total Eclipses — Number of Eclipses — Calculation of Eclipses — Lunar and Solar Eclipses — Tables of the Moon — Moon ahead of Predicted Place

V. THE SPECTROSCOPE 72–90
Newton — Wollaston — Fraunhofer — Prisms and Gratings — Wave-Lengths — Fraunhofer Lines

VI. THE SPECTROSCOPE (Continued) 91–105
Kirchhoff — Kirchhoff Laws — Rowland Gratings

VII. THE SURFACE OF THE SUN 106–127
Greek Ideas — Distance of the Sun — Size of Sun — Photosphere — Janssen — Sun-Spots — Periodicity of Spots — Is the Sun Spherical?

VIII. MODERN ECLIPSES BEFORE 1878 128–150
Tycho Brahe — Baily's Beads — Eclipse of 1842 — At Eclipse of 1860, Prominences Discovered to Be Solar in Origin — Triumph of the Spectroscope at Eclipse of 1868

IX. NINETEENTH CENTURY ECLIPSES AFTER 1878 151–166
Great Extensions of the Corona in 1878 — Intra-Mercurial Planets — Spectrum of the Corona — Flash Spectrum Discovered in 1870 — Importance of Photography — Results of 1900

X. PERSONAL EXPERIENCES IN 1900 AND 1901 167–182
Eclipse of 1900 — To Sumatra in 1901 — Scientific Results

XI. THE SPANISH ECLIPSE OF 1905 183–190
Photographing the Flash Spectrum

XII. THE AMERICAN ECLIPSE OF 1918 191–202
Threatening Weather — Painting the Corona

XIII. ECLIPSES SINCE 1923 203–225
Experiences in California in 1923 — in Connecticut in 1925 — in Norway in 1927 — on Niuafoou Island in 1930

XIV. THE STRUCTURE OF THE ATOM 226–246
Size of Universe Contrasted with the Size of the Electron — Radium and Its Rays — The Electron — Atomic Numbers — Isotopes — Quantum Theory — Spectral Series — The Bohr Atom — Spectrum of Hydrogen and Helium

XV. PHOTOGRAPHING THE FLASH SPECTRUM 247–276
Choice of Best Type of Spectrograph — Slit versus Slitless — Prisms versus Gratings — Fixed versus Moving Plates

XVI. DISCUSSING THE FLASH SPECTRUM 277–298
Chromospheric Intensities Estimated and Measured — Results of Pannekoek and Minnaert and of Menzel — Flash Spectrum Observed at Partial Eclipses and without an Eclipse — Present Problems

XVII. THE IMPORTANCE OF IONIZATION 299–318
Saha's Theory Applied to the Spectra of Sun, Chromosphere, Sun-Spots and Stars — Periodic Table of Elements — Atomic Theory — Ionization and Excitation Potentials — Multiplet Groups

XVIII. HEIGHTS AND RELATED PROBLEMS 319–337
Solar Rotation — Sun-Spot Vortices — Polarities — Sun-Spot Period

XIX. HEIGHTS AND DISTRIBUTIONS OF VAPORS 338–359
Multiplet Groups in Iron — Evershed Effect in Sun-Spots — Correlations between Heights, Intensities, Excitation Potentials, Evershed Effect and Relativity Shift — Abundances and Density Distribution

XX. THE CORONA 360–386
Paucity of Time Allowed for Investigations — Motions within the Corona — Disturbances — Wave-Lengths — Intensities — Coronium — Rotation of Corona — Unsatisfactory Nature of Knowledge

XXI. CORONAL THEORIES 387–420
Polarization — Law of Distribution of Light within the Corona — Total Light of Corona — Shape of Corona — Connection with Sun-Spot Period — Various Theories

XXII. THE EINSTEIN THEORY OF RELATIVITY 421–442
Has Newton's Law of Gravitation Been Discarded? — Space of Four Dimensions — Michelson-Morley Experiment — Relativity of Motion — What Is a Straight Line? — Events — Consequences of Einstein's Theory

XXIII. HAS THE EINSTEIN THEORY BEEN VERIFIED? 443–479
Eclipse of 1919 — Results from the Eclipses of 1922 and 1929 — The Spectroscopic Test — General Conclusions

XXIV. SHADOW BANDS 480–482

XXV. THE ECLIPSES OF 1932 AND 1934 483–511
New Information about the Corona — Eclipse Problems

INDEX 513–520

ILLUSTRATIONS

Total Eclipse of 1923, in Color (Butler) . . . . *Frontispiece*
Coronal Disturbances at 1930 Eclipse (Marriott) 8
Corona at 1930 Eclipse (Marriott) 9
Inner Corona, 1930 (Marriott) 12
The Corona, 1929 (Swarthmore) 13
An Ancient Egyptian Calendar 24
Quiescent Solar Prominence 25
Forty-Four Eclipses in a Series from 1211 to 1986 52
Swarthmore Expedition in 1929, Huge Camera 53
Tracks of Eclipses between 1919 and 1940 56
The Eclipse on Niuafoou Island 57
Total Eclipse of June 8, 1918, in Color (Butler) 60
Twenty-Four-Hour Development of Sun-Spot (Mt. Wilson) 84
60-foot Tower Telescope at Mt. Wilson Observatory 85
The Sun in the Light of Glowing Calcium Vapor 108
Spectroheliograms Showing Development of Spot 109
The Sun's Surface, September 1883 (Janssen) 116
Sun-Spot Group (Meudon Observatory) 117
Details of Prominences, June 8, 1918, in Color (Butler) 128
Outer Layers of the Chromosphere (Deslandres) 136
Test of Radial Motions in Sun (Deslandres) 137
The Corona Photographed by Lick Parties 148
Prominences in Projection against the Sun's Disk 149
Corona of May 28, 1900 (Barnard and Ritchey) 164
Prominences Photographed without an Eclipse (Yerkes) 165
Pasig River; Krakatoa (Barnard) 168
Railroad in Sumatra; Water Buffalo (Barnard) 169
Sumatran Beauty; Market Scene (Barnard) 176
Sumatrans, Male and Female (Barnard) 177
Filipino and Sumatran Homes (Barnard) 184
Scenes at Daroca, Spain (Barnard) 185
The "Heliosaurus" at 1918 Eclipse (Barnard) 196
American Eclipse of 1918, Lick Photograph 197
Mount Wilson Instruments at 1923 Eclipse 208
Disturbed Region in 1923 Corona (Swarthmore) 209
Corona of 1923 (Swarthmore) 212
The Corona of January 24, 1925 (Slocum) 213
The Corona of January 24, 1925, in Color (Butler) 224
Flash Spectrum, 1905, Region of H and K 248
Flash Spectrum with Concave Grating in Blue and Green Region . 249

PAGE
Flash Spectrum with Moving Plate (Campbell) 260
Changes in Solar Prominences (Slocum) 261
Fraunhofer and Flash Spectra Compared 268
Observers at the 1905 Eclipse in Spain 269
Prominence of May 29, 1919, Photographed without an Eclipse 272
Total Eclipse of May 29, 1919 273
Approaching Shadow at Eclipse of June 1918, in Color (Butler) 284
Calcium Flocculi (Fox) 300
Eclipse Cameras in 1905 and 1918 301
Sun-Spot and Prominence (Mt. Wilson) 320
Prominence Projected on Sun's Disk (Mt. Wilson) 321
Zeeman Effect in Sun-Spots (Mt. Wilson) 332
Bipolar Sun-Spot (Mt. Wilson) 333
Solar Vortices (Mt. Wilson) 340
Spectrum of Sun and Spot Compared (Mt. Wilson) 341
The Sun, Direct and in Glowing Calcium Vapor (Yerkes) 348
The Total Eclipse of 1918 (Lick) 349
Corona by Lick and Canadian Astronomers 364
Total Eclipse of 1922 (Lick) 365
Coronal Rings at 1930 Eclipse 372
Detailed Structure in Corona, 1930 373
Lick Photographs of Coronal Spectrum, 1918 380
General Spectrum of Corona (Lowell) 381
Corona in Polarized Light, 1908 (Lick) 388
Curves Showing Intensity in Corona (Bergstrand) 389
Inner Corona in 1926 (Swarthmore) 400
Lion-Like Prominence at 1929 Eclipse (Swarthmore) 401
Einstein Cameras at the 1919 Eclipse 424
Eclipse of 1922, Canadian Photographs 425
Corona in 1922 (Lick) 448
Fine Structure in Corona, 1929 (Swarthmore) 449
Lick Camera for Testing the Einstein Deflection 460
Australian Sailors Erecting Canadian Cameras in 1922 461
Corona in 1932 (Lick) 476
Einstein Camera in 1934 477
Clouds and Prominences in 1932 484
Moon's Shadow at Five-Second Intervals 485
Flash Spectrum in 1932 (Greenwich) 492
Flash Spectrum in 1932 (Lick) 493
Corona in 1934 (Satome) 500
Instruments at the 1934 Eclipse 501

INTRODUCTION

SCIENCE seems bent on increasing man's estimate of the span of time since creation began. It is now confidently believed that the human race has inhabited this terrestrial ball for at least five hundred thousand years, while Mother Earth herself must have been in existence for a thousand million years. On the assumption that mass is electrical and that all kinds of energy possess mass, we have found that the sun's heat can not be maintained for a space of time exceeding fifteen millions of millions of years, provided all its energy, including that of the atom, could be liberated. Surely such an interval is sufficiently great to satisfy even the most exacting! And yet what changes have taken place in our mode of life within the memory of man!

Today, as never before, our daily life follows its course surrounded by the wonders of science. The age of steam has brought the powerful locomotive drawing its fast and comfortable passenger train on its long journey across the continent, has constructed the gigantic ocean steamship whereby Europe is brought within four days of the shores of America, has been, indeed, the cause of the remarkable development experienced in all quarters of the globe. The internal combustion engine has largely revolutionized modes of transportation on the land, on the surface of and under the sea and in the air, so that the extravagantly fanciful writings of Jules Verne are common-place, every-day realities. The age of electricity has wrought the greatest changes of all in our daily life bringing as it has stupendous transformations in illumination vastly different from the feeble tallow dip; it has wrought the powerful motor; it has made possible the electric telegraph and telephone and the still more wonderful wireless, with the result that distance is eliminated and there is served fresh each morning with our coffee the very latest news garnered from every quarter of the globe. Through the agency of the cinema and the wireless telephone the inhabitant of a backwoods village can gaze upon great actors, can listen to the greatest of singers and the best of symphony orchestras quite as well as if he lived in London, Paris or New York.

The bridge separating pure from applied science has been so much shortened that the gap has practically disappeared. The refined researches of the physicist or chemist today may be applied in the arts and sciences of tomorrow. Investigations in Hertzian waves in the laboratory have made the wireless telephone possible, the discovery of Mme. Curie and the remarkable work of Rutherford and others on the penetrating powers of the rays of radium have given to the cancer specialist the means of alleviating the horrors of this dread disease; X-rays are now in every-day use by the doctor, the surgeon and the dentist. And in the quiet rest of the laboratory, the chemist has experimented with his vials and test tubes and has made noxious and poisonous gases which, liberated amidst the din and roar of the battlefield, have caused the destruction of thousands of lives.

It is less than forty years since the discovery of X-rays and radium. The marvelous researches connected with radioactivity have shown the amazing degree of refinement necessary in the work of the scientist. New forces, hitherto unknown, have been discovered, while recent researches display the amazing fact that each chemical atom takes its place in Nature as a miniature solar system. Modern investigations have revealed that one chemical element may be transformed into another, thus realizing the age-long dream of the alchemist, the transmutation of one metal into another. It has been proved that several separate series of radioactive transformations have as their end-product the metal lead — but fortunately for the economic future of the world, the realization of the alchemist's vision of changing base lead into glittering gold seems as far distant as ever. Helium is a product of almost every radioactive transformation and it has thus become a household word with every trained physicist. And yet helium was discovered — a result of solar eclipse observations — a bare half-century ago and was isolated from pitchblende only in 1895! Within two decades the demands of the Great War brought to the attention of scientists the urgent necessity of obviating the disastrous explosions that have taken place with hydrogen-filled balloons with the consequent result that helium was produced in such large quantities that balloons were filled with this inert gas.

But amongst all the wonders of all the wonderful sciences there is no science which deals with such a gorgeous spectacle as is exhibited by the queen of the sciences, astronomy, at the moment when the earth is gradually shrouded in darkness and when around the smiling orb of day there appears the matchless crown of glory, the so-called corona. Nor can any science duplicate the wonderful precision shown by the work of the astronomer in his capacity to predict hundreds of years in advance the exact hour and minute at which an eclipse will take place and the locality on the earth's surface where such an eclipse will be visible.

The great progress of science in the last fifty years is nowhere better illustrated than in the attitude of astronomers towards observations at the time of a total eclipse of the sun. Until about the middle of the nineteenth century little interest was taken in the subject although information regarding the sun was comparatively scant. The eclipse was observed only if perchance its track happened to cross the home of the observer; the only observations of value being the exact times of contact of the limbs of the sun and moon, taken for the purpose of perfecting the lunar tables. The beautiful corona was watched with awe and admiration, a few sketches were made of its form, — but these were done with such indifferent skill that they added but little to the information available at the time. Practically no expeditions were equipped and sent away from home. How different it is in the twentieth century! In the year 1901, the United States Naval Observatory financed and sent an expedition as far from home as it could possibly go — half way round the world — and all for the purpose of making observations during a few short minutes of time. At the recurrence of the eclipse in 1919, the British observers in Brazil and West Africa startled the thinking world by verification of the Einstein problem. In September, 1922, the British astronomers had spent more than six months on Christmas Island in the Indian Ocean only to have their work come to naught because of clouds at the critical moment, while American and Canadian scientists had gone to Northwest Australia in quest of information about the relativity shift. On isolated " Tin-Can Island " in the South Pacific, important discoveries were made in 1930.

To make such expensive and time-consuming expeditions worth while there must be problems connected with the sun whose solution is of vital importance to the astronomer. In this book the author will endeavor to state some of these problems and the methods devised for their solution. At the same time it will be necessary to refute one of the " twice-told tales " (See *Science* during 1921) that the moon is a more important body than the sun since the moon gives us light at night when it is dark and we need its light, whereas the sun shines during the daytime when it is bright and we could possibly get along very well without it! The light that lightens the world, the heat that gives our bodies comfort, the wind that cools our heated brows and wafts the sailing ships across the ocean, the coal that warms and illuminates our homes and generates steam and electricity to carry on the world's mighty commerce, the rain that descends from heaven and waters and fertilizes the soil and causes the flow of our mighty rivers; all these and many other benefits come from the sun. Without the sun there would be no grass, no flowers, no wheat or corn or vegetable life of any kind, and without the sun there would be no animal life, no man upon the earth. If the sun were blotted out for the space of one short month, there would not be one of us left alive to tell the tale; we should all be frozen to death! It is well then that we should endeavor to learn as much as possible of the giant luminary in which are centered our light and heat and our very life, even when the quest for knowledge carries one as far from home as does eclipse work.

The life of such an observer might be likened to that of

a hunter after big game. Many months and even years are spent in quietly investigating the problems, a costly equipment is accumulated and each piece of delicate apparatus is carefully tested at home to see that it will properly perform its designated functions far afield, a long journey is often necessary, frequently of thousands of miles, by rail and sea. Arrived at the destination, instruments, cameras and spectroscopes are erected and most carefully adjusted, and after six or eight weeks of preparation in the field the eventful day approaches. Each and everyone of the party drills constantly so that the task allotted to him may be well and carefully done, so that the photographic slides may be drawn and each camera lens may be opened at the appropriate instant. Success lies in seeing that every one of a thousand possible chances of failure are obviated. At a certain hour, minute and second, the "zero-hour," operations are due to begin. But alas! there may be no "game," the eclipse may be eclipsed by clouds, and the long months of preparation may be of no avail since it will not be possible to try again on the morrow when the clouds have rolled away.

The author has traveled ninety thousand miles to witness nine total eclipses of the sun. The total accumulation of time afforded him for scientific observations during these nine eclipses has been approximately eighteen minutes.

ECLIPSES OF THE SUN

CHAPTER I

EARLY HISTORICAL ECLIPSES

"In old Cathay, in far Cathay,
  Before the western world began,
They saw the moving font of day
  Eclipsed, as by a shadowy fan;
They stood upon their Chinese wall.
  They saw his fire to ashes fade,
And felt the deeper slumber fall
  On domes of pearl and towers of jade." — NOYES.

THE earliest recorded eclipse of the sun is one which happened more than four thousand years ago, an account of which is given in the ancient Chinese classic *Shu Ching*. According to the competent authority Oppolzer, this took place on October 22, 2137 B.C., about 1400 years more remote than that recorded by any other nation. This eclipse is celebrated not only for its great antiquity, but also for the dire fate of the two royal astronomers, Hsi and Ho, who instead of staying in the sober paths of science went and got beastly drunk, with the result that they were taken unawares and were unprepared to perform their customary rites of shooting arrows, beating drums, etc., for the purpose of delivering the sun from the monster that was devouring it. To show his great displeasure, not so much for failing to predict the eclipse, but on account of the intense confusion that prevailed, Chung K'ang, the fourth emperor of the Hsai dynasty, ordered that they should be punished and that their heads should be chopped off. And with this tragic warning in view, there is no record from that day to this that an astronomer has ever dared to follow in the steps of the unfortunates, Hsi and Ho, and been drunk at the time of an eclipse.

This eclipse is of such great importance that it will be well to give it more than a passing glance. The early eclipses, both of sun and of moon, have been of such great interest to astronomers and scholars that a very large number of investigations have been devoted to the determination of as exact dates as possible for these phenomena. Eclipses can take place only when the centers of sun, earth and moon are approximately in a straight line. These circumstances can be predicted with accuracy if we know with a sufficient degree of precision the motions of the earth about the sun, and of the moon about the earth. The times of eclipses thus give to the astronomer the means of accurately testing his calculations regarding the motions of both these heavenly bodies. The movements of the earth render no difficulties, but with the earth's child, the moon, the matter is entirely different. It may be truly said that the motions of the moon have given the mathematical astronomer more work and worry than those of all the balance of the gigantic universe put together. The earlier the eclipse that can be checked up with accuracy the more reliably can the motion of the moon be verified, since any accelerations in this motion depend on the square of the elapsed time. Valuable as these early eclipses are to the astronomer, they are equally important to the historian or chronicler in fixing the dates of remote antiquity. The astronomical problem has been well studied and is now regarded as a simple one. The crux of the whole question lies in the exactness and precision with which the event has been described by the author or historian.

The *Shu Ching*, or Book of Historical Documents, is a collection of public speeches and proclamations beginning with the reign of the legendary Emperor Yâo who lived in the twenty-fourth century B.C., and closing with the year B.C. 625. The book is not a historical, chronological narrative, nor indeed does such a book exist in the Chinese language. This ancient eclipse has been fully discussed by a number of eminent authorities. The most complete monograph is by Schlegel and Kühnert, Die Shu-King-Finsterniss, *Verhand. der Könin. Akad. van Wettenschappen, Letter-*

kunde 19, 5, 1890. Oppolzer communicated to *Monatsberichte der Kön. Preuss. Akad. der Wissens. zu Berlin,* 166, 1880. Briefer accounts may be found by S. M. Russell, *Observatory, 18,* 323, 1895; in Chambers' *The Story of Eclipses,* 65, 1900; and in the Halley Lecture delivered by Dr. J. K. Fotheringham, May 17, 1921.

In the third century B.C., Shi Hwang-Yi, a great military genius, conquered the whole of the China of those days. In 221 B.C. he proclaimed himself the "First Emperor," and he decided, on the advice of his prime minister, that every form of human progress including literature should begin with his reign. He therefore ordered the destruction of all books except those belonging to the three utilitarian branches of knowledge, agriculture, divination and medicine. To see that his orders were properly carried out, he personally examined each day a hundred pounds of books so that he might decide which were useless. Four hundred and sixty scholars were put to death for disobeying the royal commands, and others were banished for life. All copies of the *Shu Ching* seem to have perished except one incomplete copy later recovered from a receptacle where it had been walled up by a devoted scholar. The book in which the eclipse is mentioned is not included in the authentic copy. It was added later, probably to conform to a table of contents that scholars believe may have been included in the original volume, and which may possibly have been written by the great Confucius. The preface to this book is as follows:

"Hsi and Ho, sunk in wine and excess, neglected the ordering of the seasons, and allowed the days to get into confusion. The prince of Yin went to punish them. Description of this there was made, *The Punitive Expedition of Yin.*"

It should be noted in the above that there is no mention of an eclipse. Tso, a scholar and commentator of the fifth century before Christ wrote concerning the lost work, *The Punitive Expedition of Yin* in the following words, "The Sun and Moon did not meet harmoniously in Fang. The blind beat their drums; the inferior officers galloped and the common people ran about." There is further added, "That is said of the first day of this month; it was in the fourth month of Hsia, which is called the first month of summer."

In the *Annals of the Bamboo Books,* a work of the early third century before Christ, the reference to Hsi and Ho is as follows: "In the fifth year of Chung K'ang, in the autumn, in the ninth month, on the first day of the month, there was an eclipse of the Sun, when he ordered the prince of Yin to lead the imperial forces to punish Hsi and Ho."

The text does not say that the expedition was for the purpose of punishing the astronomers for failing to predict the eclipse. The scholar who restored the missing text used such flowery language that it is difficult to obtain the exact meaning, although it seems to be implied that the eclipse was the cause of the punitive expedition.

There are many uncertainties that prevent an accurate interpretation of this eclipse, chief among which may be noted: (1) The month is entirely unknown, whether it is the "first day of the last month of autumn," or, "the first day of the first month of summer." (2) The meaning of the word *Fang* is doubtful, whether it means "the order of the constellations," which seems to be the better opinion, or whether *Fang* is used in the more restricted sense as the name of a small constellation including β, δ, π and ρ Scorpii. (3) The Emperor's name is very uncertain as it is not given except in the *Bamboo Books.* Little is actually known of Chinese chronology further back than B.C. 841. We do know the names of the emperors in succession, but whether they reigned five years or fifty is unknown, and consequently all that it is possible to say about the emperor Chung K'ang is that he must have been on the throne a century or two earlier or later than the year 2000 B.C. (According to the *Encyclopaedia Britannica* the first of the kings mentioned in the Shu Ching reigned from 2357 to 2255 B.C., an interval of no less than 102 years!)

One interpretation of the above uncertainties is probably about as good as another. One date fixed by the Chinese astronomers of about the eighth century A.D., was the year

B.C. 2155. But in consequence of the fact that the eclipse of that year was invisible in China, there has arisen the well-known ditty:

> "Here lie the bones of Ho and Hi,
> Whose fate though sad was risible,
> Being hanged because they could not spy
> The eclipse which was invisible."

Taking "the last month of Autumn," and *Fang* in the restricted meaning of the word that it referred to the constellation, Oppolzer fixes the date as given above, B.C. 2137, October 22. Other authorities equally competent give other dates. The only conclusion that can be reasonably drawn is that it is not possible to identify the eclipse with any approximation to certainty. Nothing is known of the fates of the astronomers Hsi and Ho. The regulations regarding eclipses as given in the *Shu King* read: "Being before the time, the astronomers are to be killed without respite; and being behind the time, they are to be slain without reprieve." With such a fate in store, who would dare be an astronomer?

Many admirers of the Chinese, not being conversant with the true facts, have pointed out this eclipse as evidence of the great learning of the Chinese and their proficiency in astronomical knowledge twenty centuries before the beginning of the Christian era. Although not wishing to dim the halo of glory that has surrounded early China, honesty compels one to draw attention to the facts. For many centuries the Chinese were unable to predict the position of the sun accurately among the stars for determining the length of the year. In fact, they seem to have relied wholly on observation from year to year to settle their calendar. No conclusions or systematic deductions were made from their observations as is shown by the fact that their calendar was continually falling into confusion.

To those who are interested in early chronological data to be fixed by this and other remote eclipses, reference should be directed also to Delambre, *Histoire de l'Astronomie Ancienne*, Paris 1817, and to Johnson's *Historical and Future Eclipses*, London, 1896.

Other Chinese Eclipses

The history of the Chinese Empire is very closely associated with the name of Confucius, the immortal sage. Confucius is the Latinized form of K'ung tsze (meaning the philosopher K'ung, this being the family name). Through his veins flowed the best blood of China. His birth has an interesting romance connected with it. His father Heih had a family of nine daughters but only one son who was unfortunately a cripple. Realizing in his old age that the K'ung name would probably become extinct with him, he went to his neighbor of the Yen clan, and told of his plight, and asked in marriage one of Yen's three daughters. The youngest of the three became the mother of Confucius, 551 or 550 B.C., Heih then being over seventy years of age. The father died in the child's third year.

In his twenty-second year Confucius began his life as a teacher, and he put into practice principles which every college professor of the twentieth century would like to follow, namely, he would accept any pupil no matter how small the fee, but as soon as lack of capacity or diligence was manifested, the youth was sent away. "When I have presented," he is reported to have said, "one corner of a subject and the pupil cannot of himself make out the other three, I do not repeat my lesson." Would that these methods of education were prevalent today! His professed disciples numbered 3000, of whom 70 or 80 were "scholars of extraordinary ability." Most of his life was spent in the province of Lu, the modern Ho-nan and Shantung.

Among the writings ascribed to Confucius are the following:

(1) A preface to the *Shu Ching,* already referred to. The preface is little more than a table of contents, and it is very doubtful if Confucius had anything to do with it.

(2) He compiled, or edited the *Shih Ching,* or Book of Poems. Originally numbering about 3000, only 311 were retained by Confucius, of which we now possess 305. It is the most ancient book of poetry in the world. The latest of the poems has the date 585 B.C.

(3) From his own hand comes the *Ch'un Ch'in,* or "Spring and Autumn," which is best known as the *Annals of Lu,* his native state. This work has been preserved almost complete. It is a brief summary of the chief events that took place in Lu during a period of 242 years from 722 to 481 B.C. It is a model for a historical document. The facts are briefly itemized according to the seasons in which they fell. As an example, in the year B.C. 612 there are twelve entries, the fifth of which was recorded, "In the autumn, in the seventh month, there was a comet which entered Pei-ton (in Ursa Major)."

In the *Annals of Lu* are records of no less than thirty-six eclipses of the sun, the first eclipse being observed February 22, 720 B.C., and the last on July 22, 495 B.C. The first of them is described as follows: "In the 58th year of the 32nd cycle in the 51st year of the Emperor King-Wang of the Chou Dynasty, the 3rd year of Yin-Kung, Prince of Lu, in the spring, the second moon, on the day called Ke-Sze, there was an eclipse of the Sun." These ancient eclipses have been carefully investigated and many papers concerning them have been published in the *Monthly Notices of the Royal Astronomical Society.* The conclusions are that it has been possible to identify no less than thirty-two of the thirty-six. Apparently four of the eclipses, those of April, 645 B.C., June, 592 B.C., September 19, 552 B.C., and June 18, 549 B.C. are in error and did not take place. The record is a history of eclipses that were actually observed to have taken place. The inference is readily drawn that the astronomers who recorded these erroneous eclipses saw some curious phenomenon around the sun; and remembering the sad fate that befell the two negligent astronomers, Hsi and Ho, and not wishing to take any chances whatever regarding the safety of their heads, they recorded the phenomenon as an eclipse.

In addition to the works from the hand of Confucius himself we are fortunately in possession of the *Tso Chuan,* a so-called commentary. This is presumably by some one with the name of Tso who took the bare entries in the work by Confucius and enlarged upon each one to such an extent and with such remarkable genius and dramatic brilliancy that the Tso commentary reads more like a prose epic than an elaboration of a series of facts or annotations on the text of a literary work. By its means there is vividly portrayed the intrigues, the alliances, the treacheries and the jealousies of the various states which made up feudal China; we can see with the clearness almost of a photograph, assassinations, battles, heroic deeds, brilliant rescues and the torments and horrors meted out by the conquerors to their unfortunate victims. The *Annals of Lu* make the chronicles of China in the 5th, 6th and 7th centuries before the birth of Christ as full of action and as attractive as those of France and England twenty centuries later. As a matter of fact there appeared to be more of literary culture in China in these early centuries B.C. than there was in Europe a bare five hundred years ago.

The eclipse of February 22, 720 B.C. recorded in the *Annals of Lu* is not the first Chinese eclipse whose date can be accurately fixed, for fifty-six years earlier there was an eclipse of the sun preceded by an eclipse of the moon. In the *Shih Ching*, or Book of Poetry referred to above, there is an account which runs somewhat as follows: "Tenth moon, her conjunction first day, sin-mao, sun he had eclipse, also it very bad." The record ends by adding, "That moon in eclipse is a thing only common. This sun in eclipse is a thing very bad." The eclipse of the sun referred to took place on September 6, 776 B.C., the eclipse of the moon on August 21. The solar eclipse was visible only as a partial eclipse in China while that of the moon was nine-tenths total. It was during the reign of the notorius Yu-wang. It is apparent that the eclipse of the sun was regarded by the bard as a special warning to the lascivious Emperor. The evidence of this eclipse is indisputable; it fixes absolutely without controversy the first certain date in the history of China. How far before this epoch the Chinese history stretches is a matter of personal judgment and individual conjecture. As already pointed out, the records of early dates are so unreliable that one of the early kings is supposed to have reigned the impossible period of 102

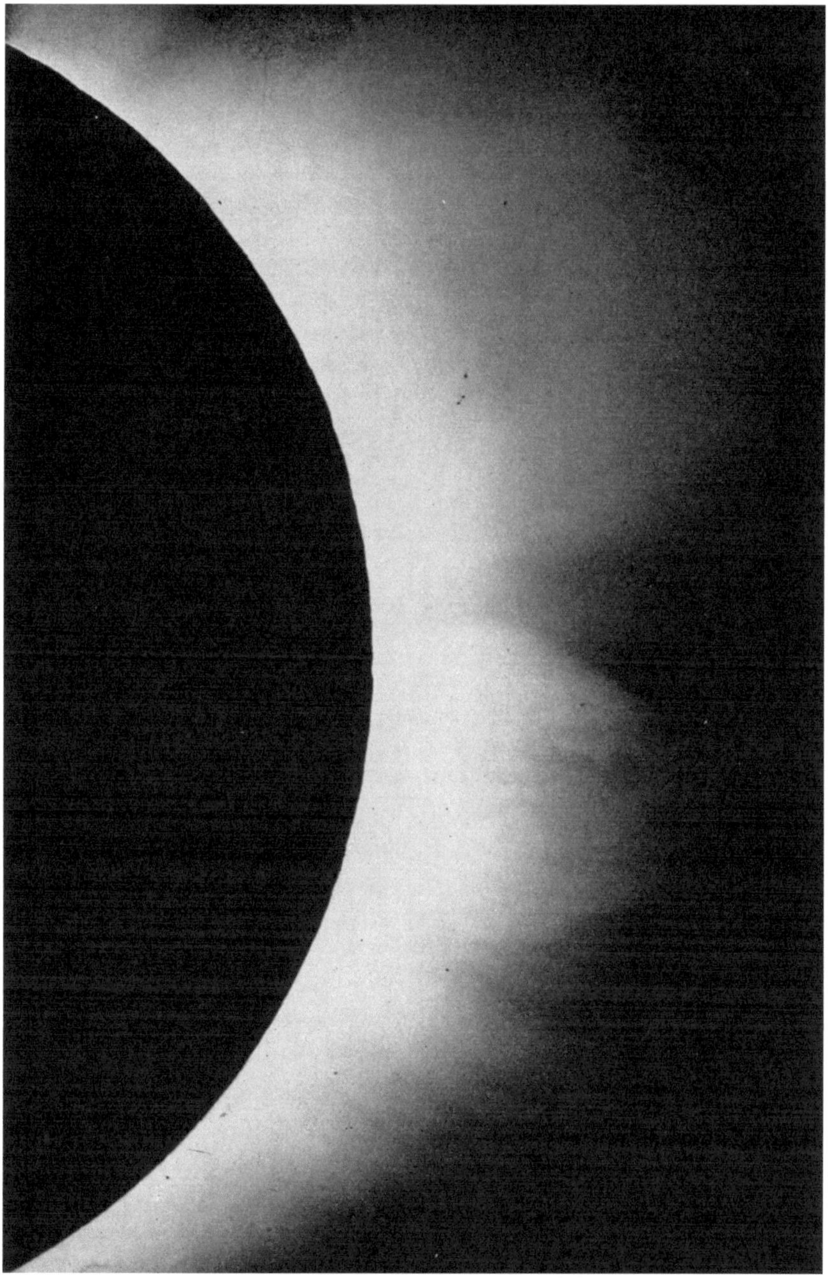

Coronal Disturbances to East of Sun in 1930, Showing "Strawberry Dome" and Longest Streamer

Photograph by Marriott with 63-foot tower telescope.

Corona at the "Tin-Can Island" Eclipse, October 21, 1930
Photographed with 15-foot horizontal camera.

years. Egypt has her pyramids and ruined temples as evidence of a much earlier civilization, but there are no such relics in China. This Chinese eclipse antedates by thirteen years the celebrated Nineveh eclipse.

Mr. John Williams, formerly Assistant Secretary of the Royal Astronomical Society, made a careful examination of the Chinese historical work called *Tung-Keen-Kang-Muh,* a record of one hundred and one volumes which contains a summary from earliest times to 1368 A.D. Information concerning these eclipses can be found in the *Monthly Notices R. A. S.* Fifty-six solar eclipses occurred between 481 B.C. and the beginning of the Christian era, and nearly one hundred additional ones before the fourth century A.D.

ECLIPSES OF BABYLON

" In Babylon, in Babylon,
  They baked their tablets of the clay;
And, year by year, inscribed thereon
  The dark eclipses of their day;
They saw the moving finger write
  Its *Mene, Mene,* on their sun.
A mightier shadow cloaks their light,
  And clay is clay in Babylon." — NOYES.

According to L. W. King, *Chronicles concerning the early Babylonian Kings, 2,* 76, 1907, it is recorded that " on the twenty-sixth day of the month Sivan, in the seventh year, the day was turned to night, and fire in the midst of heaven. . . ." Cowell[1] has concluded that this probably refers to a solar eclipse visible in Babylon on July 31, 1063 B.C. By carrying the Babylonian records backwards from this eclipse, a check is had on the acknowledged Kea-Tsze 60-day calendar of the Chinese from which comes the date of the eclipse of 2137 B.C. Unfortunately, there is no knowledge concerning the name of the king in the seventh year of whose reign the eclipse occurred. If this fact were known, the early Babylonian chronology would be fixed with fairly great certainty. It is possible to conjecture as Langdon has done that the name of this king was Nabu-shum-libur, but this name must be looked upon only as a

[1] Monthly Notices, R. A. S., *65,* 861, 1905.

suggestion. All that is known of *exact* dates in Egyptian and Babylonian chronology depends on the science of astronomy, — on the occurrence of an eclipse of the sun in 763 B.C., and on the records of the astronomer Ptolemy as given in his great work the *Almagest.*

There are no more interesting portions of the world's early history than that connected with the lands contiguous to the Persian Gulf. In Egypt, the gigantic pyramids, the construction of which is justly regarded as one of the seven wonders of the world, and the ancient temples point to a civilization that was in a high state of development five thousand years ago. Across the narrow Red Sea, more interesting even than the land of the Pharaohs, is the country of the Holy Scriptures where lived Adam and Noah, the great King David, the holy prophets and the Lord Jesus Christ. Up to the middle of the nineteenth century little was known of ancient Babylonia and Assyria other than the account given in the Bible. The country which is nearly enclosed by the two great rivers the Tigris and Euphrates from Bagdad to the Persian Gulf is bounded on the north by Mesopotamia, on the east by the plain of Elam, on the south by the Persian Gulf and on the west by the Arabian Desert. To the north, in Mesopotamia, the country is more or less mountainous, while to the south the terrain is flat and marshy. In ancient times the plain was covered with a complicated network of canals. The remarkable fertility of the soil brought such great prosperity to these peoples that they aspired to be the masters of the world. According to Herodotus (i, 193), wheat commonly returned two hundred-fold to the sower, while Pliny (*N. H.* xviii, 17) states that it was cut twice each year. The native historian Berossus remarked that wheat, barley, palms, apples and many kinds of shelled fruit grew wild. The country was studded thickly with towns. But alas, — the neglect of the canals has changed the face of the land so completely that the fertility that was once the wonder of the ancient world has departed, and in its place there is nothing but a barren and desolate waste, part of the year the land being a series of swamps and marshes.

The ancient city of Babylon was on the east bank of the Euphrates about seventy miles south of Bagdad. The plain of Babylonia was called Edin, the "Eden" of Genesis ii. The chief city of Assyria, Assud, was on the west side of the Tigris, while the three other great cities of Nineveh, Calah, and Arbela were on the eastern bank of the Tigris.

The excavations at Nineveh by Botta and Layard opened up a new world. Layard after indescribable difficulties laid bare the palace of Sennacherib (705–680 B.C.) and that of Assurbanipal (668–626 B.C.), and in the palace of the latter was found a great library of tablets which finally numbered no less than thirty thousand. These objects when not too massive were transported to the Museum of the Louvre or to the British Museum. The success of these discoveries stirred scholars throughout the world, and the different nations vied with each other in the prosecution of the interesting search. The British and French continued their investigations, and in 1886 German archaeologists began their work. Since 1888 expeditions have been organized by the University of Pennsylvania and since 1903 by the University of Chicago.

The language in which the ancient history is preserved is the so-called Babylonian-Assyrian wedge-writing, the cuneiform language of signs. No such thing as an alphabet was known. As the $ sign means much to the American, while to the resident of the British Isles the synonym of money is conveyed by a different symbol, £, so these primitive people expressed themselves in their tablets entirely by signs. A wedge was used to make an impression in clay, and the different combinations served to express a variety of ideas. The decipherment of these tablets was a difficult task. The most important progress was made through the work of an Englishman, Henry C. Rawlinson, who observed at Behistun a rock stretching up almost 1700 feet above the plain, and at a point about 350 feet above its base there was a large space carefully smoothed off. On this was found a mass of inscriptions distributed in columns of various lengths. After long years of work he succeeded in transcribing the whole of the record. Careful study and long and arduous toil revealed the fact that the writing was in three different languages, old Persian, Elamite and the Babylonian. The decipherment was excessively difficult due to the enormous number of signs employed. For simple names there were about 600 independent and distinct signs made by combinations of wedges numbering anywhere from two to thirty. But two to six of these signs might be compounded to express more complicated ideas, with the result that there are something like 20,000 different signs known to students of Assyriology. The number of inscriptions so far found number about one hundred and fifty thousand.

The earliest mention of Babylon appears to be about 3800 B.C. The first great king was Sargon of Akkad who lived about 2750 B.C. Even greater was Khammurabi the Amorite who flourished about 2100 B.C. Concerning this latter monarch, who lived over four thousand years ago, we have positive information in a group of fifty-five of his letters edited by L. W. King, and in a monument giving a code of splendid laws. From his letters we learn that the Babylonians kept their record of time entirely by the moon, a new month beginning with each new moon. Twelve lunar months made up the year. Since the synodic month is of twenty-nine and a half days in length, it became necessary to add in the reckoning of the year a thirteenth month from time to time in order that the calendar should not get astray.

This method of measuring the length of the year, which is both inaccurate and inconvenient, was followed by the Hebrews and the Persians. Even in the enlightened age of the twentieth century the Jewish and Mohammedan calendars are still based on the old custom. An intercalary month being inserted in the reign of Khammurabi, he accordingly sent out a circular letter to his governors advising them that the delinquent tax gatherers should not take advantage of the change in the calendar and that the taxes must be paid without delay. Schools flourished even at this remote period, and tablets have been found which evidently were exercises of the children as they struggled to learn the language of signs. In fact a Babylonian proverb

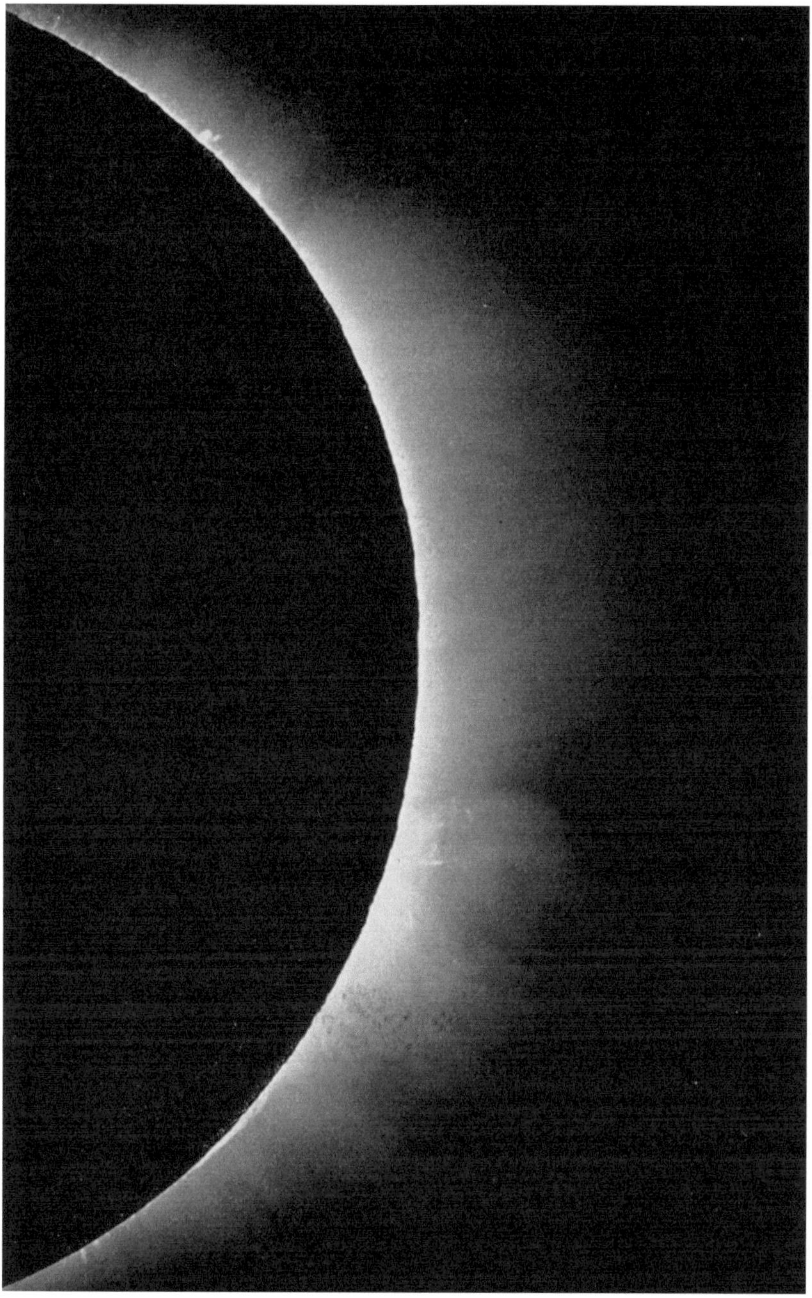

Brighter Portions of the Corona Showing Same Region as Page 8
Photograph by Marriott with 63-foot tower telescope.

THE CORONA, MAY 9, 1929
Photographed with 15-foot camera of the Swarthmore Expedition.

has been unearthed which reads, "He who shall excel in tablet-writing shall shine like the sun."

The kingdom of Assyria begins about 3000 B.C. Assyria was a warlike nation, and lived by the sword. In the eighth century B.C., their conquests had been pushed as far westward as Damascus and to the Hebrew city of Samaria. While besieging this latter city, the king was deposed and a usurper called Sargon took the throne. Under him and his son Sennacherib, Assyria was at the zenith of its military power. Campaigns were carried out in Ionia and Greece and also southward as far as Egypt, which became tributary to Assyria. In Nineveh, Sennacherib constructed the largest and most magnificent palace the world had seen up to this time. In 689 B.C. he conquered the city of Babylon, and he caused this holy city to be completely destroyed and all of its buildings razed to the ground. After many years Babylon under Nabopolassar, revolted from the Assyrian yoke and under his son Nebuchadrezzar — the Nebuchadnezzar of the Bible — construction on a great scale was carried out, and Babylon was rebuilt and became the wonder city of the ancient world. According to Herodotus, the walls were no less than 56 miles in circumference, the walls being 335 feet high and 85 feet wide. Each city had its great temple and its own god, the god of Babylon being Bel-Merodach. In the temple stood an enormous image of the god forty feet high, together with a table, a mercy seat and an altar, all constructed of gold. In this reign (604–561 B.C.), the rebuilt Babylon was at the height of her glory and the summit of Chaldean civilization had been reached.

This glimpse of the history of these early peoples of such great interest to all Christian countries is for the purpose of giving a setting for ascertaining what they had accomplished in art, literature and science — particularly in the science of astronomy. Assyria differed much from Babylonia. The former state was an armed camp and held sway by the sword, probably even in those early times believing that might is right. On the other hand, the Babylonians were peace-loving people, a land of merchants and farmers, and withal deeply religious. Consequently, Assyria had little culture of its own and the little it did possess was borrowed or acquired from Babylonia.

One of the grandest epics of ancient Babylonia was the description of Creation which is mentioned here on account of its astronomical significance and also that it may be compared with the Biblical description. In the first book an account is given of the creation of the world out of the primeval deep and the birth of the gods of light. Then comes the story of the struggle between the gods of light and the powers of darkness, and the final victory of Merodach, who clove the dragon of chaos, Tiamat, asunder, forming the heaven out of one half of her body and the earth out of the other. Merodach next arranged the stars in order, along with the sun and moon, and gave them laws which they were never to transgress. After this the plants and animals were created, and finally man. Merodach here takes the place of Ea, who appears as the creator in the older legends, and is said to have fashioned man out of clay.

Thus from very earliest times, the religion of the Babylonians was inspired by the belief that the heavenly bodies were subject to law and order and that they ruled the destinies of man. It was evident even to the simple mind of the primitive man that the sun was the cause of all life on the earth, and it was a natural conclusion that the ever-changing aspect of the sun, moon, planets and stars should be connected with the constant mutations in the lives of the individuals of the race. Hence their gods and goddesses were identified with the heavenly bodies. The priests became astrologers. To be able to read the signs of the heavens seemed to be synonymous with understanding the happenings on earth, and consequently if this knowledge were to be utilized to predict what would take place in the future it was necessary to follow with the minutest care the various motions of the sun, moon, planets and stars — and hence the priests became astronomers. These religious beliefs caused a systematic study of the heavenly bodies; and as a result astronomy, the first and parent of all of the Sciences, had its birth.

The movements of the sun and moon were symbols of law and order, and accordingly they were worshipped as gods. The motions of the planets, though more difficult to understand, seemed also the subject of order and not of caprice, and so likewise they were worshipped. Jupiter was the abode of the Babylonian god Mardok, and Venus was the home of the goddess of love, Ishtar. Saturn was identified with Ninib, Mercury with Nebo, and Mars with Nergal. In order to understand the manner in which the gods of the sun, moon and planets influenced the lives of men, it was necessary for the Babylonian priests to observe the position of these bodies, not only with respect to each other, but also with respect to the more prominent and easily recognizable fixed stars. In the case of the moon it was very necessary to note the time at which the new moon became visible as a crescent, its position in the heavens, and the angle made with the horizon by the line through the horns. (Even at the present supposedly enlightened age of the twentieth century A.D., we still hear of the "wet" moon, and the "dry" moon.) The changes of the moon were followed with meticulous care since each change afforded the opportunity of interpreting some phase of human activity. And through forty centuries of the world's history some of these superstitions have survived in the popular belief that a change in the moon means a change in the weather, or that certain crops should be planted "in the dark of the moon." These observations of the heavenly bodies being thus gradually accumulated must have reached a mass of considerable proportion. For his interpretation of these observations the astrologer rested on written records or on the recollection of what had happened under similar circumstances in the past, but sometimes merely on the vagueness of association of ideas.

The motion of the sun and moon among the stars near the ecliptic being carefully observed, it was but natural that the twelve full moons of the year should suggest the division of the circle of the year into twelve parts. We thus owe to the Babylonians the constitution and nomenclature of the twelve signs of the zodiac. Ages before Assurbanipal's great library of tablets was constructed, the eighth month was known as "the month of the star of the Scorpion," the tenth as "the star of the Goat," while the twelfth was the month of "the star of the Fish of Ea." The convenience of the duodecimal system was thus early recognized. Each day was divided into twelve "double hours." The *ner* of 600, the *sar* of 3600 was formed from the *soss*, or unit of 60. The circle was divided into 360°, or six times sixty, with further divisions by 60 into minutes and seconds of arc, the hours being likewise divided into minutes and seconds of time. According to Miss A. M. Clerke, "In the Chaldean signs of the zodiac, fragments of several distinct strata of thought appear to be subdivided. From one point of view they shadow out the great epic of the destinies of the human race; again the universal solar myth claims a share in them; hoary traditions were brought into *ex post facto* connection with them; or they served to commemorate simple meteorological and astronomical facts." Astronomy was thus of very old standing in Babylonia. The principal astronomical work, called the *Illumination of Bêl* was compiled for the library of Sargon of Akkad; it was inscribed on 70 tablets, and apparently went through numerous editions, one of the tablets being in the British Museum. It treats, among other things, on observations of comets, the pole star, the conjunction of sun and moon, and the motions of Venus and Mars.

However, since the study of orderly motions of the heavenly bodies was primarily for the purpose of forecasting human events, it is not surprising to learn from modern researches that the early astronomical knowledge was very crude, with very little perfection in the observations. It is true that as early as the days of Khammurabi (2100 B.C.) there were combinations of prominent groups of stars into outlines of animals and figures, yet there is no evidence that prior to 700 B.C. more than a small number of the constellations of the zodiac had found their place in the sky.

The Babylonians had tables of squares and cubes calculated from 1 to 60, they measured time by the sun-dial and by water clocks, were familiar with lever and pulley, and even possessed a lens turned on a lathe, but which was certainly not used as a magnifying glass. Such proficiency in this early people would lead one to expect that though their observations may have been crude they had nevertheless perfected systems of moon calculation and planetary tables of a high order of excellence. The discovery of the Saros of 223 lunations, so useful and important in the predicting of eclipses, has been attributed to the Chaldeans. They seem also to have been aware that Venus returns in almost exactly eight years to the same point in the sky and to have established similar relations of 46 years for Mercury, 59 for Saturn, 79 for Mars and 83 for Jupiter. They were thus able to fix in advance the positions of the heavenly bodies, of such great importance in astrological lore. In fact it is generally thought that Hipparchus, the first great astronomer of the world, acquired much of his information from the Chaldeans, though unquestionably he verified all of their findings. The Babylonian sage Berossus founded a school about 640 B.C. in the island of Cos, and it is not impossible that Thales of Miletus was one of his pupils.

In spite of the unbounded admiration which one must feel for the development of astronomy under the Chaldeans it seems much more plausible to believe that the greatest work of the Chaldeans was accomplished in the later years of their history, this taking place even after the fall of Babylon in 539 B.C., and in fact after the Greeks had invaded the valley of the Euphrates. This matter, and particularly their acquaintance with the Saros, will be further discussed in dealing with the famous eclipse of Thales.

The study of astrology thus securing a firm foundation under the Babylonians, it spread from them, directly or indirectly, to all quarters of the globe. It came to Greece about the middle of the fourth century B.C., and reached Rome before the opening of the Christian era. In India and China both astronomy and astrology were acquired from the Greeks, and are largely reflections of Greek theories and speculations. By the introduction of Greek culture into Egypt, astronomy and astrology were both cultivated in the land of the ancient Pharaohs during the Greek and

Roman periods. The Arabs developed astrology and also astronomy, many of the names of the brighter stars being Arabic. In Europe as late as the 14th and 15th centuries, astrologers had important positions at the royal courts and were consulted on all matters of great moment to the nation. With the revival of learning, and particularly after the development of the Copernican system which showed that the earth was but one of the planets, astrology became more and more pushed into the background, though as late as the 17th century horoscopes were cast by the great astronomer Kepler. Even in the United States of America and in England there are still many thousands of people who believe in the flat earth, and who have great confidence in the efficacy of horoscopes to correctly forecast the future. Traces of the Babylonian astrology are still found in our every-day language, for we still have such phrases as "I thank my stars," he was "born under a lucky star," or an "ill-starred undertaking."

As already stated, the Babylonians had a calendar of twelve lunar months, and a seven-day week, the seventh day being a day of rest. The early Babylonians reckoned events from some great catastrophe, such as in our day, Chicago and its great fire, but later in their history they counted time by the years of the reigning king. There were several early dynasties, but the succession of rulers is fairly certain. In Assyria, on the contrary, a plan unique among the early peoples was followed in naming the year after officers, called eponyms, whose term extended for one year only. This "Eponym Canon," as it is called, began with the year 911 B.C. In the *Almagest* of Ptolemy there is a list of Babylonian, Assyrian and Persian kings who ruled in Babylon together with the years each of them reigned beginning with the accession of Nabonassar in 747 B.C., to the conquest of Babylon in 331 B.C. by Alexander the Great. This Ptolemaic Canon is confirmed by other Babylonian chronicles and also by the Assyrian Eponym Lists.

These Assyrian tablets record three eclipses of the sun. The first, interpreted by Rawlinson in 1867, is referred to as follows: "In the Eponymy of Burgasole, Governor of

Gozan, a revolt in Assur took place in the month Sivar, and the sun was eclipsed." To call attention to the importance of the event the ancient scribe drew a line across the tablet. This eclipse has been carefully investigated by a number of astronomers, by Ginzel, Airy, Hind and others, and the general conclusion is that Nineveh, where lived the Assyrian scribe, was just outside the path of totality, and that the greatest obscuration was about 9:47 in the morning of June 15, 763 B.C. The second eclipse mentioned in the Assyrian records took place in the reign of Esar-haddon, the son and successor of Sennacherib. This eclipse was partial in Assyria, but annular farther to the east, and the date was May 27, 669 B.C. The record of the third is a little uncertain, but it is probably the eclipse of June 27, 661 B.C., during the reign of Assurbanipal.

These are not the only eclipses observed by the Chaldeans, for in the *Almagest* of Ptolemy is a record of three eclipses of the moon observed in Babylon. The first of the three was a total eclipse on March 19, 721 B.C., while the other eclipses, partial only, were observed the following year, on March 8, and September 1. As has been stated above, Ptolemy gave a list of the kings who reigned in Babylonia. These eclipses, of sun and moon, fix the dates of Eastern chronology with great exactness, the earliest date in the world's history to be thus accurately determined being the year 911 B.C. in Assyrian chronology. Earlier dates are known with increasing uncertainty. The various estimates of historians for the beginning of the first Egyptian dynasty differ as much as two thousand years!

As a result of the Babylonian eclipses, it has been necessary to alter the chronology of the Bible by lowering the dates to the extent of twenty-four years. This will be referred to later in connection with passages from the Holy Scriptures in Amos and Isaiah.

Egyptian Astronomy

In the neighboring country of Egypt excellent opportunity must have been afforded for a study of astronomy on account of the clear skies of the East and from the intercourse which must certainly have taken place between the Egyptians and the Chaldeans of Asia Minor. Egyptian records accordingly have been carefully scrutinized to find what traces there are of early astronomical knowledge in the land of the Nile. And since a study of eclipses must concern itself with the beginnings of astronomy, a brief glance will be given here to early history in Egypt.

The early Chaldean monuments are probably of an earlier date than those of Egypt, but the former are almost formless piles of sundried brick, while the tombs and pyramids of the early Egyptian dynasties are many of them in excellent preservation. The history of Egypt is generally divided into five periods:[1] (1) The Ancient Empire, 3400–2160 B.C., comprising ten dynasties of kings; (2) the Middle Empire, with Thebes as capital, two dynasties; (3) the second Thebic, or New Empire (1588–1150 B.C.), comprising dynasties XVII–XX, separated from the Middle Empire by the Shepherd Kings of Arabia; (4) the Decadence period of six dynasties, 1150–324 B.C., which includes the Persian conquest in 525 B.C.; and (5) the Ptolemaic and Roman period, 324 B.C.–300 A.D.

Little is known of exact dates before the Conquest by Alexander although repeated attempts have been made to determine them. The list of the succession of the Egyptian kings is known with certainty, but the chronology is uncertain for the following reasons: (1) The lengths of the individual reigns are not known with precision, since the record seldom reached to the end of the reign, and did not allow for co-regencies. (2) Calculations on the *probable* length of a period leads to no trustworthy information. (3) Comparisons with other records, particularly with those of Babylonia and Assyria, give the most valuable knowledge. But the dates before 911 B.C. are not known accurately even in Assyrian chronology, and it is therefore not surprising that the date of the beginning of the XIXth dynasty should be estimated by competent authorities anywhere from 1490 B.C. to 1315 B.C. (4) Of most interest to us here is the

[1] Hamlin, *Encyclopedia Americana,* 10, 12.

astronomical information. The Egyptian number system was decimal, each power of 10 up to 100,000 being represented by a different figure, on much the same principle as the Roman numerals. The day was divided into two periods, each of twelve hours, the beginning of the day being the time of sunrise. The year was divided into twelve months, each of thirty days, or a year of 360 days. As early as the Vth dynasty, the premature arrival of the seasons was noted, and as a consequence five complementary days were added making a year of 365 days. The extra days were counted either at the beginning or at the end of the year. The year was divided into three seasons, each month into three weeks, each of ten days. Since the season of agricultural growth depended more on the inundation by the Nile than on the motion of the sun, the first of the year was reckoned as the beginning of the rise of the waters of the river. As this took place with fair regularity, the Egyptians were thus furnished with a useful starting-point for their annual counting of time. It was noticed that the brilliant dog-star, Sirius, or Sothis, rose with the sun about this time (July 19), and as this star is so brilliant, refined astronomical measurements were not necessary in order to observe it. But the tropical year, the year of the seasons, is approximately a quarter of a day more than 365 days. (The tropical year is now known to be equal to 365.2422 mean solar days.) Hence the Egyptian calendar got astray one month every 121 years, or one year every 1461 years. This period, during which the New Year's day of the Egyptians traveled all round the calendar, was known to Greek and Roman writers in the first century B.C. Would that Julius Caesar when revising the Calendar — the Julian being the basis of the Gregorian, or modern calendar — had followed the simplicity of the Egyptian calendar, at least in keeping each month of a uniform number of days, and would that pride and jealously had not robbed poor February of two days! The modern calendar is very awkward and inconvenient, and there have been many attempts suggested for the purpose of revising it. Most of the modern revisions proposed call for an extra day in the year, or two extra days in leap years, which days are not to be included in the reckoning of any week or any month. Attention should be called to the fact that the five extra days of the Egyptian year — recorded as early as the Vth dynasty — were regarded as so peculiarly unlucky that there is not a single instance, among the many thousands of inscriptions brought to light, of any contract being entered into on any one of the five complementary days, nor has any event of importance ever been recorded as taking place on one of these unlucky days.

The brilliant skies of Egypt favored the development of observations. It is generally believed that the Egyptians observed the motions of the heavenly bodies for the purpose of fixing the dates of their religious festivals. These times were probably noted by observing the times at which certain of the brighter stars arose at dawn just before the sun. For accomplishing this object no instruments were necessary.

They also determined the hours of the night by observing the meridian passage of certain stars. What is thought to be the oldest type of observing instrument in the world, known to the Egyptians as *merkhet,* was probably utilized for this purpose. It consisted of a handle supporting a plumb bob and a reed which served as a sighting vane. By means of this and with the help of the ancient Egyptian *horoscopus,* or clock, the meridian could be laid out and time determined by observation. There have been preserved the titles of several temple books, which books apparently recorded the movements of the sun, moon and stars, but unfortunately, not a single one of these records themselves have survived. The Egyptians were of an intensely practical turn of mind, and they had little desire to acquire knowledge for the sake of the knowledge itself. If the information could be utilized for any practical use it was carefully cherished and preserved, or if it could assist in their religious speculations it was then regarded of great value. If we are to judge of the accomplishments of the Egyptians by the written documents alone, we must assume that they knew little and cared less for any branch of science. But then, there are the pyramids, and the great temples such as Karnak! Surely the architects and engineers who constructed such colossal monuments must have possessed knowledge of surveying vastly greater than is shown by any written record. The construction of the Great Pyramid apparently took place less than a century and a half after the elevation of the earliest piece of stone masonry known. Since these pyramids could have been erected only by means of great mechanical power, we are forced to the conclusion that the progress in the control of such power was greater at the thirtieth century B.C. than at any other age of the world's history except the present. The pyramid was placed four square to the points of the compass, and leading to the sepulchral tomb there was a passage at an angle of 26° to the horizon, which has been thought furnished a corridor for observing the star which was then the pole star, α Draconis. With this assumption, an estimation has been made of the possible age of the pyramids. According to C. Piazzi Smyth, the passage in the pyramid facing the south was used for the observation of the meridian passage of the stars and planets, and it is even possible that the moon and sun were similarly observed. Proctor thinks that if the Egyptians had utilized an opaque screen with a small round hole in it at the southern end of this gallery they could even have observed sun-spots!

It is plausible to believe that the Great Pyramid was possibly designed to serve as an astronomical observatory, but the truth is that not a single competent Egyptologist supports this view. As Breasted has pointed out, the careful orientation of the pyramid with its passage directed towards the north star is in keeping with the popular belief that the king was transmuted into a star,—hence the shaft directed towards those stars in the sky which never set.

According to E. W. Maunder, the symbol of the sun-god, so frequently found on Egyptian monuments, is sufficient evidence to prove that total eclipses of the sun were observed in Egypt. The general design of this symbol is a circle with striated wings to right and to left, and with a lesser extension in the downward direction, but with nothing above the circle. As pointed out by Maunder, the extensions to right and left call to mind and resemble the form of the solar corona when the eclipse, occurring at sun-spot minimum, has the equatorial wings. This explanation is indeed plausible, and it is not impossible that it represents the truth. To use the coronal form for the symbol of the sun would have necessitated that the corona had been frequently observed by the Egyptians, and at more than one age. Such symbols could be adopted by a people only after slow awakening and continued observation. Consequently, if eclipses had been so frequently observed it would be rational to expect that records of some of them would have been preserved. Considering the frequency with which eclipses were observed in Babylon, it is surprising in the highest degree to find not a single reference anywhere in Egyptian antiquity to an eclipse, either of the sun or of the moon. We are therefore forced to the conclusion that the Egyptians had little part in the development of the science of astronomy.

The New Empire beginning with the XVIIIth dynasty marks the golden period of Egyptian life. This was the age of conquest with an empire stretching from the Euphrates to the fourth cataract of the Nile. Egypt had become a great military empire whose world power lasted from the early sixteenth to the twelfth century B.C., or over four hundred years. The capital was the once vast city of Thebes constructed on both sides of the river Nile, the first great monumental city built by man. On the east side of the river is the temple of Karnak which is nearly a quarter of a mile long and was nearly two thousand years in the course of construction. The obelisk of Queen Hatshepsut, the first great woman queen of history, is still standing. This is about one hundred feet high and weighs 350 tons. The first of the world's great generals was Thutmose III, who ruled for fifty years beginning about 1500 B.C., and on the walls at Karnak are inscribed the stories of his great exploits.

The recent discovery of the tomb of Tut-ankh-Amen has brought into sudden prominence that portion of Egyptian history immediately following the reign of Thutmose III. While little light so far has been thrown on the actual historical facts of the lives of Tut-ankh-Amen and his im-

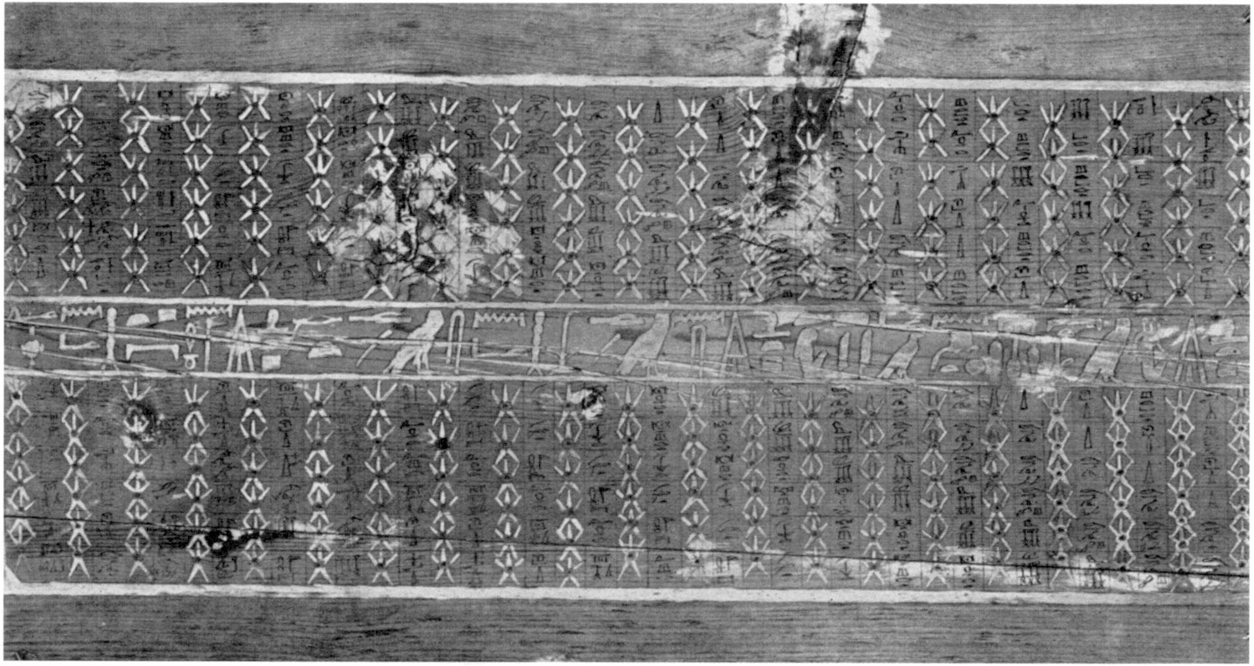

AN ANCIENT EGYPTIAN CALENDAR

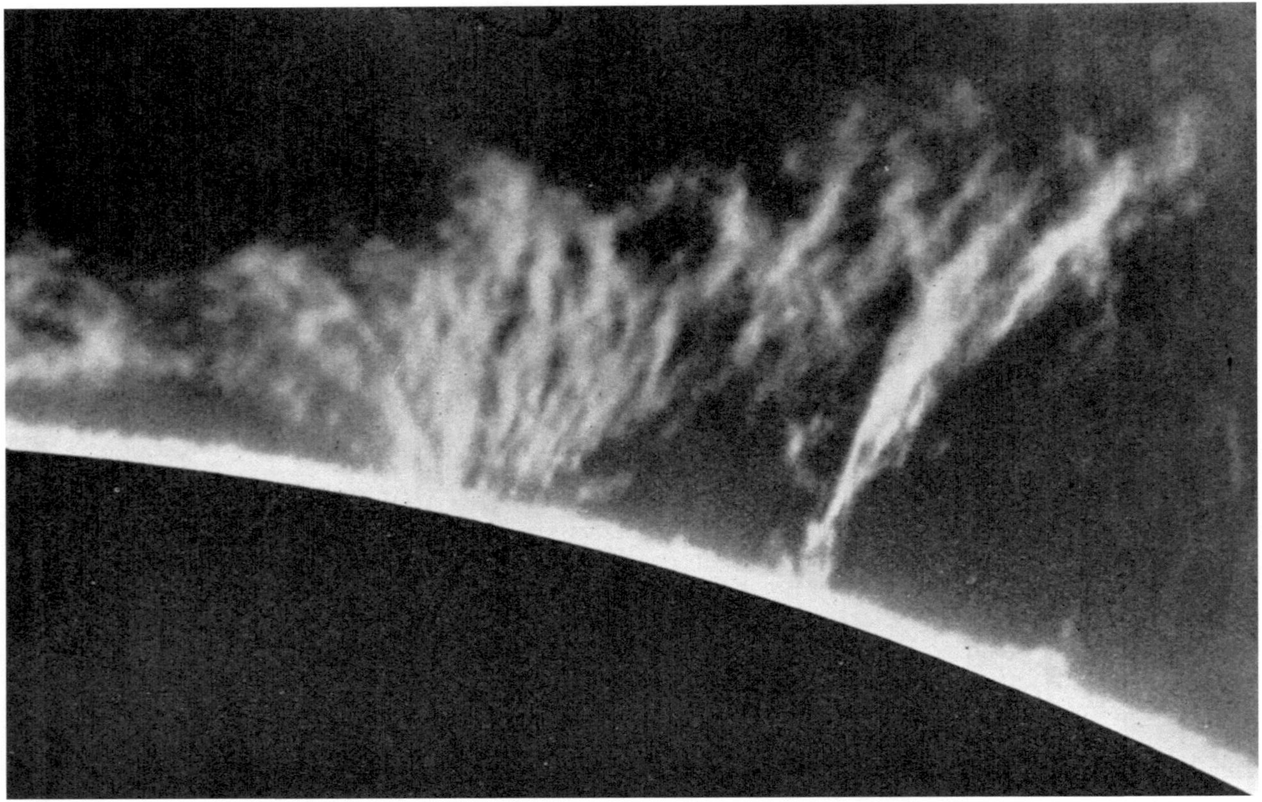

QUIESCENT SOLAR PROMINENCES 80,000 MILES HIGH
Photographed at Mt. Wilson Observatory.

mediate predecessors, the incomparable beauty of the objects of art found within the tomb bids fair to revolutionize the history of Egyptian art and literature. It is to the influence of the religious upheaval brought about by Khu-n-aten, the successor of Thutmose IV, and reputed father-in-law of Tut-ankh-Amen, that the revolution in artistic and literary expression is due. This monarch, known as the "heretic king," physically a weakling, mentally a poet and dreamer, revolted from the polytheistic religion of his forefathers and founded a new faith which deified the sun as the giver of all light and life. This worship, with Aten the sun as its visible symbol, had many points in common with Christianity and has been regarded as the most beautiful of the ancient religions. In his enthusiasm for the new faith, however, Khu-n-aten allowed the material needs of his empire to go unheeded to such an extent that at his death the confusion and distress throughout Egypt was so great that his successor was forced to give way to the strong political power of the priests of the old religion, and Amen once more became the chief Egyptian god. It was in the shadowy period between Khu-n-aten and Rameses I that Tut-ankh-Amen's brief reign occurred.

The oldest astronomical instrument in the world has recently been discovered by Breasted. The interest in this remarkable discovery is all the greater for the reason that it is believed that the instrument was actually made by no less a personage than King Tut-ankh-Amen himself. This object, a *merkhet,* consists of a rectangular strip of ebony wood 10 x 1 x $\frac{1}{2}$ inches. Along each edge, extending from end to end, is an inscription which states that the instrument was made by King Tut-ankh-Amen, "with his two hands," as a restoration of a similar monument stolen from the tomb of his ancestor, Thutmose IV. Indeed, the tomb of the latter contains a remark in ink on the wall stating that his tomb had been restored (of course after robbery) by Harmhab, who was practically the successor of Tut-ankh-Amen.

Apparently the ancient Egyptians were quite accustomed

to despoiling the tombs of their ancestors; that of Thutmose IV had already been robbed during the reign of Tut-ankh-Amen while that of the latter king was soon to follow a similar fate.

The XIXth dynasty began with Rameses I, whose reign, according to Breasted, commenced about 1315 B.C. This king reigned only two years, but he had planned and had started construction on the great colonnaded hall at Karnak, the greatest of such edifices ever constructed by man. It is 338 feet long and 170 feet wide. There are one hundred and thirty-six columns in sixteen rows, those in the nave being higher than the rest. Rameses shared his throne with his son Seti I for one year, and Seti continued the gigantic plans of his father. In addition he built a splendid temple at Abydos, and a magnificent tomb in the Valley of the Tombs of the Kings. This galleried tomb is the very finest of its kind. The decorations are wonderfully charming, full of life and color though perhaps lacking some of the strength and character of earlier art. The outlines are forceful, the postures natural and the coloring beautiful. Egyptian art had reached the pinnacle. The Metropolitan Museum of Art of New York has been conducting operations at this magnificent tomb of Seti I. The photographs of the tomb examined with care give glimpses of the glories of the tomb and show some of the decorations which include the Lion surrounded by stars and the Bull. A careful scrutiny of photographs of this and other tombs appears to show that the stars connected with the Lion and the Bull in no way refer to the well-known zodiacal constellations of Leo and Taurus. In fact, the signs of the zodiac were not acquired from the Chaldeans until fairly late in Egyptian history, for there is no record of them until after the Ptolemaic period. If the Grand Pyramid had been in Babylon instead of Egypt it is safe to say that it would have been used as an astronomical observatory. The Egyptians were not interested in systematized observations, and throughout the long years of their interesting history they contributed but little to the *science* of astronomy.

CHAPTER II

BIBLICAL AND CLASSICAL ECLIPSES

IT IS safe to say that there is certainly one allusion in the Holy Scriptures to a solar eclipse, with one or two others possible. In Amos viii, 9, appear the words, " I will cause the sun to go down at noon, and I will darken the Earth in the clear day." The language is so unmistakable and gives such a precise description of an eclipse of the sun that commentators have generally agreed that such a phenomenon must have taken place. This and other Scriptural references have been so fully and so carefully investigated by Chambers in *The Story of Eclipses,* and by Johnson, *Eclipses, Past and Future,* that a brief reference only will be given here.

The date set down in the margin of certain Bibles opposite this passage in Amos is 787 B.C. Eclipses were visible in the neighborhood of Samaria in the years 791 B.C., 771 B.C., 770 B.C. and 763 B.C. This last mentioned is the eclipse of Nineveh already described, an eclipse which was visible also in Samaria. There seems no doubt whatever that this is the eclipse predicted by Amos. It therefore becomes necessary to lower by twenty-four years the date given in Amos; and if this is done there is a ready explanation of the story in the life of King Hezekiah, as given in II Kings xx, 11. The Old Testament thus reads: " And Isaiah the prophet cried unto the Lord: and he brought the shadow ten degrees backward, by which it had gone down in the dial of Ahaz." Ahaz was the father of Hezekiah, and the " dial " was probably a sun-dial similar in construction to the ancient sun-dials of masonry still existing at Benares and Delhi in India.

The Biblical peoples at this period of their history were divided into two. In the North, the land was rich and fertile and great prosperity and wealth abounded. The people lived in large towns, they dressed and lived extravagantly. Moreover these people of Israel had adopted the gods of the Canaanites, each town in fact having its own god, or " baal." To the South, the land was poor, encroaching as it did on the desert, and the people of Judah had to struggle for their existence. The only large town was Jerusalem. The Jews were still faithful to the Hebrew God, Jehovah. The shepherd Amos being a devout follower of Jehovah made a journey to Israel, and thundered his denunciations against their many gods, and against their gaudy clothes, their licentiousness and harsh treatment of the poor.

About this time the Hebrews were beginning to learn to write, and they were abandoning the clay tablet and the cuneiform inscription of the Assyrians and Babylonians. They wrote on papyrus and with pen and ink, which method of writing they had acquired from their former masters, the Egyptians. However, they borrowed the first alphabet the world had developed, that of Phoenicia. The papyrus rolls written by Amos and other historians have descended to us as the Hebrew Scriptures. On account of the evil lives of the people of Israel, Amos predicted their destruction. Damascus was taken by the powerful Assyrians in 732 B.C., and thus being unprotected on the north, Samaria, the capital of the kingdom of Israel, was captured in 722 B.C., the Israelites being taken away captive and their nation detroyed.

Shortly before 700 B.C. arose the great prophet Isaiah. In one great oration after another he exhorted the people of Jerusalem to believe in their God, Jehovah, and not be dismayed for He would deliver them from the hosts of the powerful Assyrian, Sennacherib, who was then threatening to batter down the city. The king of Jerusalem, Hezekiah, was sick unto death, and the same tragic fate as had befallen Damascus and Samaria seemed to be awaiting the people of Jerusalem. In answer to a prayer of Hezekiah for recovery the prophet was sent to him with this message: " Thus saith the Lord, the God of David thy Father, I have heard thy prayer, I have seen thy tears: behold I will add unto thy days fifteen years . . . and I will defend this city, and this shall be a sign unto thee from the Lord, that the Lord will do this thing that He hath spoken. Behold, I will bring again the shadow of the Sun, which is gone down in the sun-dial of Ahaz ten degrees backward, by which degrees it had gone down." (Isaiah xxxviii, 5–8.)

There was a large partial eclipse of the sun visible in Jerusalem on January 11, 689 B.C. If we accept the correction of twenty-four years derived from the record of the eclipse of Nineveh, then according to Chambers (*loc. cit.*), this eclipse of Jerusalem will completely satisfy the Biblical narrative at all points. History tells that a pestilence from the marshes of the Nile caused great havoc in the army of Sennacherib, and thus was Jerusalem miraculously spared. See also in this connection, II Kings, xix, 32–37. About a century later the Hebrews rejoiced over the destruction of Nineveh in 606 B.C. But unfortunately, the days of Jerusalem were numbered. The Chaldeans followed Assyria as masters of Palestine, Jerusalem was destroyed in 586 B.C. under Nebuchadnezzar, and the people were carried away into exile in Babylon.

The dates herewith given in Babylonian and Assyrian history admit of no uncertainty since they are determined by eclipses. As a consequence, the dates appearing in the Bible must be altered to fit those of verified history. It would be interesting to follow the history of the Hebrews through the lives of Jeremiah and other prophets — but space forbids. The reader is referred to Breasted, *Ancient Times, a History of the Early World,* and also to many other books on this period of history which is of the greatest of interest to all Christian peoples.

Eclipses of Classical Literature

It will be possible to give an account here of but a few of the eclipses referred to in the Classics, and we shall begin with Homer. On the day of the slaughter of the suitors, there is a passage in the *Odyssey* (v. 351–357), which probably refers to an eclipse of the sun. The lines run:

"Ah, wretched men, what evil is this that you suffer? Shrouded in night are your heads and your faces and your knees beneath you; kindled is the sound of wailing, bathed in tears are your cheeks, and sprinkled with blood are the walls and the fair rafters. And full of ghosts is the porch and full the court, of ghosts that hasten down to Erebus beneath the darkness, and the Sun has perished out of heaven, and an evil mist hovers over all."

According to Fotheringham (*loc. cit.*), the words, "The Sun has perished out of heaven," must refer to a total eclipse, and in fact most commentators, including Plutarch and Eustathius, agree in this interpretation. Apparently, the eclipse was contained in the legend as it descended to Homer, — though naturally it is impossible to fix the exact date of the eclipse.

The next eclipse is one referred to by the Greek poet Archilochus, a portion only of whose work has come down to us. The lines are: "Zeus, the father of the Olympic Gods, turned mid-day into night, hiding the light of the dazzling Sun; and sore fear came upon men." According to the investigations of Oppolzer and Cowell, the poet must have witnessed the eclipse of April 6, 648 B.C., the eclipse being total about 10 A.M. at Thasos and in the northern part of the Aegean Sea. We know that the poet spent part of his time at Thasos, and if his statement actually gives the description of an eclipse, it thus furnishes the earliest date in Grecian chronology to be fixed with certainty. The early accepted dates must accordingly be reduced by fifty years, but no surprise need be felt as they were known only with great uncertainty.

Eclipse of Thales

The most celebrated eclipse of all history is that connected with the name of Thales of Miletus who lived from 640 to 546 B.C. He was the founder of Greek astronomy, geometry and philosophy. He was regarded by the Greeks with great veneration, and he was the chief of the seven "wise men," a reputation which rested not only on his scientific eminence but also on his political sagacity. According to the historian Herodotus (i, 74), the reference to the eclipse is as follows: "There was war between the Lydians and the Medes for five years, each won many victories from the other, and once they fought a battle by night. They were still warring with equal success, when it chanced, at an encounter which happened in the sixth year, that during the battle the day was turned into night. Thales of Miletus had foretold this loss of daylight to the Ionians, fixing it within the year in which the change did indeed happen. So when the Lydians and Medes saw the day turned to night they ceased from fighting, and both were the more zealous to make peace." The truce concluded was cemented by a double marriage, "for," adds the historian, "without some such strong bond, there is little of security to be found in men's covenants." The same eclipse is referred to by both Pliny and Cicero.

The narrative by Herodotus contains two statements, that the eclipse was total (a "night battle"), and that the eclipse was predicted by Thales. It is in regard to the second fact that the controversy has arisen among astronomers. Various dates for the eclipse have been assigned, from 625 B.C. to 583 B.C. But after careful researches of many competent authorities — Airy, Hind, Zech, Hansen, Ginzel, Newcomb, Cowell, Fotheringham and others — the date of May 28, 585 B.C., has been fixed, the eclipse taking place in the afternoon. The general consensus of opinion is that Thales was probably familiar with the Chaldean Saros of 223 lunar months or 18 years 11 days. In fact, by some it has been thought (*see,* George Smith, *Assyrian Discoveries*) that the Babylonians were actually making use of the Saros for the prediction of eclipses. Herodotus does not claim that the day and hour of the eclipse were predicted nor yet the locality where it was to be visible, these being much more difficult problems and incapable of being solved by Thales. The eclipse was predicted only "within the year." It is therefore not necessary to credit Thales with extraordinarily abnormal powers, or to assume that he was in possession of information which was known first to Hipparchus more than four hundred years later. Apparently the Greek " wise man " utilized the eclipse of May 17, 603 B.C., for the purpose of his prediction.

Probably the greatest difficulty of all concerning the identification of this eclipse of Thales has been to know just how much credence could be placed in the account of Herodotus. Some of the ancient writers go so far as to accuse him of " intentional untruthfulness." Modern critics while not going to this extreme, none the less feel that his love of effect and his loose and inaccurate habits of thought make of him an attractive writer, but a poor historian. A case in point might be cited, the eclipse which is described by Herodotus (vii, 37) as follows: " At the first approach of spring, the army quitted Sardis and marched towards Abydos; at the moment of its departure the Sun quitted its place in the heavens and disappeared, though there were no clouds in sight and the day was quite clear; day was thus turned into night." We are told that " As the King was going against Greece, and had come into the region of the Hellespont, there happened an eclipse of the Sun in the East; this sign portended to him his defeat for the sun was eclipsed in the region of his rising, and Xerxes was also marching from that quarter." The generally accepted date of the battles of Thermopylae and Salamis is 480 B.C. No eclipse occurred in the spring of that year. Many attempts have been made to find an eclipse that would harmonize with the statement of Herodotus, but all of no avail, — unless we are willing to make changes either in the date or in the narrative. It seems necessary to conclude that the eclipse incident adds greatly to the attractiveness of the story, — but that it is not history. Unfortunately, even at the present day, authors still try to embellish their stories with celestial phenomena, but do not stick too closely to the facts. One of the " best sellers " appearing recently in the field of fiction had in it no less than four astronomical blunders.

The next eclipse to be mentioned was seen in Athens in B.C. 431, August 3. It is described by Thucydides (ii, 28) who states that during the Peloponnesian War " things formerly repeated on hearsay, but very rarely confirmed by facts, became not incredible, both about earthquakes and eclipses of the Sun which came to pass more frequently than had been remembered in former times." An eclipse occurred in the first year of the war, "in the same summer, at the beginning of a new lunar month (at which time alone the phenomenon seems possible), the Sun was eclipsed after midday, and became full again after it had assumed a crescent form and after some of the stars had shone out." The account, differing from the flowery but uncertain language of Herodotus, is clear and definite, and refers to an eclipse which evidently was not total at Athens where Thucydides was supposed to be. The only difficulty in fixing the eclipse has arisen from the last part of the statement, that stars were visible. Venus was 10° distant and undoubtedly must have been readily seen, but what other star, or stars? The eclipse in B.C. 431 was seven-eighths total. To make a long story short, it sufficeth to say that at the eclipse of April 8, 1921, which was eight-ninths total at Oxford, the star Vega was readily seen. It must be concluded, therefore, that the account of Thucydides is strictly in accord with the facts and that the *stars* Venus, Vega and probably Jupiter 43° distant were seen.

This same eclipse has an interesting anecdote connected with it which has been narrated by Plutarch in his *Life of Pericles* who was the commander of the Grecian naval forces. "The whole fleet was in readiness, and Pericles on board his own galley when there happened an eclipse of the Sun. The sudden darkness was looked upon as an unfavorable omen, and threw the sailors into the greatest consternation. Pericles, observing that the pilot was much astonished and perplexed, took his cloak, and having covered his eyes with it, asked him if he found anything terrible in that, or considered it a bad presage? Upon his answering in the negative, he asked 'Where is the difference, then, between this and the other, except something bigger than my cloak causes the difference?'"

Another eclipse recorded by Thucydides (iv, 52) has been identified with the annular eclipse of March 21, 424 B.C.

The last of the eclipses described by Thucydides (vii 50)

is one of the moon, and it has a tragic story connected with it. The Athenian fleet and army were engaged in an attack on Syracuse, the fleet being under the command of Nicias, and Demosthenes had arrived with large reinforcements for the army. The latter had failed in his purpose of capturing a wall which the Syracusians had thrown across the Athenian lines, and as a result of this failure it had been decided to withdraw the whole Athenian forces. On the night of August 27, 413 B.C., the whole force had embarked and was ready to flee. Before a start was made, however, an eclipse of the moon took place which became total. Nicias was greatly terrified, for it seemed easy to understand the cause of an eclipse of the sun, but very difficult to make out "how the moon, when at the full, should suddenly lose her light and assume such a variety of colors." The soldiers and sailors added their terror to that of the commander and as a result the soothsayers were consulted. The advice was to remain for thrice nine days, — but alas, before that time the Athenians had been utterly routed, and Nicias and Demosthenes were both executed.

Plutarch records an eclipse of the sun which took place July 13, 364 B.C., and both Plutarch and Pliny give an account of a total eclipse of the moon September 20, 331 B.C., eleven days before the victory of Alexander over Darius at Arbela in Assyria.

Another interesting eclipse of the sun is connected with the name of Agathocles, the tyrant of Syracuse. The narratives of the two historians Justin and Diodorus Siculus agree so well that the astronomer has been able to fix exactly the circumstance of the eclipse. In 310 B.C. Syracuse was again besieged, this time not by Athens, but by Carthage. On the evening of August 14, Agathocles slipped out of the harbor with his whole fleet of sixty ships. He was pursued by the Carthaginians, who abandoned the chase at night-fall. According to the historical accounts, there was an eclipse of the sun on the following morning, and as a consequence day was turned into night, and stars appeared everywhere in the firmament. The soldiers were at first afraid, believing the eclipse to be an ill omen, but their fears were quickly calmed by their commander who, according to Justin, told them that eclipses "always signify a change of affairs, and therefore some change was certainly signified, either to Carthage, which was in such a flourishing condition, or to them, whose affairs were in a very ruinous state." Apparently, these words of Agathocles seemed to have inspired his men that an eclipse might even be an omen for good. At any rate, he was able to make good his escape from the fleet of the Carthaginians, and after six days and nights he succeeded in landing on the coast of Africa and in devastating the territories of proud Carthage.

The readiness with which Agathocles and Pericles quieted the fears of the sailors calls to mind the genius of Columbus who, on more than one occasion, had to quell mutinies among the members of his little fleet, they being greatly perturbed on discovering that the compass needle did not point to the true north as given by the pole star.

Mention of only one other eclipse before the beginning of the Christian era will be made here. An eclipse is generally regarded to have taken place when Julius Caesar was crossing the Rubicon on his triumphal return to Rome from Gaul. But it seems to be necessary to class this eclipse along with that connected with the name Xerxes, — for it failed to occur.

In the investigations into the authenticity and accuracy of the early eclipses, the greatest difficulty has arisen from the vagueness and uncertainties of the account of the chronicler. References to eclipses are couched in such hazy language that the account might have been equally well applied to some other rare phenomenon. Writers of literature in all ages seem to prefer flowery to straightforward language and employ imagery rather than fact, refusing to "call a spade, a spade." For the astronomer it is impossible to take one interesting eclipse and decide on the *pros* and *cons* in favor of this or of that interpretation. Eclipses depend on the motions of the sun and moon, and the history of all eclipses must make a harmonious story and not a disjointed one. And so the eclipse of Agathocles must be considered in reference to that of Thales, and that in turn to the eclipse of Nineveh in 763 B.C. The astronomer is not a dealer in necromancy nor does he acquire information about the motions of the sun and moon by mystical incantations, or by divination. The work of the scientist is as far removed as possible from the mountebank and charlatan. Scientific knowledge is acquired only by the patient study of all that has been accomplished in the past, and happy and fortunate is he who can make a correct interpretation of the past! The manner in which the story of eclipses has added to astronomical knowledge will be treated in a future chapter. Those who wish additional information regarding the early eclipses would do well to consult *Encyclopaedia Britannica,* article on *Eclipses* by Simon Newcomb, but specially the monumental work by Oppolzer, *Canon der Finsternisse,* which gives a record of no less than 13,200 eclipses, of which 5,200 are of the moon and 8,000 of the sun. These eclipses are the ones which have taken place since 1208 B.C., over three thousand years ago, or which will take place before the end of the year 2162 A.D., two hundred years in the future. In addition to the tables of the astronomical elements of each eclipse, 160 charts are furnished giving the location on the earth's surface where each total and annular eclipse of the sun will be visible. It should, however, be pointed out that the eclipse tracks could not be located by Oppolzer on the maps with the greatest of refinement, since it was manifestly impossible to calculate the location of each path for more than three positions, sunrise, noon and sunset. Consequently, it need not be a matter of surprise, particularly with the early eclipses, that the tracks may at times be incorrectly placed by as much as a hundred miles.

CHAPTER III

THE PREDICTION OF ECLIPSES

AFTER the destruction of Babylonia and Assyria, astronomy migrated from its Chaldean home to Greece. As early as the fifth century B.C., attempts were made to demonstrate some of the practical uses of astronomy in the regulation of time and particularly in the arranging of the calendar. The Greeks had inherited from the Chaldeans a calendar founded on the lunar month, but since the month is 29.5 days in length, attempts to bring the seasons and the calendar into harmony ended always in hopeless confusion. The astronomer Meton (born about 460 B.C.) made the discovery that 19 solar years were very nearly equivalent in length to 235 lunar months — the difference, in fact, being only about half an hour. The Metonic cycle, as it has since been called, has been of the very greatest service since its discovery so many centuries ago. It should not be confounded with the interval called the Saros, since the Metonic cycle is not used for the prediction of eclipses. After nineteen years, new and full moon and the various phases are repeated with a considerable degree of exactness. This cycle is still used for finding the day on which Easter will fall.

The Chaldeans had confined their energies to the making of astronomical observations, but their successors, the early Greek philosophers, made practically no observations whose record is worth keeping. On the other hand they were much interested in inquiring into the causes of things, and accordingly it may be said that the *science* of astronomy had its birth with the Greeks. We are already familiar with Thales and part of his work. To Pythagoras we owe the doctrine of the "music of the spheres," a notion which has descended even to the present day. To Plato (428–347

B.C.) astronomy owes but little, he even going so far as to decry, as degrading, the observation of the heavenly bodies. Far different, however, was the case with Aristotle (384–322 B.C.) who appears to have collected and systematized the best thought of the time regarding astronomy. He believed that all of the heavenly bodies, including sun, earth and moon, were spherical; he taught that the moon had no light of its own but shone by reflected sunlight and he explained its phases. Most important of these conclusions was the proof of the spherical shape of the earth. This, as Aristotle demonstrated, depends on two facts: first, that the earth casts a shadow and that an eclipse of the moon is due to the passing of the moon into this shadow; and second, that the shadow being circular in outline was proof that the body casting the shadow, the earth, was circular, at least in the section turned towards the moon.

After the time of Aristotle, Greek astronomy moved its home to Alexandria, and there under the protection of successive Ptolemies the great museum was erected and a library and an observatory were incorporated as important adjuncts. It is not the purpose here to trace the development of astronomy except as it progressed through the study of eclipses. The greatest astronomer of antiquity, in fact the most famous up to the time of Newton, was Hipparchus. He was not of Alexandria, though he probably visited the city and may have made some of his observations there. Little is known of his birth. This may have taken place in Bithynia or in Rhodes, though it is known that he had an observatory in the latter place and there did most of his work. We know of him mainly through the astronomer Ptolemy, since all of the original works of Hipparchus have been lost with the exception of one important book. His great fame rests upon three pedestals: (1) The invention of trigonometry, (2) the making of an extensive series of careful observations, (3) the comparison of his own with earlier observations. As a result of his exact and methodical work, he discovered the precession of the equinoxes, he made the first catalogue of the positions of stars (to the number of 1080), and he contributed greatly to the theory of the motion of the sun and moon, and, as a direct consequence, to the subject of eclipses.

Hipparchus was the first to notice that the seasons were of unequal length, the time from vernal equinox to summer solstice being 94 days, while the summer season was only 92½ days in length. Spring and summer together are seven days longer than fall and winter. Hipparchus gave the correct explanation of this fact, viz: that the earth is not always at a constant distance from the sun, being removed from the center of the orbit (which was then regarded as circular). At the present day this is readily verified by the change in the angular diameter of the sun, — but this confirmation in Hipparchus' time would have required measurements too exact for his crude instruments. By noting the time it takes the moon to pass through the earth's shadow at the time of a total eclipse of the moon, Hipparchus was able, by a method due to Aristarchus, to obtain a value of the relative sizes of earth and moon, and also to estimate the distance of the moon to be 59 times the terrestrial radius, — which is not very far from the truth.

The motion of the moon was much more difficult to investigate than that of the sun. Little observation is necessary to note that the moon moves eastward in the sky, changing her position among the stars by her own diameter in approximately an hour. A revolution of the moon from a star to the same star again is known as a *sidereal* month, and its length is 27.3 days. In this interval the earth has proceeded in its orbit about the sun, and consequently the moon, as seen from the sun, requires more than two additional days to swing into line with the earth. The interval of time from new moon to new moon, or from full moon to full moon, is the ordinary month of 29.5 days, called the *synodic* month. Further observation shows that the sun is always in the ecliptic, in fact, the sun's apparent motion determines the ecliptic, whereas the motion of the moon, although approximately in the ecliptic, is nevertheless inclined at a small angle, which Hipparchus for the first time fixed as at an angle of 5°. The moon's path thus crosses the ecliptic at two points called the nodes. If the moon's path were exactly in the plane of the ecliptic there would be two eclipses each month, once at the time of new moon when the moon would come between the earth and the sun, bringing about an eclipse of the sun; and again at full moon, when the moon would pass into the shadow cast by the earth and be itself eclipsed. But as the moon's orbit is inclined to the ecliptic, it is manifest that eclipses, either of sun or moon, can occur only when the moon is near the plane of the ecliptic, or, in other words, near its node. If the length of time it takes the moon to return to its node were the same as the sidereal month, it is evident that the moon's node would be fixed in space and would have no motion relative to the stars. The length of the month determined by the return of the moon to the node is called the *nodical,* or the *draconic* month. The meaning of the first designation is evident. But why the second? It is not difficult to trace it back to the ancient belief that the sun was swallowed by a dragon at the time of an eclipse, and this superstition is therefore the reason why the month on which eclipses depend should be called draconic. Indeed the symbols that are universally used by astronomers to denote the ascending node (☊) and the descending node (☋) are generally supposed [1] to represent the head and tail of the dragon. Hipparchus found that the moon's nodes were not fixed, but that they completed a revolution in the plane of the ecliptic from east to west in about 19 years. A fourth kind of month was likewise known to Hipparchus, the length of a revolution from the position of perigee to perigee in the moon's orbit. By making use of the eclipses observed by the Chaldeans, Hipparchus was enabled to obtain greatly improved values of the lengths of the various months. The values that follow are those furnished by the recent investigations of E. W. Brown:

Synodical	=	$29.^{d}530588$	=	29^{d}	12^{h}	44^{m}	$2.^{s}8$
Sidereal	=	27. 321661	=	27	7	43	11. 5
Anomalistic	=	27. 554550	=	27	13	18	33. 1
Nodical	=	27. 212220	=	27	5	5	35. 8

[1] See Berry, *A Short History of Astronomy,* p. 48.

The lengths of the different kinds of year will be inserted here for future reference, the values being Newcomb's:

Tropical (ordinary)	=	$365.^{d}242199$
Eclipse	=	$346.\ 620031$

Since the motion of the moon's node is in the direction to meet the oncoming sun, the interval of time from node to node (the eclipse year) is less than the interval from vernal equinox to vernal equinox, the ordinary year on which the seasons depend. From the above figures, the eclipse year is 18.62 days shorter than the tropical year. Since the sun and earth are always in the plane of the ecliptic, all that is required to permit the prediction of an eclipse of the sun is to find *at the time of the new moon* (when alone a solar eclipse can take place) whether the moon is sufficiently near enough to the plane of the ecliptic for her to come between earth and sun. And as an eclipse of the moon can happen only *at full moon,* similar investigations will furnish the means of predicting lunar eclipses. With the modern values given above, these predictions can be carried out remarkably easily. Let us see:

19 eclipse years	=	$6585.^{d}\ 7806$
223 ordinary months	=	$6585.\ 3211$
247 nodical months	=	$6585.\ 3572$
239 anomalistic months	=	$6585.\ 5374$

The integral parts of these four quantities are the same, the second being the *Saros* (meaning " repetition ") which amounts to 18 years 11 days, if only four leap years intervene, or 18 years 10 days if the 29th of February has come five times. The eclipse of June 8, 1918 was a repetition of that of May 28, 1900; the eclipse of September 10, 1923 followed that of August 30, 1905 by the interval of one Saros. The author observed the 1900 eclipse in Georgia and that of 1918 in Oregon. It was necessary for him to travel to Spain to witness the 1905 eclipse and to California for the one of 1923. Succeeding eclipses in the Saros are visible each time from a location on the earth's surface farther west. The reason for this, and the causes of further differences in eclipses which follow each other are readily explained from the fact that the numbers given above are not exactly equal to each other but differ in the decimal parts. Before going into this in detail, it will be well to take up some of the geometrical features useful in the prediction of eclipses.

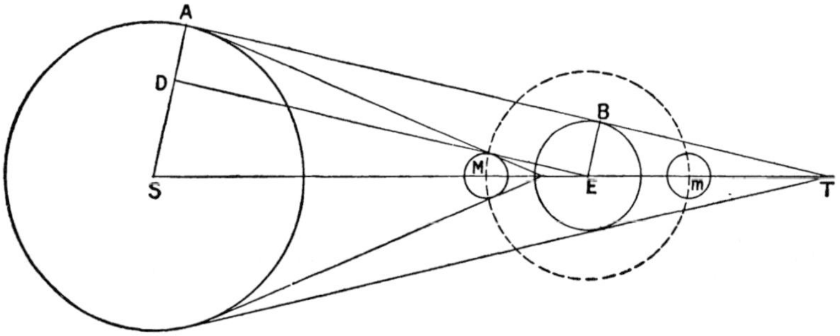

FIG. 1

The length of the shadow cast by earth or moon can be readily found. S is the center of the sun, E that of the earth, and M and *m* the moon's center at new and full moon, respectively. A shadow cone envelopes the sun and earth, its vertex being at T. If the moon in her orbit passes wholly into the shadow there is a *total eclipse of the moon.* If the line ED is drawn parallel to TBA, then it is seen that the triangle DSE and BET are similar, and hence

$$\frac{ET}{EB} = \frac{SE}{SD}$$

But $SD = SA - AD = R - r$, if R and r denote the radii of sun and earth respectively. SE is the distance of sun to earth which equals Δ. Hence the length of the shadow,

$$ET = \frac{r}{R - r}\Delta$$

But R, the radius of the sun, is equal to 109.5 times the radius of the earth, r, and accordingly, the length of the earth's shadow is equal to

$$\frac{\Delta}{108.5}$$

Assuming the average value of the distance from sun to earth as 92,900,000 miles, then the length of the earth's shadow is on the average 857,000 miles, the distance of the earth to moon being much less than this, or 238,840 miles.

In quite a similar manner it is possible to find the length of the shadow cast by the moon, by taking the moon instead of the earth, and assuming C as the length of the moon's radius, and D the distance from centers of sun to moon. The length of the shadow cast by the moon is equal

$$\frac{C}{R-C} \cdot D$$

But R and C, the radii of the sun and moon, are each constant, and substituting the values it is found that the length of the moon's shadow is $\frac{1}{399.55}$ of the moon's distance from the sun, or approximately one four-hundredth part of the distance. Under the conditions given a *total eclipse of the sun,* the distance from sun to moon, D, is equivalent to the difference between the distances from sun to earth and from earth to moon. On account of the fact that the orbits of earth and moon are both ellipses and not circles, the distance D will vary considerably and the length of the moon's shadow will change proportionally. Under *average* conditions, the length of the shadow measured from the center of the moon is 231,650 miles, but this may vary about 4000 miles each way from the mean, or from 235,700 miles as a maximum to 228,120 miles as a minimum. Since the mean distance from center of earth to center of moon is 238,840 miles, or from moon's center to earth's *surface* 234,900 miles, it is seen that under average conditions the moon's shadow is not long enough to reach the earth's surface. The distance D will have its greatest value when the earth is at its greatest distance from the sun, at aphelion, which takes place about July 1 of each year, and the moon at its least distance from the earth, or at perigee. Owing to the revolution of the moon's line of apsides, perigee may come during any month of the year. The distance D is least when the earth is at perihelion (January 1) and the moon at apogee. On account of the elliptical character of the moon's orbit, the distance from center of earth to center of moon varies, from approximately 253,000 miles as a maximum to 221,600 miles as a minimum. The latter distance is 217,650 miles from the earth's surface. Since the shadow cast by the moon may be 235,700 miles in length, its vertex therefore at times may extend 18,000 miles beyond the earth's surface. Under these conditions the area of the moon's shadow cone intercepted by the earth will be a maximum, and when all conditions are most favorable, the diameter will be 163 miles. Such conditions are possible when the total eclipse is visible about July 1, with the moon at perigee, the eclipse being visible at the earth's equator at noon. The diameter of the moon's shadow path intercepted by the earth may vary from its maximum value to a vanishing width when the shadow is just long enough to reach the earth's surface. The average width is approximately one hundred miles. Inside this shadow cone an observer will see the light of the sun totally cut off; outside of it the sun will be partially eclipsed.

Since also the distance may be as great as 253,000 miles from the earth's center, or about 249,000 miles from its surface, while the moon's shadow may be only 228,120 miles in length, the shadow may fall short of the earth's surface by more than 20,000 miles. Under these conditions an observer on the earth located on the axis of the shadow produced, would see an annular eclipse and not a total one, a ring or annulus of light appearing round the edge of the moon when the eclipse is central. The diameter of the so-called negative shadow may be as great as 230 miles. Since the maximum width of the shadow causing a total solar eclipse is only 163 miles, it is evident that the number of annular eclipses are more frequent than total — only two out of five central eclipses being total.

The work of the observational astronomer has been greatly hampered by the moon which illuminates the sky and renders it impossible to see or to photograph the faintest stars in the telescope. As already has been noted, the moon has furnished much labor and vexation of spirit to the mathematical astronomer. Yet if our satellite were banished from the sky, what a dreary old world it would be to poets and lovers without their " inconstant moon "! If the diameter of the moon were decreased a scant hundred and forty miles (less than seven percent), a total eclipse of the sun would be impossible! How thankful solar astronomers should be for the strange coincidence that the angular diameter of the moon is about equivalent to that of the sun and that the diameter of the moon is not ten percent less than it is! The greatest possible excess of the angular radius of the moon over that of the sun is only 1′ 19″.

Knowing the diameter of the shadow intercepted by the earth it is easy to find the duration of totality. For this purpose it is necessary to find the speed of the shadow over the earth. The moon advances in her orbit approximately her own diameter in an hour, or more exactly, about 2100 miles per hour. On account of the great distance of the sun, this is the speed that the moon's shadow passes across the earth regarded as a whole, but on account of the rotation of the earth on its axis the velocity of the moon's shadow over the earth's surface is far different from 2100 miles per hour. The diurnal rotation causes an observer at the equator to complete a circuit of nearly 25,000 miles in twenty-four hours which is at the rate of 1040 miles per hour. Away from the equator, farther north and south in latitude, the speed is progressively less at greater and greater distances from the equator, being in fact 1040 miles per hour multiplied by the cosine of the latitude. At 30° north and south an observer is carried 900 miles per hour, at 45° only 735 miles, while at 60° this is reduced to 500 miles. The moon moves in her orbit from west to east and her shadow in consequence traverses the earth in this direction also. This being likewise the direction that an observer is carried by the earth's rotation, the moon's shadow path travels with respect to the earth's surface the difference of the two speeds, which is 1060 miles at the equator, and 1600 miles at latitude 60°.

The above values refer only to the condition that the eclipse occurs on the meridian for the observer in question, or in other words, at noon. When the eclipse takes place near sunrise or sunset, the observer being turned not directly towards the sun and moon, the projection of the moon's orbital velocity along the earth's surface may be very large. The slowest speed that the moon's shadow can have over the earth's surface is 1060 miles per hour, when the eclipse occurs at noon and at the equator, and the speed under conditions of higher latitudes and eclipse at sunrise or sunset may be as great as 4000 or even 5000 miles per hour. With the conditions causing maximum width of shadow of 163 miles, an eclipse of the sun may have the maximum possible duration of the relatively brief period of 7 mins. 31 secs. Even a six-minute eclipse, such as the Sumatra eclipse of May 18, 1901 and its repetition eighteen years later in Brazil and Africa, is considered by astronomers unusually long.

The discussion above regarding the conditions of a solar eclipse has been made with reference to the *umbra* of the moon's shadow. An observer on the earth situated within the confines of this shadow would find the sun's light totally obscured. Inside the *penumbra* of the moon's shadow, the sun is not entirely covered up by the moon and a partial solar eclipse will result. The diameter of the penumbra intercepted by the earth is readily determined. Since the angular diameters of sun and moon are nearly equal, the diameter of the penumbra measured at right angles to the line joining sun and moon is about twice the diameter of the moon, or roughly 4400 miles. Consequently, for a distance of 2200 miles along the earth's surface on either side of the moon's umbra a partial eclipse may be visible. And by a process of reasoning similar to that followed out in the case of total eclipses, it is readily perceived that in latitudes away from the equator and when the eclipse is at sunrise or sunset, the distance along the earth's surface when a partial solar eclipse is visible may be increased to 3000 miles from the central line of totality.

Again, since the sun and moon's angular diameters are about equal, and the moon moves the extent of her diameter in an hour, it will take about one hour from the first beginning of an eclipse, either solar or lunar, to the beginning of totality, or, as the astronomer expresses it, from first to second contact. Similarly an hour will elapse between third and fourth contacts, or between the ending of totality and the end of the eclipse.

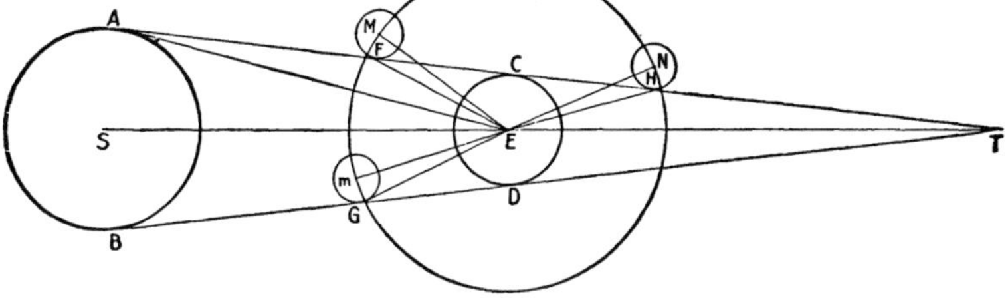

FIG. 2

For a *partial* eclipse of the *sun* to occur, it is necessary for the moon at M to encroach on the shadow-cone $ACDB$ enveloping sun and earth. The angle between centers of sun and moon under these conditions is readily determined. The angle SEM in Fig. 2 is the sum of three angles, SEA, AEF and FEM. SEA and FEM are the angular semi-diameters of sun and moon respectively, while the angle AEF is equal $CFE - CAE$. But CAE is the angle subtended by the earth's radius at the distance of the sun, and is accordingly the sun's horizontal parallax. If S and s represent the angular semi-diameter of sun and moon, and P and p the horizontal parallaxes of sun and moon respectively, then SEM is equal to $S + s - P + p$.

For the solar eclipse to be *total*, the moon must be entirely within the shadow at m, and for these conditions the angle SEm is found to be $S - s - P + p$.

For an eclipse of the *moon*, with moon at N, the angle between the moon and the center of the earth's shadow, the angle NET is equal $NEH + HET = NEH + CHE - HTE$. But $SEA = EAT + HTE$, hence $NET = s + p - S + P$.

Since the centers of the sun and earth are always in the ecliptic, the angles just determined give the amount in angle that the moon may be distant from the ecliptic at the time of new or full moon in order that there may be an eclipse of sun or moon. Angular distances measured at right angles to the ecliptic are called celestial latitudes, and hence in order that a partial eclipse may take place, the latitude of the moon must be less than

$$p + s + (S-P) \quad \text{for a } \textit{solar} \text{ eclipse,}$$
$$\text{and } p + s - (S-P) \quad \text{for a } \textit{lunar} \text{ eclipse.}$$

Since S has a mean value of 32′, while P is never greater than 9″, it is evident that a solar eclipse can take place with the moon at a greater angular distance from the ecliptic than is possible for an eclipse of the moon. This is patent by referring to Fig. 2 and noting that a section across the shadow-cone enveloping sun and earth is greater through the moon at new moon, i.e., for a solar eclipse, than at full moon, or for a lunar eclipse.

For the conditions under which total eclipses may occur, we find that the celestial latitudes of the moon at the time of conjunction of sun and moon are (by changing $+ s$ to $- s$ in the above formula)

$$p - s + (S-P) \quad \text{for } \textit{total solar} \text{ eclipse,}$$
$$\text{and } p - s + (S-P) \quad \text{for } \textit{total lunar} \text{ eclipse.}$$

On account of the varying distances of sun and moon from the earth, the values of the semi-diameters and parallaxes of sun and moon are changing proportionally. In the table below the values are taken from the *American Ephemeris:*

		Greatest	*Least*	*Mean*
Semi-diameter of sun	$= S$	16′ 18″	15′ 46″	15′ 59″.6
Semi-diameter of moon	$= s$	16′ 46″	14′ 43″	15′ 32″.6
Horizontal parallax of sun	$= P$	8″.9	8″.7	8″.8
Horizontal parallax of moon	$= p$	61′ 28″	53′ 55″	57′ 2″.7
Inclination of moon's orbit	$= i$	5° 19′	4° 57′	5° 8′ 43″

An eclipse, either of sun or moon, might be predicted by finding from the above formulas the moon's celestial latitude. For purposes of computation it is more convenient to find the angular distance that the sun or moon may be from the moon's node, but measured in the plane of the ecliptic. These angles, being celestial longitudes, determined at the time of new moon give what is known as the *solar ecliptic limit,* and at full moon furnish the *lunar ecliptic limit.* In Fig. 3, *NS* is the ecliptic, *MN* the moon's path, *N* the moon's ascending node. *M* is the moon, and the circle at *S,* for a solar eclipse, represents a section of the shadow cone at *M* (in Fig. 2), and, for a lunar eclipse, a section through *N* (in Fig. 2), The angle *SNM* represents the angle, *i,* of inclination of the moon's orbit to the plane of the ecliptic.

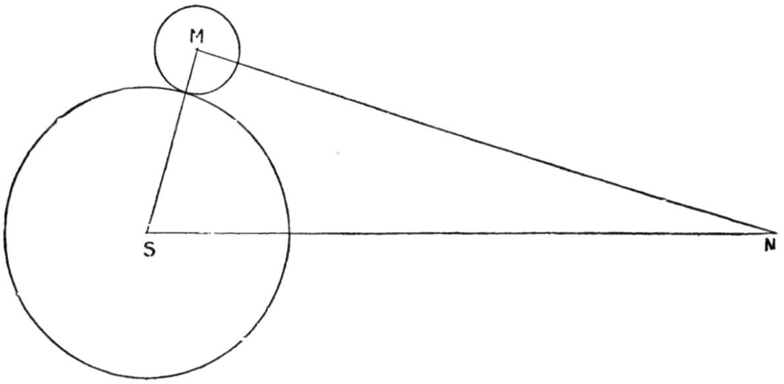

FIG. 3

For a solar eclipse, the angular distance *SM* is known from above as equal to $p + s + S—P$. The angle *i* is known, and if the angle at *S* is assumed as a right angle the triangle *SNM* can be solved, and the distance *SN* determined. An approximate method is all that is required, and since the mean value of *i* is nearly one-eleventh of a radian, hence *SN*, the ecliptic limit is approximately eleven times *SM*. But as the value of *SM* depends on the semi-diameters and parallaxes of sun and moon, and as we have seen these quantities continually are varying, the value of the ecliptic limit must also vary proportionally, and also because the angle *i* changes likewise. Hence it is necessary to distinguish between the maximum and minimum values of the ecliptic limit, for solar and lunar eclipses and for total and partial eclipses. Using the appropriate maximum and minimum values, it is easy to find the following values of eclipse limits.

	Major	*Minor*
Solar eclipse	18° 31′	15° 21′
Lunar eclipse	12° 15′	9° 30′
Total solar eclipse	11° 50′	9° 55′
Total lunar eclipse	6° 0′	3° 45′

As a simple exercise in mathematical astronomy these quantities might be used for predicting eclipses. All of the additional information is readily obtainable from one of the government publications like the *American Ephemeris and Nautical Almanac.* As already stated, an eclipse of the sun can take place only at new moon. For this instant of time look up in the *Ephemeris* the longitude of the sun and that of the moon's node. The difference is the angular distance of the sun from the node. If this longitude is greater than the major ecliptic limit of 18°31′, an eclipse of the sun cannot possibly take place; if the quantity is less than the minor limit of 15°21′, an eclipse must certainly take place. If the difference in longitudes is less than 18°31′, but greater than 15°21′, it is impossible to state whether an eclipse will take place or not, and recourse must then be had to a calculation of the ecliptic limit using the particular values for the semi-diameters, parallaxes and inclination for the day in question at the time of new moon. The conditions for a *total* eclipse are found by using the major and minor ecliptic limits of 11°50′ and 9°55′. Similarly, for an eclipse of the moon, which occurs only at full moon, by utilizing the values of the lunar ecliptic limits, it is possible to predict whether an eclipse of the moon will happen or not.

In practice, however, in predicting the number of eclipses to occur in any one year, the eclipse limits defined above are not used. It is much simpler to use the *Saros* to bring forward the data of the *Ephemeris* of 18 years before. All of the various series of eclipses, lunar and solar, that are taking place at the present time are known to the astronomer. When no great accuracy is needed the Saros provides a simple method. The information so obtained should be checked by the magnificent book by Oppolzer, *Canon der Finsternisse* where are given the elements of all of the eclipses (8000 solar and 5200 lunar) between the dates 1208 B.C. and 2162 A.D. Various lists of total solar eclipses have been published. (See for instance, *Encyclopaedia Britannica,* article on *Eclipses*).

Referring now to the lengths of the different kinds of months on p. 40, we recall that 19 returns of the sun to the moon's node are equal to 6585.7806 days, while 223 synodic months consume the slightly different amount of 6585.3211 days. If therefore a new moon fell exactly at the node, then after 18 years 11 days, the new moon would take place before the node was reached. The difference of the two quantities above is 0.4595 days, in which time the sun moves 28′. Accordingly, at each succeeding Saros the new moon is found 28′ farther and farther west of the node. It is now possible to trace the total number of eclipses and the progressive changes in them as they pass through the Saros. According to the value of the ecliptic limit already found, we know that if the new moon happens within 18° of the node, an eclipse of the sun may take place. If the node is the ascending one, the conditions will be as represented in Figs. 3 and 2, the moon being north of the shadow-cone. An eclipse of the sun will accordingly be visible in high northern latitudes on the earth. After 18 years 11 days the conditions will be nearly identical, but the new moon will take place 28′ nearer the node, and as a result, the eclipse will be visible on the earth a little farther south of the preceding position. With each succeeding return, the new moon moves nearer and nearer the node and the eclipse track shifts farther and farther south on the earth. When the moon is within about 11° of the node the solar eclipse becomes central, and the eclipse may be total or annular. As before, the central eclipse track will first touch only high northern latitudes, each succeeding eclipse moving farther and farther to the south. Total or annular eclipses will now take place at each return of the Saros until the new moon takes place 11° west of the node, when the central eclipse track passes off the earth at its south pole, and a series of partial eclipses ensues until the moon is 18° west of the node, when these particular eclipses cease. In such a series there are from 68 to 75 solar eclipses, depending on conditions, extending over some 1200 years. In each series there are approximately 25 partial eclipses and 45 central eclipses. The numbers vary for different series of eclipses. Of the central eclipses, total eclipses follow total eclipses with about the same duration of totality, and annular eclipses follow annular. If the eclipses had taken place at the descending node of the moon's orbit instead of at the ascending node, eclipses would have come on the earth at its south pole and gradually moved north, going off the earth at the north pole, as shown in the illustration facing this page.

On account of the ecliptic limits for lunar eclipses being smaller in value than for solar eclipses, there will be fewer repetitions in a lunar series, there being 48 or 49 altogether. Of this number there will be 22 or 23 total, with 13 or 14 partial eclipses, both before and after the total eclipses. The interval for a series of lunar eclipses consumes about 870 years.

There is one other very important relation brought out by the data on page 41. It is there found that 239 returns of the moon to the line of apsides, or to perigee, amount as a total to 6585.5374 days. For predicting the nature of the eclipse that will be found at the next return this quantity is almost as valuable as the Saros itself. At the end of 223 lunations the moon not only returns very closely to its original position with respect to the sun and the node, but also with respect to the line of apsides, with the necessary result that the distance from earth to moon is very closely repeated. This fact in turn brings two consequences; firstly that the duration of the eclipse, or at least that part of it which depends on the moon's distance, is altered but little; but secondly, that the perturbations of the moon's orbit which otherwise might have displaced by several hours the time of the eclipse will now be repeated almost as they were before, and with no consequent relative effect on the time of the eclipse.

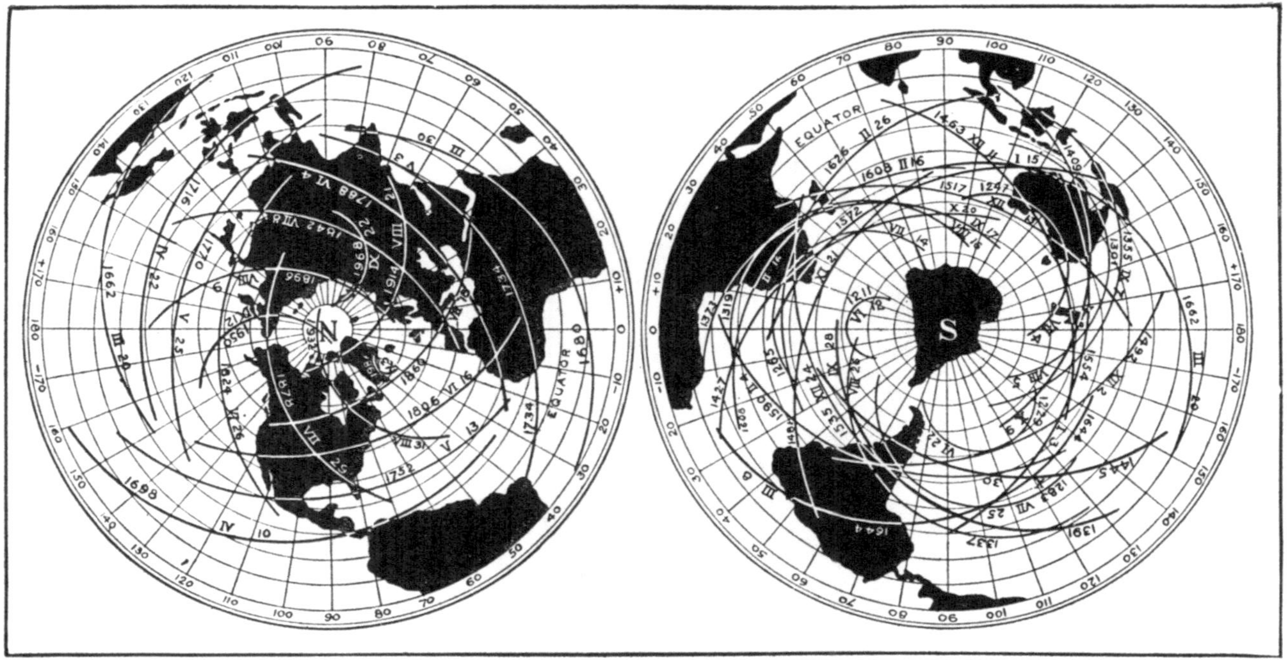

A Series of Forty-Four Total Eclipses Beginning near the South Pole July 12, 1211 and Ending October 3, 1986 near the North Pole

The eclipse of August 31, 1932 belongs to this series.

(Drawing by S. W. Balch.)

63-FOOT TOWER TELESCOPE OF THE SWARTHMORE EXPEDITION INSTALLED IN SUMATRA FOR THE 1929 ECLIPSE

CHAPTER IV

THE VERIFICATION OF ECLIPSES

IN THE last chapter it was shown that the times of eclipses can be foretold with moderate accuracy by means of the Saros. Equally important with the actual prediction of the times is the fact that the circumstances attending the eclipses following each other in the Saros will be very closely repeated. A large partial eclipse of the sun will be followed by a large partial eclipse of the sun, an annular eclipse by an annular eclipse, a total eclipse of short duration by a similar short eclipse. And so also with lunar eclipses. There remains still to explain the signification of the decimal portion of the time of the Saros, 6585.3211 days. If the sun had been on the meridian at the middle of the eclipse, the eclipse therefore occurring at noon, the next following eclipse will be repeated at a place 0.3211 of a revolution of the earth, or in other words 7h. 42m. of longitude farther west. After the return of three Saroses, or 54 years, the eclipse tracks will have gone almost around the world and will have returned again to nearly the same longitude. If the eclipses belong to a series that is taking place at the moon's ascending node, the later eclipse track will be found farther south than the earlier one, and if at the descending node, farther north than the eclipse fifty-four years earlier.

Those who wish to amuse themselves by playing with figures may find other remarkable coincidences by experimenting with the lengths of the various years and months on p. 41. Newcomb has found a very interesting period at the end of 358 lunations.

358 synodic months	=	10571.95 days
30.5 eclipse years	=	10571.91 days

This period amounts to 29 Julian years, less 20.3 days. But 358 lunations are equal in length to 383.673 anomalistic months, and hence, as explained in the preceding chapter, the eclipses which follow each other with this period will differ greatly in their characteristics, since the conjunctions between sun and moon take place at different positions with respect to perigee. Also, such eclipses must follow each other alternately at ascending and descending nodes. Three such periods, however, will equal 1169.019 anomalistic months, and the time of conjunction has accordingly moved very close to perigee (if it had previously been found there) and a total eclipse will thus be followed by a total eclipse. Three of these periods equal 87 years less 61 days, and 18 periods equal 521 years, actually within a day or two. Thus with the Saros of 18 years, and with the addition of the 29-year, the 87-year and the 521-year periods, the astronomer is enabled to engage in predicting at long range. Take for instance the first eclipse verified with certainty, the Nineveh eclipse of June 15, 763 B.C. (763 B.C. may be expressed also as the year —762). Applying the 521-year period we have the years 242 B.C. and A.D. 280, 801, 1322 and 1843. These eclipses each fell on June 15, O. S., or by the Julian calendar. The date of June 15, 1843, O. S., is the same as June 27, 1843, by the Gregorian or present-day calendar. By means of the 29-year period we obtain the eclipses of June 6, 1872, and May 18, 1901. (The eclipses of 1843 and 1872 were not total while that of 1901 was the six-minute total eclipse observed in Sumatra.)

Facing page 56 will be seen a map of the tracks of all of the eclipses, total and annular, taking place between the years 1919 and 1940. The map is from Oppolzer's *Canon der Finsternisse.*

In the following table are given one hundred years of total eclipses of the sun. The article on "Eclipses" in *Encyclopaedia Britannica,* 14th edition, does not contain the eclipses of 1944, 1945, 1950, and 1968. As shown in the diagram facing page 52, the last two are unimportant eclipses which occur in the north polar regions.

Date	*Series*	*Node*	*Duration in minutes*	*Where visible*
1875, April 6	1	A	4.7	Indian Ocean, Siam, Pacific.
1876, Sept. 17	2	D	1.8	Pacific Ocean.
1878, July 29	3	D	3.2	Canada and the United States.
1880, Jan. 11	4	A	2.1	Pacific Ocean, California.
1882, May 17	5	D	1.8	Egypt, Central Asia, China.
1883, May 6	6	D	6.0	Pacific Ocean, Caroline Islands.
1886, Aug. 29	7	A	6.6	South America, Central Africa.
1887, Aug. 19	8	A	3.8	Northern Europe, Siberia, Japan.
1889, Jan. 1	9	D	2.2	California, Oregon, Canada.
1889, Dec. 22	10	D	4.2	South America and Central Africa.
1893, April 16	1	A	4.8	Venezuela to West Africa.
1894, Sept. 29	2	D	1.8	East Africa, Indian Ocean.
1896, Aug. 9	3	D	2.7	North Europe, Siberia, Japan.
1898, Jan. 22	4	A	2.3	East Africa, India, China.
1900, May 28	5	D	2.1	United States, Spain, North Africa.
1901, May 18	6	D	6.5	Sumatra, Borneo.
1904, Sept. 9	7	A	6.4	Pacific Ocean.
1905, Aug. 30	8	A	3.8	Canada, Spain, North Africa.
1907, Jan. 14	9	D	2.3	Russia, Central Asia.
1908, Jan. 3	10	D	4.2	Pacific Ocean.
1911, April 28	1	A	5.0	Australia, Polynesia.
1912, Oct. 10	2	D	1.8	Colombia, Ecuador, Brazil.
1914, Aug. 21	3	D	2.1	Scandinavia, Russia, Asia Minor.
1916, Feb. 3	4	A	2.5	Pacific Ocean, Venezuela.
1918, June 8	5	D	2.4	The United States.
1919, May 29	6	D	6.9	Peru, Brazil, Central Africa.
1922, Sept. 21	7	A	6.1	East Africa, Australia.
1923, Sept. 10	8	A	3.6	California, Mexico, Central America.
1925, Jan. 24	9	D	2.4	Northeastern United States.
1926, Jan. 14	10	D	4.2	East Africa, Sumatra, Philippines.
1927, June 29	11	A	0.7	England, Scandinavia.
1929, May 9	1	A	5.1	Sumatra, Siam, Philippines.
1930, Oct. 21	2	D	1.9	Pacific Ocean, Patagonia.
1932, Aug. 31	3	D	1.5	Canada, New England.
1934, Feb. 14	4	A	2.7	Borneo, Celebes, Caroline Islands.
1936, June 19	5	D	2.5	Greece to Central Asia and Japan.
1937, June 8	6	D	7.1	Pacific Ocean, Peru.
1940, Oct. 1	7	A	5.7	Colombia, Brazil, South Africa.
1941, Sept. 21	8	A	3.3	Central Asia, China, Pacific Ocean.
1943, Feb. 4	9	D	2.5	China, Pacific Ocean, Alaska.
1944, Jan. 25	10	D	4.1	South America, West Africa.
1945, July 9	11	A	1.1	Canada, Greenland, Scandinavia, Russia.
1947, May 20	1	A	5.2	South America, Africa.
1948, Nov. 1	2	D	1.9	Central Africa, Madagascar.
1950, Sept. 12	3	D	0.9	Northeastern Siberia.
1952, Feb. 25	4	A	3.0	Africa, Persia, Central Asia.
1954, June 30	5	D	2.5	Canada, Scandinavia, Russia.

Date	*Series*	*Node*	*Duration in minutes*	*Where visible*
1955, June 20	6	D	7.2	Ceylon, Siam, Philippines.
1958, Oct. 12	7	A	5.2	Chile, Argentina.
1959, Oct. 2	8	A	3.0	Canaries, Central Africa.
1961, Feb. 15	9	D	2.6	France, Italy, Austria, Siberia.
1962, Feb. 5	10	D	4.1	New Guinea.
1963, July 20	11	A	1.5	Alaska, Canada, Maine.
1965, May 30	1	A	5.3	Pacific Ocean.
1966, Nov. 12	2	D	1.9	Bolivia, Argentina, Brazil.
1968, Sept. 22	3	D	0.5	Northern Siberia.
1970, Mar. 7	4	A	3.3	Mexico, Florida, Georgia.
1972, July 10	5	D	2.7	Northeast Asia, Northeast America and Atlantic Ocean.
1973, June 30	6	D	7.2	South America, Africa.
1974, June 20	12	D	5.3	Southwest Australia and Indian Ocean.

From the table it will be seen that the total eclipses which have been or will be observed in the United States are: 1878, July 29; 1880, January 11; 1889, January 1; 1900, May 28; 1918, June 8; 1923, September 10; 1925, January 24; 1932, August 31; 1963, July 20; 1970, March 7. So many eclipses in the past, so few in the coming years! In the future, the next eclipses to visit the United States will be those of 1945 and 1954, both of which begin at sunrise about on the international boundary and their tracks will go northwards over inaccessible spots in Canada. To observe either of these eclipses it will probably be necessary to travel to Scandinavia. The next eclipse to be observed scientifically in the United States is that of July 20, 1963, the path crossing Maine not very far northeast of the 1932 eclipse. The eclipse of March 7, 1970, will be visible in Florida and along the coast of Georgia. The next American eclipse after that will be on February 26, 1979, visible in northwest United States and in Canada across Hudson Bay. What hardy astronomer will be willing to brave the probabilities of a blizzard and 40° F. below zero in order to add to scientific knowledge by observing this total eclipse?

In the immediate future, the next total eclipse to be observed will be on June 19, 1936, with a duration of two and a half minutes. At sunrise the total eclipse begins near Greece, the path of totality later passing through Siberia and Japan.

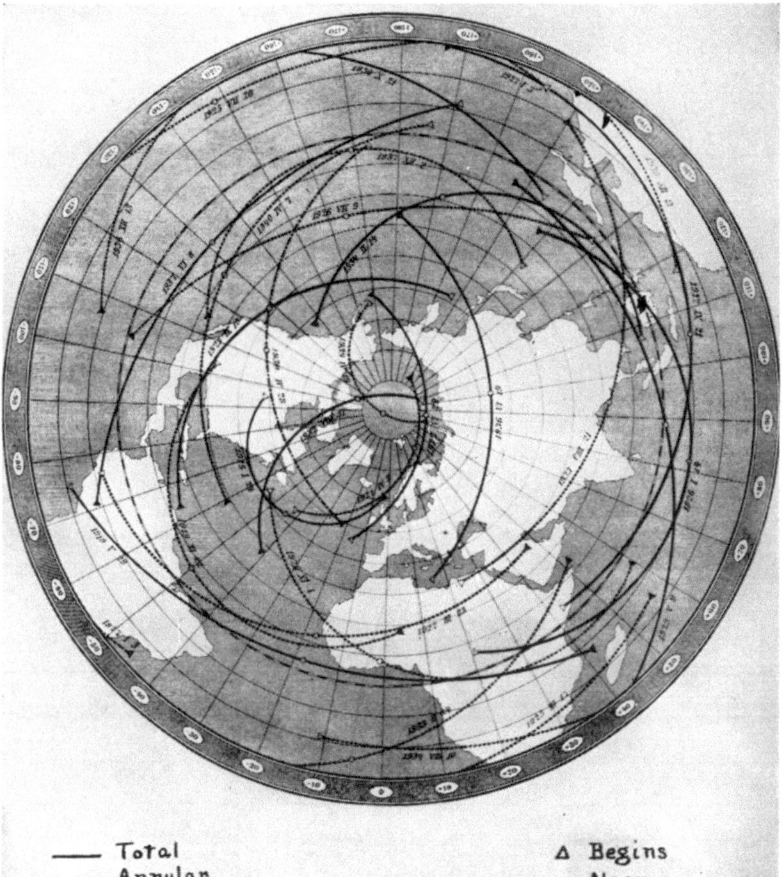

TRACKS OF ECLIPSES BETWEEN THE YEARS 1919 AND 1940
From Oppolzer's *Canon der Finsternisse.*

Desolation on Niuafoou Island after the Eruption in 1929

Huge Cameras Installed for the 1930 Eclipse

The most favored sites for observation will be at the cities of Omsk and Tomsk on the Trans-Siberian Railroad. The astronomers from the United States will probably locate in Japan. The following eclipse is June 8, 1937, with a duration of totality of more than seven minutes. The track passes across the wide expanse of waters of the Pacific Ocean and touches land only near sunset in Peru. With the sun at a very low altitude it appears probable that no important scientific observations will be possible. The next eclipse, that of October 1, 1940, will probably be extensively observed in South Africa. Evidently the coming generation of American astronomers will be forced to make long trips away from home if they expect to continue the important discoveries made at eclipses. At the present time many eclipse problems, discussed in the final chapter, still await solution. It is confidently expected, however, that the coming generation will find methods of investigating chromosphere and corona with such success that observations at eclipses will no longer be necessary. When science shall have progressed to this extent, future astronomers will be able to point to the long trips taken by their forefathers, sometimes half way round the globe, in order to observe a total eclipse for a few brief minutes of time. In the twenty-first century the great god of Efficiency may be worshipped even more profoundly than is the case in the early part of the twentieth century. Under such circumstances a total eclipse will still be viewed with interest as a fascinating phenomenon, which, however, will afford ample evidence of the crude methods of investigation of the astronomers a century before.

If so few eclipses are visible in such a large country as the United States, how about those in a relatively small territory like the British Isles? The eclipse of June 29, 1927 was enthusiastically observed in the northern portion of England in spite of heavy clouds; that of August 11, 1999, grazes the south of Ireland and Land's End, and on August 12, 2026, the eclipse track may cross the southwestern tip of Ireland but will not be seen in Great Britain.

Since the sixth century the following total eclipses have touched parts of the British Isles:

594 July 23	1185 May 1
603 Aug. 12	1330 July 16
639 Sept. 3	1424 June 26
664 May 1	1433 June 17
878 Oct. 29	1598 Mar. 6
885 June 15	1652 April 8
1023 Jan. 24	1715 May 2
1133 Aug. 1	1724 May 22
1140 Mar. 20	1927 June 29

Eclipses have included London twice, Dublin twice, and Edinburgh five times.

On the other hand, some parts of the globe are visited by eclipses in rapid succession. Spain witnessed the eclipses of 1842, 1860, 1870, 1900 and 1905. In the vicinity of the East Indies the eclipses of May 18, 1901, January 14, 1926, May 9, 1929 and February 14, 1934 have been observed, while the track of the eclipse of September 21, 1922, lay but a few hundred miles south.

Assuming that the moon's shadow path intercepted by the earth averages 100 miles in width, it is easy to find that on the average a total eclipse of the sun will be visible in any one locality once every 360 years. A stay-at-home astronomer might have little opportunity during his lifetime to witness many such eclipses!

With a knowledge of the ecliptic limits found on page 50, it is possible to find the number of eclipses that will occur in any given time, say a year. If at the time of the new moon, the moon (and also the sun) is more than 18° 31′ from the moon's node there cannot possibly be an eclipse of the sun. Since the angle of 18° 31′ may be east or west of the node, there is thus a zone of 37°, within which an eclipse of the sun *may* take place provided that a new moon also occurs somewhere within the zone. The minor ecliptic limit for a solar eclipse is 15° 21′, and similarly if new moon is found anywhere in the zone 30° 42′ in length an eclipse of the sun *must* happen, there being no possibility of the moon slipping by without intercepting the light from the sun as it comes in the direction towards the earth. For an eclipse of the moon, the ecliptic limits are smaller. It is only when full moon comes within 9° 30′ of the node that a lunar eclipse is certain to take place, though an eclipse may possibly happen if the limit is as great as 12° 15′. In a synodic month the earth moves along the ecliptic to an average extent of 29° 6′. But the moon's node is moving in the opposite direction by an amount of 1° 31′ each month. *Relative to the node,* therefore, new and full moon are found farther east each synodic month by the sum of 29° 6′ and 1° 31′, or 30° 37′, and it is this motion relative to the node on which the circumstances of an eclipse depend. By an analogy one can see how easily the question of eclipses is thought out. Suppose a man walking around a circular track of any diameter in which there are two mud holes four feet in width diametrically opposite each other. If he is taking a constant pace of three feet, it will be impossible for him to walk around the track without stepping at least once into each mud hole during each circuit. If when he comes to the muddy spot he happens to plant his foot near the middle he will have only one wet foot, but if the first foot gets in the hole near the edge, the second foot may be planted in the mud near the other side of the puddle. If the mud hole were only two feet in width, his standard pace of three feet might carry him across it without either foot getting into the mud. And so it is with eclipses of sun and moon. The moon's step is 30° 37′ in angular length, the danger zone for a solar eclipse is 30° 42′, or a quantity which is larger than the moon's monthly motion with respect to the node. There *must,* therefore, be at least one solar eclipse at each nodal point, or two such eclipses without fail during the course of each and every year. On the other hand, the danger zone for lunar eclipses being much smaller, it is possible that the full moon may take place so far from the node each time that there will not be a single lunar eclipse during the year. There can, moreover, be only one eclipse of the moon at each node, but never two. Eclipses do not occur in the same months each year, but follow each other at the beginning and end of the "eclipse year" of 346.6 days. This being 18 days less than the calendar year it is possible for there to be three eclipses of the moon within the calendar year, if the node is passed early in January, or late in December (which comes to the same thing). Two solar eclipses may possibly take place at each node, and a fifth one within the calendar year. If the *new* moon take place near the node bringing an eclipse of the sun, the preceding or following *full* moon may be too far from the node to cause an eclipse of the moon. Under these conditions there will be only two eclipses during the year, each of the sun, and as these occur near the node, they will be central eclipses, namely, either total or annular. This frequently happens, as, for instance, in the years 1922, 1926, 1929, 1933, 1940, 1944, 1951 and 1962. If there are two solar eclipses at each node, a full moon must take place near the node and a total eclipse of the moon will result. There is thus the possibility of two partial solar and one total lunar eclipse at each node, or a total of two lunar and four solar eclipses during the year. If the node is passed about the middle of January, there may be a fifth solar eclipse within the calendar year, making a maximum of seven eclipses in a single year. There may also be three eclipses of the moon and four eclipses of the sun within the year, an event which will happen next in 1982, the three lunar eclipses each being total, and the four solar, partial. In 1935 there will be two lunar and five solar eclipses. This combination occurs but seldom, and will not happen again until the year 2485. The maximum number of eclipses falling within a calendar year is thus seven and the minimum two.

According to Oppolzer, there are 237.5 solar eclipses on the average in a century. Of these 83.8 are partial, 77.3 annular, 10.5 annular and total (the vertex of the moon's shadow just reaching the earth) and 65.9 total eclipses. Twenty-seven percent of all solar eclipses are total. On the average, therefore, two total eclipses of the sun are visible every three years, but fully half of these are inaccessible for the reason that the eclipse tracks are in high northern or southern latitudes, or pass entirely across the oceans, or lie in localities where the probability of good weather is unpromising. Consequently, on the average, about once in every three years a total eclipse somewhere on the earth is available for observation. As the duration of totality averages less than three minutes, we thus see that an eclipse observer may have an average of a minute per year for

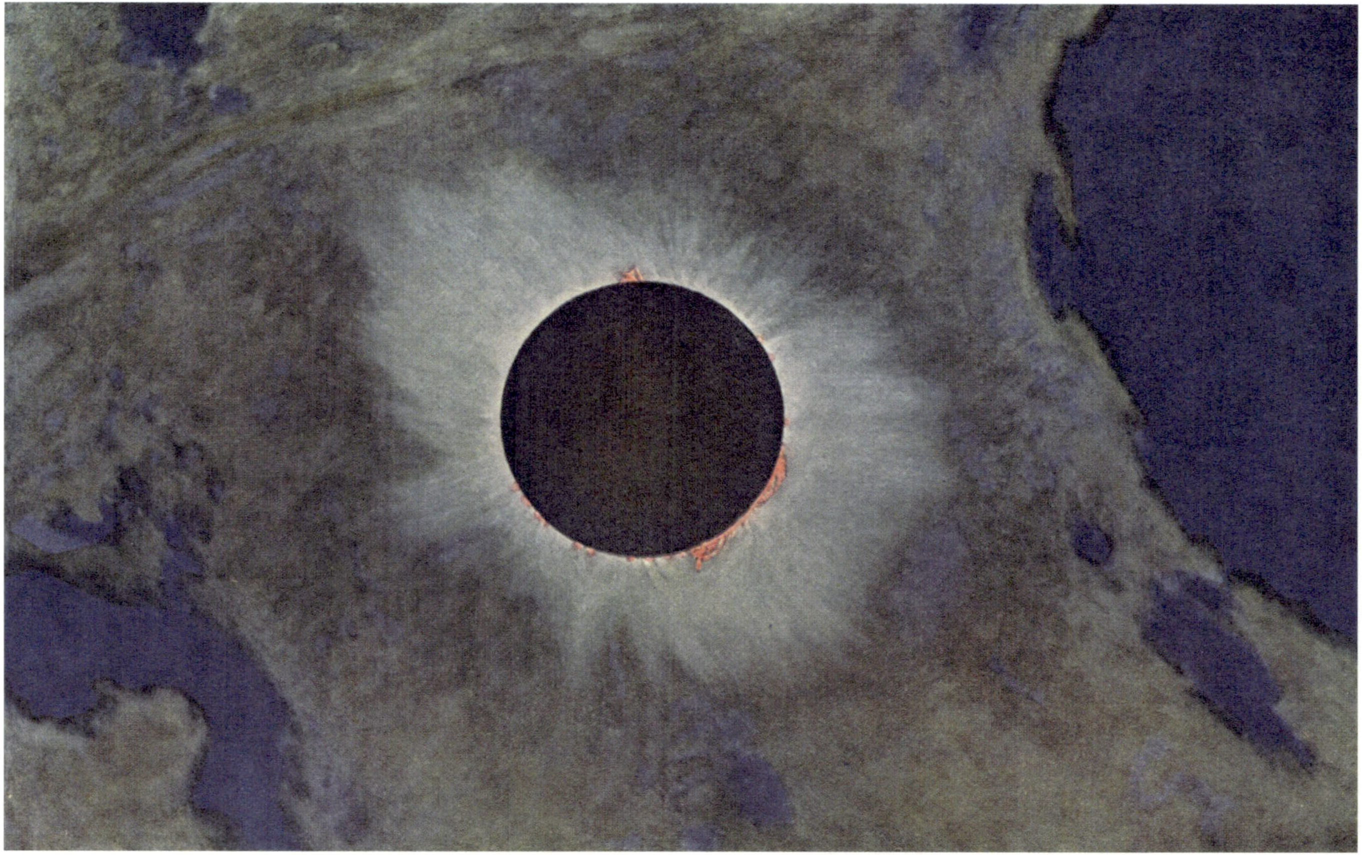

Total Solar Eclipse of June 8, 1918
From the painting by Howard Russell Butler, N. A.

scientific work, provided he has sufficient financial backing to permit him to travel long distances for every available eclipse. The author has had the great good fortune of witnessing nine total eclipses in 1900, 1901, 1905, 1918, 1923, 1925, 1927, 1930 (Oct. 21) and 1932. In 1936 he hopes to be able to observe his tenth eclipse when he will have accomplished the rare feat of observing three total eclipses in the same series separated by the Saros interval of 18 years 11 days.

The Lick Observatory has set up an enviable reputation for itself through the observation of fifteen total eclipses in the following years: 1889 (Jan. 1 and Dec. 22), 1893, 1896, 1898, 1900, 1901, 1905 (three parties in Labrador, Spain and Egypt), 1908, 1914, 1918, 1922, 1923, 1930 (April 28) and 1932. In the past there have been but three complete failures through clouds, in 1896, 1914 and 1923. Excellent observations have been secured by the British Joint Permanent Eclipse Committee under the auspices of the Royal Society and the Royal Astronomical Society. Through the coöperation of astronomers with eclipse experience working under the auspices of the Commission on Eclipses of the International Astronomical Union, problems are now carefully studied before each eclipse to the end that there may be little wasted effort.

Taking into account the *earth as a whole,* it is evident that there are more solar eclipses than lunar, in the approximate ratio of four to three. But for eclipses visible from any one locality, London or New York or Timbuctoo, the question is an entirely different one. A partial solar eclipse at best may be seen from a limited portion of the globe, while a total eclipse is observable only in a very restricted path. An eclipse of the moon, being caused by its passage into the shadow of the earth, must be visible to every locality on the earth where the moon itself is visible. At any one instant of time the moon is above the horizon to half of the world, and since the eclipse of the moon, from beginning to ending of the partial eclipse, lasts for some hours it will be plainly visible to over half the globe, unless clouds interfere. There are probably very few people living who love Nature, who have not witnessed at least one total eclipse of the moon, but probably not more than one in a thousand of the world's inhabitants has caught a glimpse of the matchless glory of the corona, seen only during the fleeting moments of a total eclipse of the sun.

Calculation of an Eclipse

The seemingly uncanny and almost miraculous power of the astronomer to predict the coming of an eclipse hundreds of years in advance has always possessed a powerful fascination for the uninitiated. By means of the Saros and of the methods outlined in the foregoing chapter, it is readily possible to foretell the happening of an eclipse of the sun or moon and the general circumstances surrounding these events, provided no very great accuracy of time or place is required. To know the *exact* times for any given locality more accurate methods must be used. The problem of eclipse calculations resolves itself into three stages. First, the accuracy of the whole problem depends mainly on the completeness of our knowledge regarding the motions of the moon furnished by the lunar theory, and also upon the motion of the earth about the sun which gives the apparent motion of the sun about the earth, the latter theory being comparatively simple. From the motions of sun and moon thus determined, it is necessary to compute their positions at equidistant intervals, and as seen from a standard place, namely, the center of the earth. Second, from these positions it is necessary to compute certain "elements" on which eclipses depend; and third, for a given latitude and longitude of any place on the earth's surface, to calculate the times, etc., of the eclipse. The first part of the problem is taken care of by the various governmental publications which appear some three years in advance. By this means it is readily possible to find the exact positions of sun and moon at any given instant of time. Elements of eclipses needed for the second step of the process have been calculated, from the earliest historic times up to the twenty-second century, by Oppolzer and Simon Newcomb. The third step in computing the details of the eclipses for a given year is now done by the *American Ephemeris* alone, but the results are published by the other nautical almanacs, thus eliminating duplication of efforts. For the 1932 eclipse Oppolzer's chart was in error by 200 miles in giving the eclipse path; his elements however furnished the position of the central line within an error of 10 or 15 miles and the times to an accuracy of about one minute. The *American Ephemeris* aims at an accuracy of one-quarter mile in position and two seconds for the times of contacts.

The calculation of an eclipse of the moon is one of great simplicity. The time at which the moon passes into the shadow cast by the earth is the same absolute instant of time no matter where the observer is on the earth. Even the inhabitant of Mars (who may or who may not exist), who observes the earth for signs of life with the gigantic telescope which his supposed advanced civilization must have furnished him, would see the eclipse of the moon at identically the same absolute second of time as that noted by the Astronomer Royal at Greenwich. The local time of the eclipse recorded by the Martian clock would not be Greenwich mean time,—but that is not part of the problem. Eclipses of Jupiter's satellites, caused by the passing of these moons into the shadows cast by the planet are seen at the same instant of time by all observers on the earth, provided of course that the terrestrial observers are on the side of the earth turned towards Jupiter. If the various observers take the value of the Greenwich mean time of such an eclipse (which can be obtained from the nautical almanacs) and compare this with the local mean time of the individual observation, the difference in times will give the longitude of the observer from Greenwich. This is a method much in vogue for the determination of longitude when no great accuracy is necessary, for such an eclipse is a gradual phenomenon and not one which comes with great suddenness. The Greenwich mean times of the phases of an eclipse of the moon are the same for every inhabitant of the earth, and consequently, in each of the annual nautical almanacs it is possible to publish the Greenwich times of the various portions of the eclipse, such as the beginning of the partial eclipse, the beginning of the total eclipse, etc. If the person observing the lunar eclipse lives in Western Europe, in Great Britain, France, Spain, Portugal, Holland, Belgium or any of the other countries that now keep Greenwich time as standard, the times of the eclipse will be that recorded by their clocks, if these keep correct time. If the observer lives in New York or Washington or anywhere in United States or Canada where Eastern Standard time is used, the times of the eclipse will be obtained by subtracting 5 h., 0 m., 0 s. from the Greenwich time, while if he lives in San Francisco it will be necessary to subtract 8 h. from the Greenwich time.

The method of calculating the times of a lunar eclipse, as given in the ephemerides, is a beautifully attractive problem which may be solved quite simply by anyone even though he is not an expert computer, and this may be accomplished either by graphical methods or by logarithmic calculation. Since the earth has an atmosphere, the outline of its shadow is not sharp and well-defined, but hazy when examined by a telescope. On account of the indefinite outline of the earth's shadow, its theoretical diameter must be augmented. Different computers utilize different values for this increase, varying from one-fiftieth to one-seventy-fifth of the theoretical value of the diameter of the earth's shadow. Accordingly, a total lunar eclipse is not a sudden phenomenon like an eclipse of the sun and there is, therefore, no need to employ very refined methods of computation, or to carry the calculations to fractions of a second of time. The *American Ephemeris* gives the times to the decimals of a minute, — an accuracy sufficiently great for the purpose. The method of calculating a lunar eclipse is given in Chauvenet, *Spherical and Practical Astronomy*, Vol. 1, p. 589. For graphical methods of determining the times of eclipses of moon and sun, and also of occultations, see a series of articles by Rev. W. F. Rigge in *Popular Astronomy*, volume III.

As just stated, an eclipse of the moon does not permit of an exact calculation owing to the ill-defined nature of the earth's shadow caused by its atmosphere. Due to the refraction of light by the earth's atmosphere, the sun's light, or rather the rays at the red end of the spectrum that are not absorbed by the atmosphere, reach the moon, even at the middle of totality when the earth is directly between the moon and the sun. If the " man in the moon " were to view the phenomenon that would form, to him, an eclipse of the sun, he would see the earth surrounded by a great ring of light, its own illuminated atmosphere. If the moon had even a rare atmosphere, there would, for this same reason, be a ring of light around the moon as it encroached upon the face of the sun at the time of a solar eclipse. No such light is visible, which is a sure proof that any atmosphere the moon may possess must be rarer than can be made by the best vacuum pump ever constructed by man. The coming of totality at the time of an eclipse of the sun is a sudden phenomenon and as a consequence the approximate methods utilized in calculating a lunar eclipse are not sufficiently precise. The best method to follow in determining the times of a total eclipse is that of Chauvenet, *Spherical and Practical Astronomy*, I, 436. Another excellent guide is Buchanan, *Theory of Eclipses*.

The calculation of a solar eclipse, in fact, cannot be treated in the simple manner of that of a lunar eclipse owing to the large size of the moon's parallax. In other words, the moon is so comparatively near the earth that it is projected on the face of the sun differently for every separate place on the earth's surface. As a result, the times of beginning and ending of the partial and total phases of the eclipse are different at every station.

The accuracy with which the times of solar eclipses can be predicted depends on the reliability of the work of the astronomers of all ages, and on the manner in which the torch has been passed on from one generation to the next. The chief cause of concern is found in the motions of our unruly neighbor, the moon. The position of the moon is furnished from observations of the times of contact of the limbs of the sun and moon, four different contacts being recognized. First contact is the instant that the moon begins to creep on the face of the sun, the eclipse beginning on the western edge of the sun. Second contact is signaled by the beginning of totality, third contact by the ending of totality and fourth contact is the ending of the eclipse, the moon passing off the face of the sun, the last contact being found on its eastern edge. The four contacts are generally observed visually by a pair of field glasses or by a telescope of moderate power. The time of first contact is difficult to observe with accuracy since nothing is to be seen at the edge of the sun until the moon is actually projected on to the face of the sun, — and first contact has already actually taken place. In other words, the observer is always too late in noting the time of first contact, the amount of tardiness depending on the size of the telescope, the state of the seeing, but especially on the skill and experience of the observer, which in turn depend on the number of eclipses witnessed. Fourth contact is easier to observe than first since the moon can be followed in the telescope until it leaves the face of the sun. The beginning and ending of totality can be more accurately observed than the two other contacts, but they are subject to some uncertainties on account of irregularities in the profile of the moon at the points of contact. By means of photographs taken just before and just after totality, when the crescent of the sun is small and changing rapidly, the times of second and third contacts can be determined with much greater degree of precision than is attainable in visual work. It goes without saying that the error of the chronometer should be known as accurately as possible (probably obtained by wireless signals), that the observed times should be recorded on the chronograph, and that the latitude and longitude be known.

The photographs for contacts will unquestionably be taken with the same instrument used during totality for the corona. It will be necessary, therefore, to utilize a different method of exposure and different technique from that employed on the corona. The brilliancy of the crescent sun compared with the corona will be very great. To diminish the brightness of the crescent images it will be necessary to use a rapid exposing shutter. The best results photographically will be secured by using slow, fine-grained plates, and they should be specially "backed" to prevent halation caused by reflection from the glass side of the plate. (See Chapter VII.)

At the eclipse of 1905, totality came ahead[1] of its predicted time, the beginning of totality being 17 seconds earlier, while the ending came 23 seconds earlier than the times predicted by the *American Ephemeris*. The middle of totality was thus 20 seconds ahead of that calculated, while the duration of totality was some six seconds less than was expected from the computations. The time predicted for the middle of totality by the British *Nautical Almanac* was identical with that furnished by the *American Ephemeris* but the duration of the former was $1^s.7$ less than that of the latter, while the duration calculated from the *Connaissance des Temps* was five seconds greater than that of the *American Ephemeris.* All observers at the Spanish eclipse had their program of observation greatly interfered with by having the moon so far in advance of its predicted place. Before the eclipse of June, 1918, the observers of the U. S. Naval Observatory party in Oregon were furnished by the Washington authorities with a correction of 12.5 seconds to be applied to the times of contacts calculated from the *American Ephemeris*. The observed times were about fourteen seconds ahead of those predicted by the *Ephemeris*. At the eclipse of 1922, the Lick Observatory party observed the beginning of totality some sixteen seconds earlier than the time predicted, while the end of totality came twenty seconds earlier than the Almanac prediction.

Apparently, therefore, after having made due allowance for all known possible sources of error, the moon has strayed from the path mapped out for it by the mathematical astronomers by an amount which is not inconsiderable. What is the cause of the moon being away from its predicted place?

The first of the great modern authorities dealing with the

[1] *Lick Observatory Bulletin*, *4*, 118, 1905.

motion of the moon was Hansen, who, in 1857, published his *Tables de la Lune*. By the help of these tables the unexplained fluctuations in the moon's motion were reduced to a small fraction of their former amount. Hansen derived the inequalities in the moon's motion largely from the gravitational theory, but in order to satisfy the observations made with great care at Greenwich between 1750 and 1850, it was necessary for him to apply two empirical terms supposed to come from Venus, the larger of the two having a long period of 239 years.

The next advance was made by Simon Newcomb, who published his *Researches on the Motion of the Moon* in 1878. In addition to the observational material utilized by Hansen, Newcomb discussed all of the lunar eclipses recorded by Ptolemy in the *Almagest* and in addition a large number of mediaeval lunar eclipses. He discussed a few ancient solar eclipses, but excluded them from his calculations on account of their unreliability. The chief value of Newcomb's great work lay in collecting and discussing observations of occultations and solar eclipses in the century and a quarter previous to 1750. Newcomb had thus 250 years of observations to discuss in place of the hundred years of observations available to Hansen. The whole material was utilized to secure the value of the moon's motion and acceleration, and clearly showed the presence of an unexplained term of long period. Newcomb did not deduce the period of this term from the observations themselves but assumed it identical with one of Hansen's terms (which is now known to be faulty). " It is no small tribute to the thoroughness of Newcomb's work that while the observations from 1750 onwards have been thoroughly examined by Cowell, Radau, and Brown, the reductions of the observations from 1621 to 1747 have never been revised by anyone but Newcomb himself." [1]

The third great investigation of the lunar theory is found in E. W. Brown's *Tables of the Moon's Motion*, published in 1920. These tables are more complete than those of Hansen and Newcomb, and take account of every term of

[1] Fotheringham, *Monthly Notices, R. A. S., 80*, 289, 1920.

appreciable significance. In addition to these three great monumental works on the Theory of the Moon by Hansen, Newcomb and Brown, exhaustive investigations have been made by many competent authorities. After making allowance for the gravitational attraction of every conceivable form of disturbance, it is unmistakable that the moon departs from her theoretical place in a very irregular manner. An error in the assumed value of the acceleration of the moon's mean longitude, even though this error is a very small one, will cause errors in the predicted place of the moon which increase in size for the reason that the errors depend on the square of the elapsed time.

Tables of the Moon are used for predicting the place of the moon in the different nautical almanacs which appear some two or three years in advance. For these predictions it is desirable to keep the theory of the moon as free as possible from arbitrary empirical terms so that the theory may not be cluttered up by too many additions. When observations are secured the results can then be directly compared with theory. Between the theoretical position of the moon given by the almanac, and the observed place, there may be a difference of several seconds of arc.

For guiding the work of the eclipse astronomer it is now deemed necessary to know the times of beginning and ending of a total solar eclipse within an error of a few seconds. Hence it has become customary in recent years to secure from Washington, or Greenwich, a month or more before the eclipse takes place, corrections to the almanac positions of the moon. These corrections derived from observations in the three-year interval since the almanac was prepared permit increased accuracy in computing the eclipse.

On account of the difficulty of seeing the moon when near the time of new moon, there is ordinarily a gap in the regular lunar observations, whether these are made by occultations or by meridian circle. An intense interest was aroused in the United States by the total eclipse of 1925 and hence a special attempt was made to utilize this interest in order to secure as many observers as possible. From large numbers of careful observations it was hoped that it might be possible to detect any short-period deviations in the motion of the moon, as well as to compare the results obtained by different methods of observation in order to see whether there are systematic errors of observation peculiar to any of them.

At the time of the 1925 eclipse it was found that the differences in the observed and tabular positions of the moon had been nearly constant during the interval since the *Ephemeris* was published. For predicting the eclipse, it was necessary to apply a correction of $+7''.0$ to the tabular mean longitude of the moon as given in the *American Ephemeris* and $-0''.50$ to the lunar declination. After applying these changes to the moon's place, the observations of the beginning and ending of totality gave the following corrections to be applied to the predicted times of second contact and duration of totality, the observations coming from trained astronomers with every facility for obtaining accurate positions of their stations and accurate time signals on the day of the eclipse.

Place	*Second Contact*	*Duration*
Poughkeepsie, N. Y.	$+3.^s3$	$-0.^s8$
Beacon, N. Y.	$+1.8$	-2.9
New Haven, Conn.	$+6.0$	$+0.8$
Middletown, Conn.	$+3.6$	0.0
Martha's Vineyard, Mass.	$+6.0$	$-6.$
Nantucket, Mass.	$+4.9$	-3.7
Dirigible, *Los Angeles*	$+6.3$	-0.9

A summary of the results of the observations discussed by Brown[1] for the 1925 eclipse gives corrections to be added to the mean longitude of the moon $\delta\lambda$, and to the mean latitude $\delta\beta$ as follows, where *n* represents the number of observations:

Method	$\delta\lambda - 7.''00$	n	$\delta\beta$	n
Occultations	$+0.''38 \pm .''10$	79	$+0.''49 \pm .''17$	79
Photographs	$-0.14 \pm .25$	11	$+0.34 \pm .26$	11
Greenwich, meridian circle	$-0.35 \pm .16$	16	$-0.85 \pm .11$	16
Washington, 9-inch transit	$-0.34 \pm .15$	16	$-0.60 \pm .16$	16
Washington, 6-inch transit	$+0.70 \pm .19$	22	$-0.13 \pm .18$	21
Cape, meridian circle	$+0.42 \pm .28$	13	$-0.34 \pm .46$	10
Eclipse	$+0.40 \pm .14$		$+0.80 \pm .12$	

1 *Astronomical Journal*, 37, 9, 1926.

The weighted mean of the results gives a correction to the moon's longitude and latitude of $+0.''28 \pm.''06$ and $+0.''32 \pm.''06$, respectively.

The comparison of the results made by the different methods is not without interest. The meridian observations differ rather widely. The occultations and eclipse results give corrections which agree within their probable errors. All the non-meridian observations agree in giving a positive correction to the lunar latitude and all the meridian results give a negative correction, and hence systematic differences of over $1.''0$ are shown. These discrepancies are partially explained by an assumption originally made by Hansen that there is a difference between the center of mass and center of figure of the moon. Brown has shown that the northern half of the moon is probably the denser half.

The ranges in the times of second contact and the duration of the eclipse, as exhibited in the table above, based on observations of skilled astronomers, appear much larger than one would expect, but the observations of contacts during the eclipse of 1914 contained in *Meddelande fran Lunds Astron. Obs.*, No. 20, show about the same differences. However, it must not be forgotten that the outline of the moon departs from a perfect circle and that the time of the beginning and ending of a total solar eclipse depends very largely on the character of the lunar surface at the point of contact, the beginning of the total eclipse not taking place until the last Baily bead has disappeared.

The marvelous accuracy in the prediction of eclipses shows the wonderful precision and perfection of the science of astronomy. An observation of an eclipse of the sun made three thousand years ago has an important bearing on the most recent refined researches on the motion of the moon!

CHAPTER V

THE SPECTROSCOPE

ASTROPHYSICS, called the "new astronomy," has revealed in a remarkable manner through its discoveries during the past half century the wonderful ability and resourcefulness of the human brain, making it evident that man is gifted with almost infinite powers. From this earth of ours, which astronomy teaches us to be but an insignificant speck among the countless orbs of the universe, man has been able to reach out and ascertain the physical constitution of the sun, and to acquire this information with almost the certainty of a chemist who could make a qualitative analysis provided that an average specimen of the sun's matter could be furnished him. And across millions and millions of miles of space, the far distant suns, the stars, shining by their own feeble light, are made to give up the secrets of their construction. Not only can we learn of the constitution of the sun and stars, but we can also ascertain their effective temperatures as well, and with information thus garnered we can arrange the stars in an orderly sequence, tracing their evolution from the swollen red stars of very minute density, the so-called "giants," through the successive stages of yellow and white to that of the stars of class B, and then on in the descending branch of development as the stars become cooler and more dense. The astronomer believes in evolution and recent researches make evident that our gigantic sun is but a yellow "dwarf," well advanced towards the state of old age and final obscurity. It is by means of the spectroscope that such information is gathered, and by the spectroscope it has become possible to investigate motions, not athwart the sky as the older astronomy was able to do, but towards us or away from us in the line of sight, and to measure these motions in miles per second. The shift of the lines of the spectrum due to motion in the line of sight, which has been confirmed experimentally in the laboratory, has given rise to many interesting developments in astrophysics: the discovery of an entirely new class of bodies, spectroscopic binaries, the measurement of the axial rotation of the sun and Jupiter, as well as providing a magnificent confirmation of the meteoric constitution of Saturn's rings. And quite recently the greatest triumph of the spectroscope has been achieved by Adams and his co-workers at Mount Wilson Observatory being able to determine the distances of the stars and thus find their luminosities compared with that of the sun. Since its birth in 1859, when Kirchhoff discovered the principles of spectrum analysis, astrophysics has advanced by leaps and bounds. In no branch of astronomy has the new instrument of research shown its outstanding value quite as clearly as in the development of the subject of eclipses of the sun. For a clear understanding of the matter it will therefore be necessary to give a brief account of the history of the new astronomy.

The scientific world owes much to John Kepler for handing down to us the three great laws of planetary motion, but his activities extended beyond the realm of mathematical astronomy into the domain of physics. He was the first to show — as he did in his "Dioptrics" — that if a beam of sunlight be allowed to fall upon a prism in a certain way it passes out as a colored beam, giving the colors of the rainbow. And as the great Sir Isaac Newton took Kepler's Laws and used them for finding the law of gravitation, so likewise he extended the physical work of Kepler and by reason and experimentation greatly increased our knowledge. In fact, it is from Newton's labors that the science of spectrum analysis virtually had its birth.

In 1666, Newton, by allowing sunlight to pass through a hole in a shutter and to fall on a glass prism, found a colored ribbon of light, an impure spectrum. His description of the effect observed is so clear that the following is copied from *Optics*, Third edition, page 21.[1]

[1] See also, *The Spectroscope and its Work*, Newall, 1911. For greater details consult *Chemistry of the Sun*, Lockyer, 1887, and *Handbuch der Spectroscopie*, Kayser, Vol. 1.

"In a very dark chamber at a round hole about one-third part ot an inch broad made in the shut of a window I placed a glass prism, whereby the beam of the sun's light which came in at that hole might be refracted upwards toward the opposite wall of the chamber, and there form a colour'd Image of the Sun. The axis of the prism (that is the line passing through the middle of the prism from one end of it to the other end parallel to the edge of the refracting angle) was in this and the following experiments perpendicular to the incident rays. About this axis I turned the prism slowly, and saw the refracted light on the wall or coloured image of the sun first to descend, and then to ascend. Between the descent and ascent when the image seemed stationary, I stopped the prism and fix'd it in that posture, that it should be moved no more. For in that posture the refractions of the light at the two sides of the refracting angle, that is at the entrance of it, were equal to one another. So also in other experiments, as often as I would have the refractions on both sides of the prism to be equal to one another, I noted the place where the image of the sun formed by the refracted light stood still between its two contrary motions, in the common period of its progress and regress; and when the image fell upon that place, I made fast the prism. And in this posture, as the most convenient, it is to be understood that all the prisms are placed in the following experiments, unless where some other posture is described. The prism therefore being placed in this posture, I let the refracted light fall perpendicularly upon a sheet of white paper at the opposite wall of the chamber, and observed the figure and dimensions of the solar image formed on the paper by the light. This image was oblong and not oval, but terminated with two rectilinear and parallel sides, and two semicircular ends. On its sides it was bounded pretty distinctly, but on its ends very confusedly and indistinctly, the light there decaying and vanishing by degrees."

Newton concluded that white light was made up of separate colored rays; by passing the light through the prism, the different rays suffered different amounts of refraction; they were in fact dispersed, and as a result the spectrum consisted of an infinite number of colored images of the round hole lying side by side. In the middle of the spectrum the different rays overlapped, and white light resulted, but the ends remained colored. Newton found that by the use of a slit " an inch or two long, and a tenth or a twentieth of an inch in width," the spectrum became purer; and he even tried a triangular hole, observing greater and greater purity as the vertex was approached. He proved conclusively that the colors came from the sun's light itself and not from the prisms, for he produced spectra with a variety of different prisms, and afterwards combined the prismatic colors together to make white light. He further showed that each ray of light consisted of a single color and possessed a certain definite refrangibility.

Newton's books on optics (published in 1704) are marvels of clearness of exposition, and his fundamental experiments have come down to us almost unchanged. The significant result of Newton's work is that the pure colors, and not white light, occupy the primordial position of importance, since it is possible to form all conceivable colors including white light from a mixture of the pure colors. A determined color in the spectrum is defined through its index of refraction, and since this quantity continually varies, there is an infinite number of spectral colors.

Newton did not see any of the dark lines of the solar spectrum, now known as Fraunhofer lines, though he used a narrow slit and should have been able to see them, — but his prisms were poor (as he himself said). The authority of his great name discouraged further experimentation, and no advances were made for one hundred years. Unfortunately he made some mistakes, first in dividing the colors into seven, the quantitative number of such great attraction and " perfection " to the early scientists, thereby perhaps preventing the discovery of the Fraunhofer lines, and second, in not seeing that various media dispersed differently. As is well known, Newton said that the case of the refracting telescope was a deplorable one, and following his time these telescopes were made of enormous lengths, up to two hundred or even three hundred feet, in order to minimize the color difficulty. However, if Newton had been in a position to make use of different kinds of glass of crown and flint as Fraunhofer was later, it is altogether likely that many other discoveries of great importance would have been made by him.

The long period of arrested progress was broken in 1802 when Wollaston, making use of a slit, repeated Newton's experiments. He then found,[1] "The colours into which a beam of white light is separated by refraction appears to me to be neither seven, as they are usually seen in the rainbow, nor reducible by any means to three, as some persons have conceived; but that, by employing a very narrow pencil of light, four primary divisions of the prismatic spectrum may be seen with a degree of distinctness that, I believe, has not been described nor observed before. . . . The four colors are, red, yellowish-green, blue and violet." He then goes on to describe the dark markings that were seen in his spectrum, these being ill-defined in appearance and evidently taken as the natural divisions between the colors. It is readily apparent that the prisms used by Wollaston were likewise poor in quality, for his experiments were made in a manner that should have brought to view the thousands of dark lines in the solar spectrum. However, a very great advance was made over the work of Newton, for the spectrum of a candle flame and of an electric light were examined. Spectra were found which not only differed from each other in appearance but each of which was entirely unlike that furnished by the sun. The candle and the electric light each gave a spectrum of bright lines, a discontinuous spectrum. The fundamental importance of these experiments was not at the time recognized.

The life of Joseph Fraunhofer[2] had an almost tragic beginning. At the age of fourteen he lived in a dilapidated house in an alley in Munich which tumbled down and buried its occupants in the ruins. The other residents were

[1] *Phil. Trans.*, part 1, 378, 1802.
[2] Clerke, *History of Astronomy during the Nineteenth Century.*

killed, but the boy, who was an orphan, was dragged out, more dead than alive and seriously injured. The Elector Maximilian Joseph was a witness of the accident and to show his commiseration made him a present of eighteen ducats. Besides the purchase of books and of a glass-grinding machine the money permitted his release from apprenticeship to a looking-glass maker. Through study and toil and privation he increased his knowledge of the optician's art and at the age of nineteen entered the glass-making firm of Von Reichenbach and Utzschneider. He devoted himself now with great avidity to a study of lenses for the purpose of improving the refracting telescope. After Newton's time, Dollond had discovered that it was possible to banish most of the color which so interfered with the action of the refracting telescope, by combining a lens of crown glass with one of flint. By means of many experiments with prisms of glass of different varieties, Fraunhofer was able to investigate the best combination of two lenses that would give the most perfect definition with freedom from the disturbing color. In 1817 there was finished the great "Dorpat refractor" with the then unprecedented aperture of nine and a half inches. This telescope in the skillful hands of Struve became one of the most famous telescopes ever in existence. To Fraunhofer the astronomical world is also indebted for the first really serviceable heliometer, that at Königsberg, an instrument which was to play an important part in extending our knowledge of the sidereal universe by permitting the measurement of stellar distances, or, as they are technically called, stellar parallaxes.

In 1814, Fraunhofer not only extended Wollaston's work but introduced great improvements in the method of observing. The slit was retained but placed at a great distance from the prism. Instead of allowing the refracted beam of light to fall on a screen he placed a small telescope directly behind the prism and by this means a magnified view of the spectrum was obtained.

"Into a dark room, and through a vertical aperture in the window-shutter, about 15″ broad and 36″ high, I introduced the rays of the sun upon a prism of flint glass placed upon the theodolite; this instrument was 24 feet from the window, and the angle of the prism was nearly 60°. The prism was placed before the object glass of the telescope, so that the angles of incidence and emergence were equal. In looking at this spectrum for the bright line which I had found in the spectrum of the artificial light, I discovered, instead of this line an infinite number of vertical lines of different thickness. These lines are darker than the rest of the spectrum, and some of them appear entirely black."[1]

The interrelation of these lines and streaks appears to be the same no matter what refracting substance is employed so that, for instance, a particular band is found in each case only in the blue, another is found only in the red, and one can therefore learn to recognize a particular line in the spectrum by noting its position with respect to the prominent lines. Fraunhofer observed that the strong lines did not mark the edges of the colors as Wollaston had supposed and further that the same color is found on both sides of the line, the colors grading by imperceptible degrees from one color to the next. Starting from the violet end of the spectrum, the colors are given the following names: violet, indigo, blue, green, yellow, orange and red. As Lockyer has pointed out, the first letters of these color names make the combination VIBGYOR, — an aid to memory that the beginner may find useful.

Fraunhofer constructed a map of the lines of the solar spectrum, measuring by means of the circle of the theodolite the accurate positions of over 350 lines, though he counted no less than 754. Starting at the red end of the spectrum he called the more prominent lines by the letters of the alphabet. Thus A, B and C denote lines in the red part of the spectrum, D the prominent double lines in the yellow part, which we now know to be due to sodium, while H and K in the violet are the very broad lines caused by calcium. In addition, the small letters of the alphabet were made use of, *b*, for instance, denoting a group of lines in the green due to the element magnesium. Not only are the

[1] *Denkschriften der K. Akad. der Wissen. zu München*, 5, 193, 1814.

prominent lines of the solar spectrum to which Fraunhofer gave names called after their discoverer, but all dark lines in the spectrum whether prominent or faint are known as Fraunhofer lines.

He plotted his lines not according to their wave-lengths as in the manner of all modern maps, but according to a rather arbitrary scale. Now that the solar spectrum has been more fully investigated with the perfected apparatus of the twentieth century, we can go back and gauge at its true value the worth of Fraunhofer's map. Thus we have the opinion of Hartmann of Potsdam to the effect that Fraunhofer made his map with the greatest degree of refinement and secured a precision which has warranted spectroscopists of the past hundred years placing in it the great confidence that they have felt. Fraunhofer's life and work is a splendid example of what one man by patient and careful work can accomplish for the cause of science. Any one who has ever looked into a spectroscope will realize what a colossal work the making of this map must have been.

Since the lines and bands in the color image have only a very small width, it is evident that the apparatus must be most perfect to avoid all aberrations which could either render the lines indistinct or entirely scatter them. The faces of the prism must therefore be perfectly plane; the glass to be used in such prisms should be entirely free from waves, streaks and striae; and the greatest care should be exercised in their grinding and polishing. These and other considerations, such as, that the slit must be parallel to the edge of the prism, Fraunhofer found out by careful experimenting.

Fraunhofer, however, was not content with this work. He wanted to know something of the origin of the lines, and he soon came to a conclusion on this point. It occurred to him that they might possibly be attributed to some illusion caused by the narrow aperture through which the light was admitted. We know that the shape of the slit has something to do with the forms of these dark spaces in the spectrum, but with their simple existence as spaces the slit has nothing to do, the mere shape of the lines being quite a trivial matter. To settle this question beyond doubt he changed the slit which he was using in order to ascertain if this would change his spectrum. He passed light through a small round hole of 15″ in diameter, and allowed it to fall upon the prism placed in front of the objective of the theodolite. It is clear that the color-image seen with the telescope can have only an inappreciable width, and therefore will form only a line; but in this narrow colored width no fine cross lines can be seen. In order to widen this narrow stretch of light into a band wide enough to see, Fraunhofer made use of a cylindrical lens, or a lens which is plane on one side and curved on the other resembling a portion of a cylinder of large diameter. The axis of the cylinder was placed parallel to the base of the prism and hence parallel to the line of light. Consequently, the width of the spectrum would be changed without in any way altering its length. With this arrangement the lines were observed to be exactly the same as when the light comes through a long narrow opening. Hence the bands and lines in the solar spectrum are not caused by diffraction and interference by the light passing through the slit, nor are they produced by any peculiarities in the apparatus. There is, therefore, only one other possible cause for these lines, and that is that they somehow or other belong to the light given out by the sun. Hence the solar spectrum is not a continuous spectrum having light of all colors and all wave-lengths, but is, on the contrary, a discontinuous one in which vibrations of certain lengths are missing from the sum-total which goes to make up white light.

If the sun is remarkable for this discontinuous spectrum, what types of spectra do the other heavenly bodies show? If Fraunhofer could examine the light from a small round opening at a short distance from his instrument, there is no reason, if the light were sufficient, why he should not examine the light from a round body made apparently small by the fact of its situation at a great distance. He examined, therefore, the light from Venus directly *without making the light pass through a small opening,* and he found, after spreading out the light by means of the cylindrical lens, that the same lines appeared in the light of Venus as appear in sunlight. Since, however, the light from the planet is very feeble in comparison with the light received from the sun, the intensity of the violet and red colors of its spectrum are very weak, and on this account even the stronger lines in both these colors are seen with difficulty, though in the other colors they are very easily distinguished. Fraunhofer was able to see D, E, *b* and F perfectly defined and he even recognized in the triplet *b* in the green, two lines, one weak and one strong, although he was unable to see that the stronger of these two lines was really a double line. This, of course, was due to the weakness of the spectrum, and for this same reason the other finer lines could not be distinguished satisfactorily. By measuring the arcs DE and EF it was made certain that the light from Venus contained, as far as could be analyzed, just the same lines in its spectrum as did the light of the sun.

With this same apparatus observations were made on the light of some fixed stars of first magnitude, but since the light of these stars is much weaker than that of Venus, it is natural that the brightness of the spectrum should be much less. In spite, however, of this comparative lack of brilliancy, Fraunhofer was able to recognize with certainty in the spectrum of Sirius three broad lines which appear to have no connection with those of sunlight: one of these is in the red, one in the green, and the other in the blue. In the spectra of other fixed stars of the first magnitude, lines were actually recognized by him, and it seemed certain that these spectra though very faint differed amongst themselves. The observations were made with a telescope of an aperture of about one inch, so that it was impossible for the distinguished pioneer to do more than to point out the way. In such a manner as this was the science of stellar spectroscopy born.

As we have seen, Fraunhofer examined the stellar spectra by allowing the light from the star to fall directly on the prism and after refraction to examine this light by the telescope. This same method of observation is still of the greatest scientific value and it is in this way that Harvard College Observatory has been able to accomplish such an enormous amount of sound research by methods involving the use of the spectroscope. The combination of prism and object glass is called an " objective prism " — or, when used for photographing, the " prismatic camera." If we point such an instrument to any place in the sky, we can view by our eye, or photograph on a plate, the spectrum not of one star only, but of all the stars that are in the field of the telescope.

It is not to be wondered at that a man who had thus brought sun, planet and star within the grasp of a new instrument should not rest content with these observations.

Fraunhofer next investigated at considerable length the spectra of artificial light. In an early part of his paper he states that, on examining the spectra of flames, he found that flames such as that of a lamp and candle, and, indeed, in general the light produced by the flame of a fire, exhibit between the red and yellow of the spectrum a clear and well marked line which occupies the same place in all the spectra. Returning to this subject later, he notes that in transmitting the light of a lamp through the same aperture employed for the examination of the solar spectrum, a line appears which corresponds exactly to the position of the D line in the solar spectrum. In fact, the resemblance to the solar D line is so close that both the bright artificial line and the solar D line are each a fine double line. This was the first step towards the true explanation of the dark lines in the solar spectrum; but it took many years before the true meaning was arrived at, mainly on account of the presence of this bright D pair in practically every flame and under all conceivable sets of conditions.

According to Agnes M. Clerke,[1] " the ubiquity and conspicuousness of the sodium-line long impeded progress. It was elicited by the combustion of a surprising variety of substances — sulphur, alcohol, ivory, wood, paper; its persistent visibility suggesting the accomplishment of some universal process of nature rather than the presence of one individual kind of matter. But if spectrum analysis was to exist as a science at all, it could only be by attaining cer-

[1] *History of Astronomy during the Nineteenth Century.*

tainty as to the unvarying association of one special substance with each special quality of light. It appeared, indeed, without fail where sodium *was;* but it also appeared where it might be thought only reasonable to conclude that sodium *was not.* Nor was it until thirty years later that William Swan, by pointing out the extreme delicacy of the spectral test, and the singularly wide dispersion of sodium, made it appear probable (but even then only probable) that the questionable yellow line was really due invariably to that substance. Common salt (chloride of sodium) is, in fact, the most diffusive of solids. It floats in the air; it flows with water; every grain of dust has its attendant particle; its absolute exclusion approaches the impossible. And withal the light that it gives in burning is so intense and concentrated, that if a single grain be divided into 180 million parts, and one alone of such inconceivably minute fragments be present in a source of light, the spectroscope will show unmistakably its characteristic beam."

The advance made by Fraunhofer in studying the refraction of light through prisms placed the new science on a very firm foundation. This remarkable progress, due to the very marked improvement in the quality of the glass used in the prisms, was rendered possible only by the employment of a telescope of greatly increased defining power. The utilization of the telescope permitted advances in the study of diffraction of light through a narrow aperture or slit, as important and as epoch-making as the investigations concerning the refraction of light. The instrument used by Fraunhofer was essentially a twelve-inch repeating theodolite whose verniers read to 4″. In the middle of the circle and about it there is a plane horizontal disk six inches in diameter which turns on its axis, and whose center lies exactly on the axis of the theodolite. On this disk, the slit to be investigated was placed. The width of the opening in the slit was measured by a micrometer devised for the purpose which could read to 0.0001 inches. Light passing through a narrow opening at the heliostat and falling on the screen with its slit is examined by the theodolite telescope. Fraunhofer discovered the diffraction pattern which consists of a bright strip in the center having symmetrically on each side a series of bright and dark bands, if the incident light is monochromatic. If the incident light is sunlight, instead of having bright and dark bands, there will be found a series of colored bands in which, however, the transitions from one color to another are not sharply defined. The series of bands gradually decrease in brightness in passing outwards from the central beam of light until they are finally rendered invisible. Instead of employing a long, narrow opening at the heliostat, he used a small circular opening, in front of which he placed a cylindrical lens and as a result obtained bands identical with those produced by the other method.

In the process of development of this subject, the natural course after trying diffraction through a single opening, would be to try the action of two, three and more parallel openings. This was exactly the plan followed by Fraunhofer. In order to study the diffraction through a great number of openings he stretched upon a rectangular frame a great many wires of the same thickness, parallel to each other and at the same distance apart. The light was then diffracted through the intervening spaces. In order to be sure that the wires were exactly parallel and at exactly equal distances apart, he made two very good micrometer-screws, and putting these on opposite sides of a frame, he wound on this frame very fine wire, being careful to stretch the wire at a constant tension. If now he soldered along the length of the screws, each wire was thus securely fastened, and by sawing each screw in halves, two similar wire gratings were obtained. By this method gratings were manufactured consisting of wires 0.002 inches thick, and separated by spaces of 0.004 inches. Using a grating of 260 turns of wire, and examining the light which first passed through a narrow opening at the heliostat and then fell on the grating placed in front of the theodolite objective, he found, much to his surprise, phenomena which were entirely different from those observed with a single opening. The aperture at the heliostat was seen exactly as if no grating intervened, but on both sides were seen a series of spectra as perfect as he had hitherto obtained with a good prism.

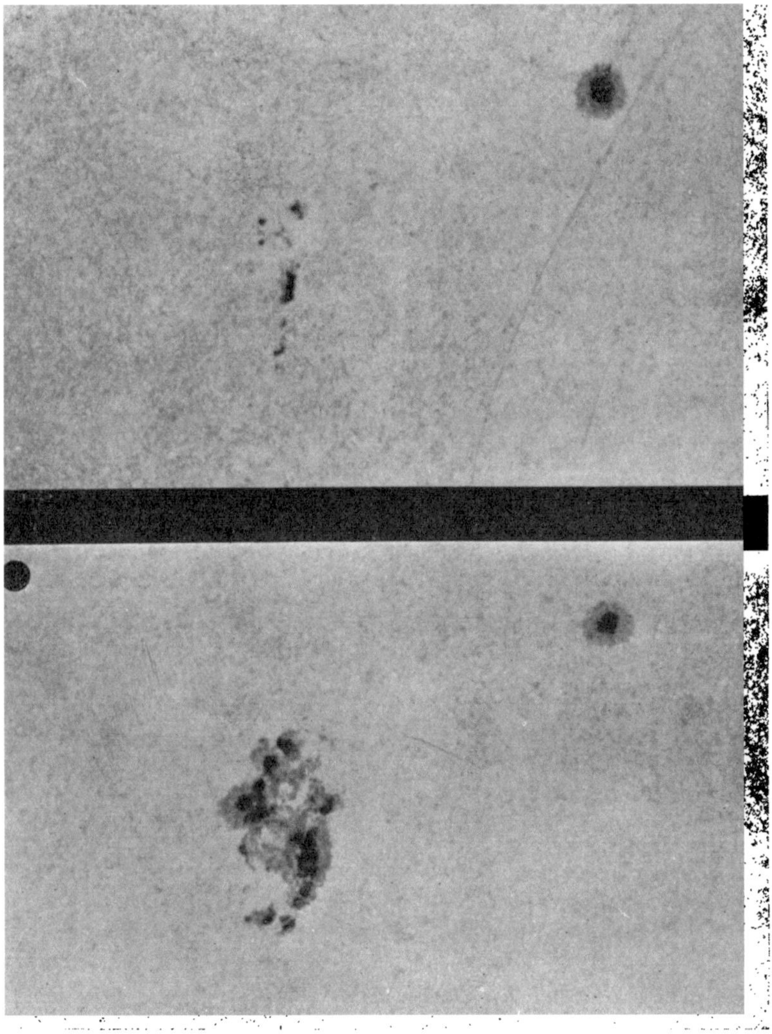

SUN-SPOT WHICH DEVELOPED IN TWENTY-FOUR HOURS
Photographed at Mt. Wilson. The black circle at left of lower picture shows the size of the earth.

The 60-foot Tower Telescope and the Snow Horizontal Telescope of the Mt. Wilson Observatory

The series of spectra gradually increased in length but decreased in intensity, and with his apparatus he was able to count thirteen spectra on both sides of the middle. In order to vary the conditions as much as possible, gratings were made of different thickness of wire and with different spaces. Wire was wound on a screw having as many as 343 threads to the inch. Gratings were also made by scratching parallel lines on a piece of glass covered with goldfoil, through which spectra were observed exactly similar to those observed with wire gratings. Fraunhofer quickly found that the size of the spectra produced did not depend upon the width of the spaces nor upon the thickness of the wires, but upon the sum of these two quantities, or the distance apart of the centers of the wires. Consequently, the finer the screw in whose grooves the wires were stretched, the longer would be the spectra, and it became immaterial of what thickness the wire was or how wide the opening. The quality and accuracy of a grating depends on the precision attained in the attempt to arrange wires of the same width throughout, and in making the wires perfectly parallel, with their centers equally distant.

Although the same dark lines were seen in the spectrum of sunlight when produced by the grating as were found when a prism was employed, one point of difference was revealed in the relative distances apart of the lines in the two spectra. In the diffraction spectrum of the grating, each of the different colors, the red, orange and so on through the blue and violet, is about equal in extent, while in the prismatic spectrum on the other hand the colors at the red end grow more and more crowded together, with the result that the violet reaches to a much greater extent in prismatic spectra than the red. Consequently, the appearance of a diffraction spectrum differs very much from that of a prismatic spectrum, and it is well to be familiar with the two different types of spectra.

In a normal spectrum produced by a grating, the dispersion from the C line in the red to the D lines in the orange is approximately twice the distance from G to the H-line in the violet; whereas with a flint prism of 27° angle the ex-tents of the red and violet are changed in such a remarkable manner that the distance from C to D is only one-half that from G to H.

There is also another marked difference between prismatic and diffraction spectra. In using a prism to form a spectrum, the red rays are the least bent from their incident direction, and as a result we speak of the red as the least refrangible. How is it with the diffraction spectrum with the bright beam in the center and spectra on both sides? Is the violet or the red end towards the bright patch? The red is the least refrangible end in prismatic spectra, but not in diffraction spectra, for the violet is nearest to the bright patch with the red end bent more from the original direction. However, since spectrum analysis was first developed from the use of prisms, a nomenclature has been adopted which, though applying essentially only to prismatic spectra, is used indiscriminately in respect to spectra of both kinds. Thus in referring to the least refrangible rays, the red end of the spectrum is always meant and never the violet, although in the diffraction spectrum the violet is the least refrangible end.

After measuring the angles by means of the theodolite, Fraunhofer was able to formulate two laws: first, the size of the spectra, and their distances from the center (or the dispersion) vary inversely as the distance between the centers of the lines in the grating; and second, the dispersions in the different spectra for any ray form an arithmetical series.

This second law states in other words that the angle of deflection of the same colored beams in the series of spectra formed by the grating are in the ratio of the numbers 1, 2, 3, etc. The experiments from which these results were deduced gave, however, such small angles that the sine, the tangent and the arc do not sensibly differ. If the angles were larger, that is, if the gratings had greater dispersion, it might be possible to determine whether it was the arcs themselves that form an arithmetical series, or some function of these angles. Having this in view, Fraunhofer made further experiments on gratings to find whether it would not be possible to get gratings of greater dispersion. As it was almost impossible to evolve a screw, for the manufacture of his wire gratings, with a smaller pitch than the one he had already employed of 343 threads to the inch, he constructed a machine for scratching parallel lines upon a piece of plane glass covered with gold-foil, and he succeeded in scratching them so closely together that he was able to rule a grating with about 900 lines per inch. If more lines than this were scratched, practically no gold-foil remained. With this grating and with others, he measured the deflection for different rays in different orders of spectra and he found that the sines of the angles of deflection increased uniformly in the different orders of spectra, or in other words $Sin\ \theta = \frac{n\lambda}{\omega}$, where θ is the angle of deflection, n is the order of spectrum, λ is the wave-length and ω the grating space.

Using this equation, Fraunhofer was able to astonish the world by telling them that he had been able to measure the infinitesimal length of a light-wave, for the D line of 0.00005888 cm., which is very close to modern determinations.

To sum up the work of Fraunhofer: he was a telescope maker, and in order to improve his lenses entered into an investigation of prisms in order to study the action of light in passing through them. Using a telescope in connection with his apparatus, he found the spectrum of sunlight filled with lines, now called " Fraunhofer lines." Investigating the spectra of flames he found the bright D line doubled and in the same position exactly as the dark D in the solar spectrum. He examined the spectra of planets and stars. Taking up diffraction, he studied the action of light passing through a single opening, and formulated the laws governing it. He then studied the action of light through parallel openings side by side. Next he attempted to rule lines in a layer of grease spread over a glass plate so thinly that the film could scarcely be recognized by the eye alone. In this grease parallel lines were scratched which were only half as far apart as the lines ruled in gold-foil. After many experiments it was found impossible to rule lines in any layer of grease or varnish much finer than the gold-foil grating. A diamond point is the only method yet known in modern engineering that will provide sharp and sufficiently clean cut lines to permit the construction of finer gratings.

Let us stop a moment and think what it means to rule a grating with a diamond and try to visualize the total lengths of the lines the diamond point must trace, practically without variation. The finest gratings ruled by Rowland have 20,000 lines per inch, and if the grating possesses a ruled surface of 3 x 6 inches, the diamond evidently rules 3 x 6 x 20,000 or 360,000 inches, a distance of nearly six miles. If in ruling this distance the diamond point wears appreciably or breaks down, the previous rulings are useless. Altogether it takes from five to six days continuous working to rule such a grating. What a task Fraunhofer must have had in ruling his gratings! Indeed it must have consumed an enormous amount of time and patience, with numerous vexations caused by imperfect rulings and the fracture of the diamond points. It was only after many trials that he succeeded finally in getting a grating with about 7500 lines to the inch. In order to obtain good definition with his little telescope it was necessary for the slit to be at some considerable distance (as much as 642 feet in one set of experiments) in front of the prism. A great advance in instrumental equipment was later made by Simms, of the celebrated optical firm of Troughton and Simms, who rendered, by the introduction of a lens between slit and prism, the incident rays parallel in falling on the prism. Thus was introduced the collimating lens which increased the compactness of the spectroscope and rendered it almost as we now use it.

In England, Brewster and Herschel undertook a great variety of experiments by means of which they examined the absorbing effects of various colored substances. A piece of red glass permits only the red part of the spectrum to pass through it, the balance of the spectrum being absorbed by the glass, while blue glass allows only the blue end of the spectrum to pass unobstructed. It consequently became evident that the absorbing effect of the terrestrial atmosphere should be carefully investigated before any certain information could be obtained regarding the cause of the dark lines in the solar spectrum. The water vapor in our atmosphere exerts a powerful absorbing action on the light of the sun as it passes through the air on its way to reach the slit of the spectroscope. The absorption commences at the blue end of the spectrum and becomes gradually greater and greater as the sun sinks towards the horizon. For very dense layers of atmosphere, when the sun is near the horizon either in rising or setting, the absorbent action is so great that little of the solar spectrum penetrates except the yellow and red parts, the violet end being entirely absorbed. For this reason the sun is always red when rising or setting, and the color becomes a deeper red when dust or haze in the atmosphere causes a greater absorbing effect. At the time of a total eclipse of the moon, the light from the sun passes through what might be termed a double layer of terrestrial atmosphere with the consequence that there is an increased absorption, the moon thus shining with a dull copper colored hue, and this in spite of the fact that the moon is immersed in the shadow cast by the earth!

By 1833, Brewster announced that he had examined the lines of the solar spectrum with various optical combinations, and had made a map of the solar spectrum on a scale four times that of Fraunhofer. Indeed for some portions of the spectrum the scale was twelve times Fraunhofer's. By observing the absorbing effect of various substances he soon found that some of these substances exerted a general darkening action on the spectrum, whereas other materials produced an absorption in a limited portion of the spectrum. At times the effect was so limited in action that bands or even lines, sharp and distinct, were added to the solar spectrum. These phenomena were so clearly defined that the conviction was borne in upon Brewster that he had obtained "the discovery of a general principle of chemical analysis, in which simple or compound bodies might be characterized by their action on definite parts of the spectrum."

In the course of his investigations, Fraunhofer varied the conditions under which he observed, using different kinds of slits to see if the dark lines in the solar spectrum were caused by some action at the slit itself; but when he found that the form of the slit had nothing to do with the presence of the lines, he concluded that the lines must be truly solar in their origin. Brewster endorsed this view and inferred that if a tube of nitrous oxide gas gave lines identical in character with the solar lines, these lines must then be caused by absorption at the surface of the sun. When in addition he found that many of the lines of nitrous acid gas appeared to be identical in position with some of the Fraunhofer lines, the verification seemed to be complete. Brewster made another important discovery which is described in his own words as follows: "When the sun descends towards the horizon and shines through a rapidly increasing depth of air, certain lines which before were little, if at all, visible, become black and well defined, and dark lines appear even in what were formerly the most luminous parts of the spectrum." As these lines appear both at sunrise and sunset, Brewster announced the discovery that these bands and lines were caused by the absorbent effect of the earth's atmosphere. Since the majority of the lines in the sun, however, appeared without change, he concluded that, "the apparent body of the sun is not a flame in the ordinary sense of the word, but a solid body or coating raised by intense heat to a state of brilliant incandescence."

To quote from Lockyer, *Chemistry of the Sun,* page 41, "It will be seen, then, that the study of the sun was now (1833) in full swing. We had at length, after waiting some centuries, a method of observing a spectrum; we had, further, the fact that there were dark lines in the solar spectrum; that colored flames gave us bright lines; that certain substances stopped some of the light which passed through them, thus producing dark lines. Hence that the solar lines might be produced in the same way."

CHAPTER VI

THE SPECTROSCOPE (Continued)

THE solar eclipse of 1836 played a very important rôle in the history of the development of the new science of astrophysics. The eclipse was visible at Edinburgh. It was not a total eclipse, however, but merely an annular one. At such an eclipse the angular diameter of the moon is smaller than that of the sun, with the result that at the middle of the eclipse there is a ring or annulus of sunlight visible around the edge of the dark moon. This eclipse was observed by Forbes. Referring to Brewster's discovery of atmospheric lines in the solar spectrum, Forbes clearly pointed out that the "telluric" lines were comparatively few in number and relatively unimportant compared with the very great number of solar lines. Moreover, he emphasized the fact that the Fraunhofer lines could not have been caused by the absorbing effect in our terrestrial atmosphere, for if this were true the spectra of the stars should be identical with that of the sun. As the spectra of the stars differed among themselves and were not always identical with the spectrum of the sun, it was manifest that it was not possible to imagine a terrestrial origin for all of the solar lines. Consequently, if the absorbing action of the earth's atmosphere could not be invoked to give an adequate explanation, there remained no other cause than that the origin of the lines must take place within the sun's own atmosphere. If, therefore, this was the true explanation, then an annular eclipse of the sun should furnish a crucial test. The terrestrial lines of the solar spectrum become more and more intensified as the sun's light reaches us through greater and greater layers of our earth's atmosphere, and in like fashion it appeared evident to Forbes that a similar effect must be visible in the lines of truly solar origin. And since the light from the limb of the sun must pass through a much greater thickness of solar atmosphere than that from the sun's center, the lines from the sun's edge, observable at the time of the annular eclipse, should be much intensified in comparison with the Fraunhofer lines ordinarily visible from the sun's center. When the eclipse took place no change whatever was observed by Forbes in the number, position or intensity of the lines, and the obvious conclusion was drawn by him that, "This result proves conclusively that the sun's atmosphere has nothing to do with the production of this singular phenomenon."

To us living in the twentieth century, it seems passing strange that Forbes did not try the experiment of comparing the limb and the center of the sun by forming an image of the sun on the slit of his spectroscope by means of a projecting lens. This plan was undoubtedly in his mind as the following shows: "Had the weather proved unfavorable for viewing the eclipse, I intended to have tried the experiment by forming an image of the sun by using a lens of long focus, stopping alternately by means of a screen the interior and central moiety of his rays, and restoring the remainder to parallelism by means of a second lens, then suffering these to fall on the slit as before. The result of my experiment during the eclipse seemed, however, so decisive as to no marked change being produced at the sun's edges that I have thought it unnecessary to repeat it."

The scientific world in the first half of the nineteenth century knew little of laboratory methods and it was not at that time a habit of mind to test any theoretical conclusions by means of experimentation. In fact it was not until the year 1866 that the idea of Forbes was carried into execution and a projecting lens utilized for the examination of local phenomena such as the spectra of the limb or of sun spots.

In 1845, there was performed in England by W. H. Miller, the very experiment that later in Kirchhoff's hands was the crucial proof of the cause of the Fraunhofer lines. This experiment consisted in passing sunlight through various vapors heated to incandescence in a flame and noting the changes in the solar spectrum. He observed that in passing sunlight through glowing sodium vapor, the D lines were intensified. It is surprising that he appears to have sought no explanation of this amazing fact, and the surprise is all the greater since Miller's experiments were carried out with the express purpose of testing the theory that the Fraunhofer lines were actually produced by absorption in the sun's atmosphere.

This line of investigation was continued in 1849 by Foucault in Paris who was able to use a new method of heating salts and metals to the glowing point by the use of the electric arc. He focused an image of the sun on the arc itself and this procedure allowed him to observe at the same time the spectra of the arc and of sunlight. Incidentally he was surprised to discover the extreme transparency of the arc which caused only a faint shadow in the sunlight. This experiment manifested to Foucault that when the two spectra were exactly superimposed, the D line of sunlight was made considerably darker, proving that the arc absorbed the D rays; but that when, on the contrary, the two spectra jutted out one beyond the other, the D line appeared darker than usual in sunlight, yet stood out bright in the electric spectrum, thus demonstrating the perfect coincidence in position ot the dark and bright rays. "Thus the arc presents us with a medium which emits the rays D on its own account, and which at the same time absorbs them when they come from another quarter." It was difficult to explain how it was possible for a glowing flame to furnish at the same time both bright and dark lines, — and the riddle of the solar lines seemed a hard one to solve. As early as 1850, Stokes seemed to have clearly grasped the solution of the problem, and he inserted a discussion of these matters in his university lectures at Cambridge.[1] Moreover, he seems to have been the first to localize the cause of the yellow D lines since he observed that the bright line was absent from a candle flame when the wick was snuffed clean, and from an alcohol flame when the spirit was burned in a watch-glass. Although he so clearly saw that the D lines,

[1] See Lockyer, *Chemistry of the Sun*, p. 51.

whether bright or dark, were caused by sodium vapor, and therefore correctly concluded that the absorbing effect of the sodium was in the neighborhood of the sun, he made no further experiments to test his deductions; and the honor for the great discovery waited another nine years. Thus again was demonstrated — which has frequently happened in the history of science — that though many investigators have converged towards the same goal of discovery, yet the prize has awaited the fortunate one who should make the critical experiment, which in itself at times has been one of little difficulty. Thus in 1855 Angström, and in 1859 Balfour Stewart by their experiments came very close to the true solution of the problem.

In 1859, Kirchhoff showed [1] for the first time that in order that sodium should be in a condition to absorb from the light of other sources it must itself be at a cooler temperature. " Fraunhofer has remarked that in the spectrum of a candle flame two bright lines appear which coincide with the two dark lines D of the solar spectrum. These bright lines can be easily intensified in a flame into which some common salt is put. I formed a solar spectrum by projection and I allowed the solar rays thus formed to pass through a strong salt flame before falling on the slit. If the sunlight were sufficiently subdued, then in place of the two dark lines D two bright lines appeared; if the intensity increased beyond a certain amount then the two dark D lines showed in much greater intensity than without the presence of the salt flame. The spectrum of the Drummond light contains as a rule the two bright sodium lines if the illuminating spot of the calcium cylinder has not long since passed the glowing point; if the cylinder remains undisturbed then these lines become weaker and finally completely vanish. If they have disappeared or are faintly visible, an alcohol flame into which cooking salt has been placed and which is brought between the calcium cylinder and the slit, causes two dark lines of exceptional blackness and sharpness, which in that respect agree with the lines D of the solar spectrum, to show themselves in their place. In this manner the D lines of the solar

[1] For complete details see Kayser, *Handbuch der Spectroscopie,* Vol. I, p. 81.

spectrum are artificially produced in a spectrum in which they are naturally not present. I conclude from these observations that colored flames, in the spectra of which bright sharp lines are found, so weaken rays of the color of these lines when such rays pass through the flames, that in place of the bright lines, dark ones appear just as soon as there is brought behind the flame a source of light of sufficient intensity in the spectrum of which these lines are otherwise lacking. I conclude further that the dark lines of the solar spectrum, which do not find their origin in the earth's atmosphere, are caused in the glowing solar atmosphere by the action of those substances which in the spectrum of a flame produced bright lines at the same place. We thus assume that the bright lines coinciding with D in the spectrum of a flame always arise from sodium contained in it; the dark D lines in the solar spectrum therefore allows us to conclude that sodium is found in the atmosphere of the sun. . . . In order that the D lines should come out dark in the spectrum of the Drummond light it is necessary to use a salt flame of lower temperature."

In these experiments, the salt flame was kept constant, the intensity of the sunlight being weakened or strengthened at leisure. In other words, this may be expressed by saying that the Fraunhofer lines in the spectrum of the sun are dark only in contrast with the more brilliant background of the sun itself. If this dazzling surface could be removed, and the comparatively dark lines of the solar spectrum could then be viewed against a background still darker, then by contrast the spectrum lines would appear as bright lines on a dark background where formerly they had existed as dark lines on a bright background. This change in the sun's spectrum, as we shall see later, takes place at the time of a total eclipse of the sun. The student of spectroscopy will save himself needless worry if he will remember that bright and dark are always to be considered as *relative* terms only.

A spectrum of bright lines on a dark background is said to be a bright-line, or an emission spectrum. On the other hand, a spectrum of dark lines on a bright background is called a dark-line, or an absorption spectrum. If a chemical element is heated to the point of vaporization, its spectrum consists of a bright-line spectrum. Sodium gives a very simple spectrum, consisting mainly of two very strong lines in the yellow part of the spectrum, the well-known D lines. If pure metallic sodium alone is used, the spectrum consists of these D lines. If the sodium is in chemical combination with chlorine, and the sodium chloride, or common salt, is heated to incandescence, the same D-lines due to sodium are shown in the bright-line spectrum. Or if any other compound of sodium is heated, the same D-lines result. The manner of the heating is of no consequence; a pinch of common cooking salt may be placed on the wick of an alcohol flame, a paper soaked in a saline solution may be placed about the burner of a Bunsen lamp, or a grain of salt may be put on the carbon of the electric arc or on one of the poles of an electric spark — the lines of sodium will always appear, the color of the lines will be exactly the same, and the wave-lengths of the lines will be unaltered no matter what the chemical compound in which the sodium is found or the manner of heating that salt to incandescence. Since the total light of the sodium consists mainly of two lines in the yellow, then when sodium is burning, it will give off yellow light only and the color of the flame will appear yellow to the eye.

If another element like lithium is examined, whether pure lithium or lithium in compound with some other element or elements, it will give its own peculiar spectrum of bright lines, and these bright lines will be found not to coincide with the D lines due to sodium. Since lithium burns with a red light, the prominent lines in its spectrum will be found at the red end of the spectrum. Some metals like sodium show a very simple spectrum, with very few lines; other metals show more lines, the greatest number of lines appearing for any one element being due to the presence of iron. It makes no difference how the iron is heated to incandescence, it makes no difference whether the iron is a piece of scrap or of polished steel, the spectrum will consist of thousands of bright lines in all the colors of the spectrum. Each of these many thousands of lines has its own particular wave-length. It is the business of the spectroscopist to find the value of the wave-length of each and every line. Some of the lines are faint, some strong, some are narrow, some broader, some are very sharp, others more fuzzy in appearance, but no matter what the quality of the iron or how vaporized, the spectrum is the same with lines of practically identical wave-lengths. Although there exist a great many chemical elements, the spectrum of each of which consists of many lines, while others of the elements have even thousands of lines in their spectra, it may almost be said that no line in any one spectrum coincides *precisely* with a line in any other spectrum. If, therefore, it is possible to determine the *exact* wave-length of a line in a spectrum, though the chemical origin of this line may be unknown, we shall have a ready means of identifying the elemental source of this line.

The principles upon which spectrum analysis depends are found in Young's *General Astronomy,* page 213, as follows:

1. A *continuous spectrum* is given by every incandescent body, the molecules of which so interfere with each other as to prevent their free, independent, luminous vibration; that is, by bodies which are either *solid* or *liquid,* or if gaseous, are *under high pressure.*

2. The spectrum of a gaseous element, *under low pressure,* is discontinuous, or in other words made up of *bright lines,* these lines being characteristic, that is, the same substance under similar conditions always gives the same set of lines, and generally does so even under widely different conditions.

3. A gaseous substance *absorbs* from white light passing through it *precisely those rays of which its own spectrum consists.* The spectrum of white light which has been transmitted through it then exhibits a " reversed " spectrum of the gas; that is, one which shows dark lines instead of the characteristic bright lines.

The third law, the great discovery of Kirchhoff, may be stated in other words as follows: The relation between the emissive power for each wave-length and the absorptive power for the same wave-length at the same temperature is identical for all bodies, and is in fact equal to the emissive power of an absolutely black body at the same wave-length and temperature. We are quite familiar with a similar effect in the realm of sound. If a voice singing or speaking sounds a note of a certain pitch in a room where there is a piano, one string of the piano will vibrate in unison with the voice, the particular piano string taking its motion from the oscillations of the air. Similarly, a tuning fork in vibration will set in motion another tuning fork nearby which is tuned to the same pitch.

The work of Kirchhoff, therefore, in connecting the emission of light with absorption gives the means of determining the chemical composition of the sun. According to our present ideas, the photosphere of the sun, the portion of the sun we *see,* consists of gases under such very high pressure that the molecules cannot vibrate independently; and in consequence the spectrum of the photosphere must be *continuous,* a ribbon of light without breaks from red to violet. The photosphere is surrounded by a cooler layer of gases under low pressure, the so-called " reversing layer." If the spectrum of these gases could be examined entirely separated from the bright photospheric background, they would exhibit the gaseous, or bright-line spectrum. Under ordinary conditions, the light from the photosphere shines through the cooler gases of the reversing layer, and certain wave-lengths are there absorbed by the gases of the reversing layer, so that the spectrum of the sun comes to us as a reversed spectrum, of dark lines on a bright background; these dark lines, however, as stated above, are dark only in contrast with the much brighter photospheric background. To determine the constitution of the sun, it becomes therefore necessary to compare the bright line spectra of the various elements with the dark line spectrum of the sun. This comparison may be made by two different methods, either by viewing or photographing with a suitable instrument the spectrum of the sun and the comparison spectrum side by side, or by an exact determination of wave-lengths in the solar and in the comparison spectrum.

But the spectrum of the sun consists of many thousands of lines. It is evidently quite possible, and even highly probable, that there should be very close agreement between some of the many lines in the sun and an equal number of lines in the spectrum of the element under investigation. These concurrences might be the result of pure accident. Kirchhoff investigated this possibility. A particular line in a comparison spectrum may exactly match in position a line in the solar spectrum. If the agreement is due entirely to chance, then by the laws of probability, it is equally probable that the line in the spectrum under consideration may or may not match a line in the sun's spectrum, or speaking mathematically the chance of an exact match taking place fortuitously is one out of two. If two lines agree in each spectrum, then the possibility of this happening by chance is but one out of four. Kirchhoff found sixty lines of iron to agree with sixty lines in the sun. The chance that this coincidence of all sixty lines is purely accidental is expressed by the number $\frac{1}{2}$ raised to the sixtieth power. At the present time over two thousand lines due to iron have been identified in the solar spectrum. If the coincidence of this large number of lines were the result of pure accident it would represent a chance of one in 2 raised to the 2000th power. This number is about equivalent to 100 followed by no less than 600 ciphers! (If one has nothing better to do, one might take a large piece of paper and put down the number one and follow it by six hundred and two zeros. Then one could divide it off into millions, billions, trillions, etc., and invent a name for this huge number!) The chance that the lines of iron and the Fraunhofer lines in the sun should agree in position entirely by accident is therefore infinitesimally small. But when in addition, we compare the appearance of the lines in the two spectra, and find that a strong line in the spectrum of the sun is matched by a strong line in the spectrum of iron, and a weak Fraunhofer line is matched by a weak iron line, then we see the utter impossibility of the coincidences being the result of mere chance.

What is true of iron is equally true of the other elements investigated. It accordingly seems perfectly certain that we are able to ascertain the chemical constitution of the sun by means of the spectroscope even though we are looking at the sun across a space of ninety-three millions of miles.

Astrophysics thus being placed on a very firm foundation, the infant science was immediately recognized throughout the scientific world to be of the very greatest importance to physicists, chemists and astronomers. But in spite of the almost universal recognition, there were a few doubting Thomases. In Chambers's very excellent *Descriptive Astronomy,* edition of 1867, page 27, is found the following: "Spectrum analysis has taken a start within the last two or three years, chiefly owing to the assertions made that it enables us to ascertain something about the physical condition of the sun. The subject is too purely a physical one, and also in too infantine a state to require notices in these pages at present, though the time *may* come."

It is remarkable to read the early history of spectroscopy and learn of the great opposition of certain scientists to the acceptance of Kirchhoff's proof; and it was but natural that there should be many claims to priority. By 1867, however, Pritchard of Oxford summarized the general feeling of the scientific world in the following words: "It may safely be asserted of Foucault in 1849, of Stokes in 1850, of Angström in 1855, and of Balfour Stewart in 1859, that each of them was in possession of an enunciated truth, which, had they traced to their natural and inevitable consequences, must have led to that grand generalization which will immortalize the name of Kirchhoff, and which forms one of the happiest and most remarkable discoveries of modern times."

With the trail so surely blazed, the path pointed out the direction of the researches to be taken by Kirchhoff's successors. The quest had a two-fold interest, for not only did the new method serve as an infallible chemical test of terrestrial substances, but it gave also a ready means of determining the constitution of the sun and also of the more distant suns, the stars. Physicists, astronomers and chemists vied with each other in pushing forward the researches as rapidly and thoroughly as possible, and opticians and instrument makers came to the assistance of the scientists by furnishing improved forms of apparatus. It was manifestly necessary to investigate the solar spectrum and the spectra of the various chemical elements. These investigations had necessarily to be carried out with the spectra, produced on as large a scale as possible, and the measurements of the positions of the lines required the very highest degree of precision attainable. To denote the position of a line, some more accurate method was necessary than that of placing it in the red or blue of the spectrum. Fraunhofer and also Kirchhoff used a rather arbitrary scale. Newton proved that a difference in color meant a difference in refrangibility. Fraunhofer went a step further and demonstrated that a difference in color meant a difference in the length of the wave causing the light. Since the time of Thomas Young, 1802, it has been known that light is a wave phenomenon, the waves, somewhat similar to water waves, moving transversely to the direction of motion. The length of the wave from crest to crest, or from trough to trough, is known as the wave-length, and this length might be measured in fractions of an inch, or foot, or meter. All scientists, whether they live in America or in Germany, or whether they speak English, Japanese or Russian, now use the meter as the unit for measuring the wave-length of light. A meter divided into ten thousand million parts, or 10^{10} parts, is called a "tenth-meter." This very small distance is known as the "Ångström Unit," or more simply as the "Angström," and it is the unit for measuring wave-lengths. The position of a line is known by its wave-length, and the more precise the investigation the more accurately do we need to know this quantity. The K-line in the solar spectrum has a wave-length according to Rowland of 3933.-826. This is printed either as λ 3933.826, or 3933.826 A. (We shall adopt the latter notation.)

Knowing the wave-length of light of a certain color, it is a very simple matter to calculate the number of waves that enter into the eye in a single second of time. All that is required is to know the velocity at which light travels. It is now known that light of all colors, whether red, blue or violet, travels at the same rate of speed, viz., the almost incredible velocity of 186,300 miles, or in round numbers 300,000 kilometers per second. To simplify the calculation, suppose the light is violet, of wave-length 4000 A. The length of these waves from crest to crest is 4000×10^{-10}, which is 4.10^{-7} meters (i.e. 4 divided by ten million). 300,000 kilometers per second is 300,000,000 meters or 3.10^{8} meters per second. If therefore we divide the distance that light travels in one second by the wave-length, we will find the number of waves. For light of 4000 A, seven hundred and fifty millions of millions (750,000,000,000,000) of waves enter into the eye in one second of time. If the light under consideration is red instead of violet, inasmuch as the wave-length of the red is longer than that of the violet, fewer red waves will consequently enter into the eye in a given time. These tiny waves impinge on the retina of the eye, creating motions which when telegraphed to the brain cause the sensation of violet or red light. The mechanism by means of which the minute motions produced in the eye by light waves cause the sensation of light has never been completely discovered. Professor John Joly in *Philosophical Magazine, 42,* 289, 1921, gives a very plausible explanation based on the quantum theory. He assumes that the origin of vision and color perception is to be sought in the liberation of electrons under light stimulus within a photoelectric substance or substances existing in the retina. In the case of the rods, rhodopsin, being such a photosensitive substance, acts as the basis of vision, and it is assumed that the same substance in the cones is responsible for the color vision. The sensitivity of the eye to faint light is extraordinary. Henri and des Baucels have found that the retina is sensitive to a minute amount of light energy which, when expressed in physical units, amounts to 5×10^{-12} erg. The quantum for green light is 4×10^{-12} erg, and hence it is assumed that one quantum, by the liberation of a single electron, is sufficient to cause the sensation of light. The action taking place in the eye seems to be quite analogous to that occurring in the light-sensitive film of the photographic plate, the latent image being caused by the movement of electrons.

The new method of research in the hands of Kirchhoff soon resulted (1861) in the discovery of two new chemical elements, caesium and rubidium. The spectroscope manufactured by Steinheil consisted of four prisms, three of 45° and one of 60°. The collimator and telescope had apertures of one and a half inches, with a focal length of eighteen inches. One half of the slit was covered by a totally reflecting prism by the aid of which two spectra could be examined side by side and direct comparisons made. Unfortunately, it was necessary to set to minimum deviation by hand, a very slow process, and as a consequence Kirchhoff shifted the prisms only occasionally, a procedure which greatly impaired his measures. He investigated the spectra of a large number of elements, and also measured the positions of the lines in the solar spectrum from A to G, though at first he only published the region from D to F, since his eyes could not stand the strain of such continuous measurement and failed him. His measures being referred to an arbitrary scale, it was necessary to reduce them to wavelengths, and this was done later by Airy, Gibbs, Watts and Hasselberg.

It has been said with great verity that these were splendid days for the laboratory scientist for the reason that each and every observation, no matter how trivial, was almost certain to prove to be a new discovery. Without attempting to trace the details in the further development of the new science, we shall try to give only the more important names in the honor roll of fame: Plücker, Hittorf, Crookes, Miller, Huggins, Rutherfurd, Angström, Secchi, Janssen, Lockyer, Young, Vogel, Cornu, Liveing and Dewar. In the twenty years following Kirchhoff, no less than ten new elements were found by the aid of the spectroscope. Naturally many mistakes were made and wrong conclusions drawn. The chief cause of the mistakes was the presence of many impurities in the elements investigated and hasty identification of lines through insufficient accuracy in wavelength determinations.

A new epoch in the history of spectrum analysis was inaugurated in 1882 by the work of Henry A. Rowland, whose gratings,[1] plane and concave, permitted a hundred-fold increase in accuracy in the determination of wave-lengths. Rowland's success came through the construction of a long screw, almost free from errors, mounted in a dividing engine in such a manner that it was practically possible to eliminate the few remaining errors of the screw. The precision that must be attained in the manufacture of gratings of the very first quality may be stated as one requiring that the average line of the grating shall be correctly placed to about one one-thousandth part of the grating space. This, for the finest Rowland gratings of 20,000 lines per inch, means that each line ruled on the grating must not on the average differ from its true position by so much as the minute quantity of one twenty-millionth part of an inch!

The highest degree of success was attained by ruling with a diamond point on speculum metal, the incident light thus being reflected from the grating surface. Gratings were ruled both on plane and on spherically concave surfaces. The largest Rowland gratings were six inches in diameter, and ordinarily the greatest radius of curvature for the concave gratings was twenty-one and a half feet. The concave gratings reduced the spectroscope to the greatest simplicity of slit, grating and photographic plate. No lenses of any kind were necessary to bring the light to a focus, and hence all aberrations introduced by the lenses and all absorption of light by the glass were eliminated, with a consequent great increase in the extent of the ultra-violet region. The concave grating is generally used in the laboratory with the "Rowland mounting" possessing two tracks for carrying grating and photographic plate perpendicular to each other, the slit being placed accurately at the intersection of the two tracks. If grating and photographic plate are each perpendicular to the arm joining the two, there then results a "normal" spectrum, or one in which the distances between the lines are directly proportional to the wave-length.

Compared with prisms, concave gratings have the following advantages: (1), An enormous increase in dispersion, definition and resolving power. To equal Rowland's grat-

[1] See Kayser, *Handbuch der Spectroscopie,* Vol. 1, 121 and 397. See also article Screw, *Encyclopaedia Britannica.*

ings in these respects, in the neighborhood of the D lines, it would be necessary to have prisms of the very first quality added to prisms with a total prism base of fifty inches of glass. (2), The spectrum produced by the grating is "normal" and not "prismatic," thus permitting wave-lengths to be determined with much greater facility and much greater accuracy. (3), A much greater extent of the ultra-violet is secured. (4), The astigmatism of the grating increases the length of the lines, and this principle, combined with the overlapping of images from the spectra of different orders, not only permits a great increase in accuracy, but also a more ready determination of absolute wave-lengths. The grating, however, has some disadvantages, chief among which is that the incident light is divided between the central beam and many different orders of spectra, with the consequent result that there is a great weakening of light in any one spectrum. In the investigation of objects giving little light, like the stars, prisms are almost universally used. In most laboratory researches and in work on the sun, gratings are generally used. Although Rowland's method of grinding the screw was not a secret, it is only in comparatively recent times that the excellence of the Rowland gratings has been equalled by others, by Michelson of the University of Chicago, by the Mount Wilson Observatory, and by Lyle and Merfield of Australia.

Coincident with Rowland's manufacture of the grating came the discovery of the modern photographic dry plate with its great increase in sensitiveness. The two most important pieces of work in this new epoch of discovery in astrophysics have been Rowland's great map of the solar spectrum, and the publication in the *Astrophysical Journal* of the wave-lengths of the lines in the solar spectrum with the tracing of as many lines as possible to their chemical origins. The more prominent names connected with investigations in the laboratory and on the sun itself are: Rowland, Jewell, Kayser, Runge, Paschen, Rydberg, Eder and Valenta, Exner and Haschek, Langley, Abbot, Schumann, Lyman, Humphreys, Zeeman, Hale, Deslandres, St. John, King, Meggers, Russell, Saunders and Fowler.

CHAPTER VII

THE SURFACE OF THE SUN

AT THE time of the Greeks how simple it was to explain all of the known facts about the sun! No supposition was necessary other than that the sun was a ball of fire. It was not at all known what fire was beyond the fact that heat was manifested, — but this deficiency in knowledge seemed of little importance. As a further illustration of the elemental beliefs of primitive peoples there might be given the following legend regarding the origin and motion of the sun which is found among the Yuki tribe of American Indians.[1]

"In the beginning there was no land; all was water. Darkness prevailed everywhere. Over this chaos of dark water hovered On-coye-to who appeared in the form of a beautiful white feather, hence the love of the Yukis for feathers. In time the spirit became weary of his incessant flight through the murky space and lighted down upon the face of the water. Where he came in contact there was a whirlpool that spun his body round and round. So rapid became the motion that a heavy foam gathered about him. This became more dense and expanded in width and length. It gathered up the passing bubbles until it was a huge floating island. On the bosom of this rested the snowy form of On-coye-to. As he lay upon this island for an almost endless flight through the dark space, the idea of a permanent resting place came into his mind. So he made the land and divided it from the water. From the form of a feather he assumed that of a man, and rested upon the land. Still there was no light, and his spirit was troubled. On-coye-to saw afar off in the firmament a star, 'po-ko-lil-ey,' and resolved to visit it and learn how it emitted its sparkling

[1] *Smithsonian National Mus. Report,* 326, 1902.

light. After a long journey, he arrived there and found a large and beautifully lighted world, inhabited by a numerous, hospitable people. Still, he saw not whence came the light. He was allowed free access to all habitations save one, the 'sweat house.' This was guarded night and day, and was accessible only to sick persons. Finally a great hunt was planned, and as time drew near all was prepared for the occasion. But On-coye-to feigned sickness that he might investigate the sweat house. When the morning arrived for the hunt he was too ill to acompany the hunters. A council was held to determine whether this stranger should be admitted to the sweat house, which is even now a sacred place with the Yuki tribe, and it was decided to give him the benefit of this house of medicine. A few old men were left to administer to his wants and to see that all went well. As he entered the sweat house he was almost blinded by the light that flashed upon him, but as he became accustomed to it, he looked around him and discovered its origin. Hanging high over his head in several baskets were as many beautiful suns. Having found the fountain of light he waited patiently until the old men were all asleep, then climbing cautiously to what seemed the brightest of the suns, he seized it, slipped from the sweat house and made his way rapidly towards his own world. He was hotly pursued by the indignant warriors, but he arrived safely after many adventures. He hung the sun in its basket in the far east, then surveyed it. It did not light up to suit him, and he moved it a little higher. Still it did not suit him, so he continued to move it, on and on. And he is moving it to the present day." Thus the Indian accounts for the moving of the sun, and thinks not that the earth moves.

As knowledge has gradually been accumulated regarding our central luminary, and as information is secured about the laws, physical and chemical, to which the sun is subjected, the more and more difficult has become the problem of finding an explanation adequate to satisfy all of the facts. And now in the twentieth century, it is discovered that each and every one of the countless billions of chemical atoms that form the sun is a solar system in miniature, with the result that solar theories must be revised and hypotheses revamped in order to take account of electrons and protons. The sun is a typical star, but owing to its proximity it may be examined in detail, its surface, the spots that are of such great interest, the reversing layer, the chromosphere and the far-flung corona. The stars are so far distant that they appear practically as points of light even in our largest and best telescopes, and consequently little more can be learned of them than what depends on their surface brightness. On account of the closeness of the sun, its surface may be covered up by the interposing moon, and as a result, envelopes of chromosphere and corona are shown. Although similar envelopes unquestionably exist on the stars, they can never be made manifest to us. By the study of eclipses a wealth of knowledge is acquired concerning the sun, but to gain an adequate idea of what additional information is thus secured, it will be well to give a brief résumé of the salient points of solar research.

Compared with terrestrial standards, the distance to the sun is colossal and its diameter enormous. The problem of finding the distance of the sun is one of the most important as well as one of the most difficult in the whole of the science of astronomy. The importance lies in the fact that the distance from earth to sun is the unit for measuring all celestial distances, except that to the moon, the solar distance being called the *astronomical unit*. The yard, or the meter, is the standard of length for all measurements in civilized countries. The " standard yard," or the " standard meter," is a bar of certain composition, of definite shape, whose length can be determined by precise measurements at known temperatures. The standards are kept in London or Paris, but various prototypes are widely distributed. Certain advantages would result from the employment of but one standard of length, the meter, which is now almost exclusively used in all scientific measurements. The reason is not because the length of the meter is more valuable than that of the yard, but rather that the decimal system employed with the meter is simpler than the

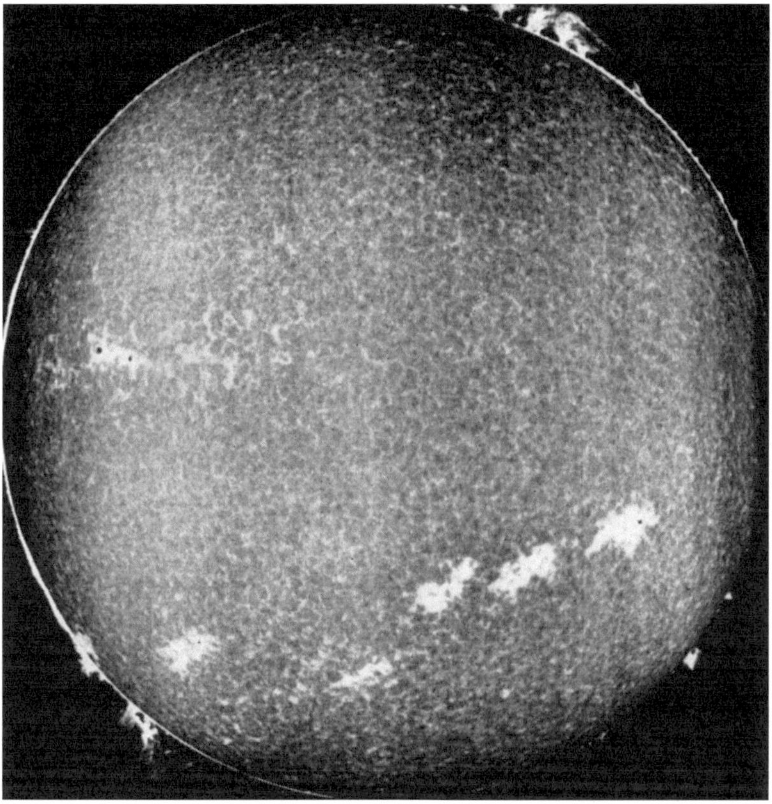

THE SUN IN THE LIGHT OF GLOWING CALCIUM VAPOR
Two spectroheliograms showing unusual solar activity. Yerkes Observatory.

N

May 6, 3^h 40^m | May 7, 3^h 37^m | May 8, 3^h 23^m

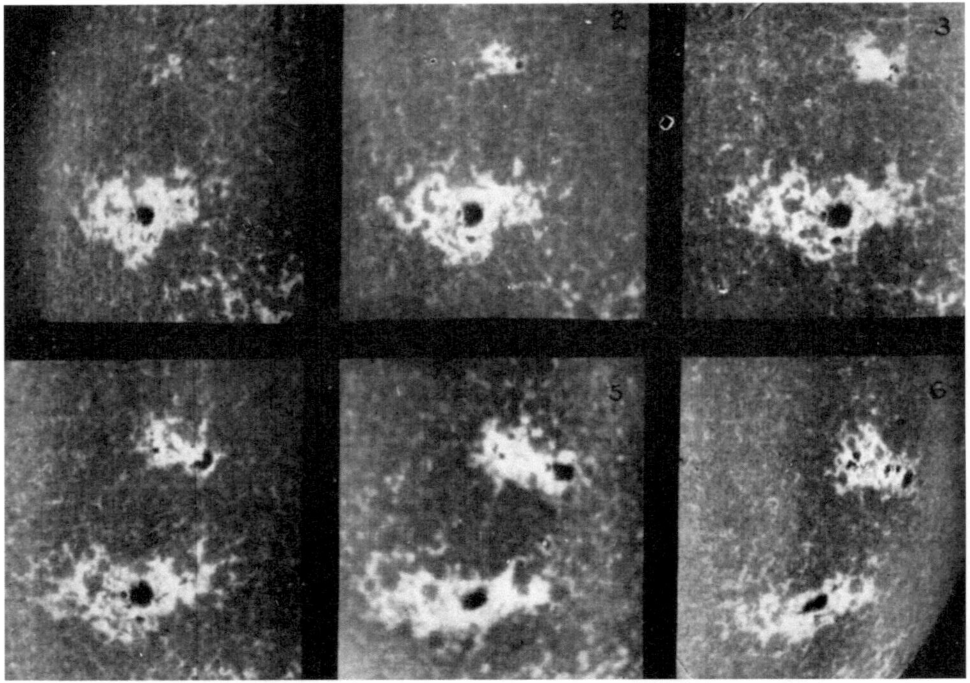

May 9, 3^h 10^m | May 11, 4^h 19^m | May 13, 4^h 19^m

SPECTROHELIOGRAMS SHOWING DEVELOPMENT OF COMPANION SPOT

Photographed May 1907 by Fox and Abetti with Yerkes refractor.

more cumbrous division of the yard. At the end of the eighteenth century the meter was designed by the French to represent the ten-millionth part of the quadrant of the earth, so that its circumference should be exactly forty million meters. At that time, however, the size of the earth was known with little accuracy. If, therefore, the meter were actually to represent a certain definite and fixed fraction of the earth's size, it would be impossible to use it as a standard for the reason that its length would change with every revision of the earth's measurement as new geodetic operations were carried out, and would alter with any variations in the earth itself, which changes geodesy and geology tell us are continually taking place.

By means of the careful researches of mathematical astronomy stretching back over hundreds, and even thousands of years, an accurate plot can be drawn to scale of the orbits of all the planets and satellites of the solar system. The periods of each have been determined accurately, the shapes of their orbits, their inclinations to the ecliptic, etc. All of the planetary distances have been found by the astronomer by referring them to the astronomical unit, the distance from earth to sun. To know the complete scale of the astronomical plan in miles, it is necessary to know accurately at least one distance, either that of the earth to the sun, or to one of the planets. Manifestly, the nearer the planet, the more accurately can the distance from the earth be determined. Celestial distances are usually found by the methods of the surveyor who wishes to ascertain the width of a river which he cannot traverse. A base-line is measured with as great precision as possible, and from each end of this base-line, angles are measured to some well-defined object on the other side of the river. A civil engineer would be in a quandary if a base-line of only three inches in length was available for measurement on his side of the river and it was necessary to determine, with the precision necessary for planning a cantilever bridge, the distance across the river. This is exactly the problem that confronts the astronomer in attempting to measure directly the distance of the sun, since the only base-line available is a chord of the earth. Instead of expressing the distance to the sun in miles or kilometers, the astronomer knows this unit ordinarily by means of the small angle at the sun subtended by the earth's radius, and this angle is called the "solar parallax." As there are many different terrestrial radii, the earth not being a sphere, the equatorial radius is assumed. The astronomical unit when expressed in miles must therefore be subject, not only to all of the errors of the astronomer in carrying out his measurements, but also to those due to the work of the geodesist in determining the shape and size of the earth and in referring these measures to the standard yard or meter.

Although this is not the place to discuss the details of the determination of the solar parallax, brief references will be made to the more promising methods. It is impossible to secure the distance to the sun by direct measures and recourse must be had to the determination of this distance indirectly by the measurement of other distances in the solar system. Great were the expectations aroused by the transits of Venus in 1761 and 1769, and again in 1874 and 1882. By the year 1882, the dry plate had been invented and it was anticipated that photography would revolutionize our knowledge of the fundamental unit. But alas! the "black drop," and the atmosphere of Venus caused the observations to be practically a dismal failure. Another attempt by this method will not be possible until the year 2004, the year of the next transit of Venus.

Great advances in precision were made by Gill in his measures of Mars and some of the minor planets by means of the heliometer. The discovery in 1898 of the planetoid Eros, which at perihelion comes closer to the earth even than Mars, gave to the astronomer a splendid opportunity of determining the solar parallax for the reason that photography could be applied to the problem. Although much was accomplished at the opposition of 1900–01, when Eros at its nearest approach was 30,000,000 miles from the earth, more will be effected from observations made in 1931 when the planet was within half this distance.

Methods based on the law of gravitation furnish the

means of determining the distance of the sun. E. W. Brown's magnificent investigation of the motion of the moon gives 8.″778 as the value of the solar parallax. Other gravitational methods furnish the mean value of 8.″780.

The velocity of light can be utilized in three different manners: (a) by the constant of aberration; (b) by the eclipse of Jupiter's satellites; and (c) by the velocity of the earth in its orbit by utilizing the Doppler principle of measuring the motion in the line of sight from spectra of stars or of planets. St. John and Nicholson at Mount Wilson, using large dispersion on the spectrum of Venus, find the solar parallax of 8.″813. (*Publications A. S. P.*, *32*, 332, 1920.) They give also the values by other observers.

A summary of the best values[1] of the solar parallax (see Abbot, *The Sun*) are:

From heliometer work on minor planets	8.″807
From the Eros campaign	8. 807
From all gravitational methods	8. 780
From the eclipses of Jupiter's satellites	8. 799
From the velocity of light and constant of aberration (20″.47)	8. 803

The present accepted value of the solar parallax is 8.″80, corresponding to a distance of the sun of 92,900,000 miles, with an uncertainty of about 50,000 miles, or an error of about one part in 2000. To walk the distance to the sun at four miles per hour and ten hours per day, 68 years would be necessary for the first million miles, or 6300 years for the total distance. An express train going at sixty miles per hour would take 175 years, while light which travels at the great speed of 186,300 miles per second takes 499 seconds.

If we know the angular diameter of the sun we can readily find its linear diameter of 865,000 miles, which is 109.5 times the diameter of the earth. Perhaps the best method of visualizing the huge size of the sun is to compare it with the distance to the moon, which in round numbers is 239,000 miles. If the earth could be placed at the center of the sun and the moon were allowed to revolve in her orbit about the

[1] See also, Newcomb-Engelmann, *Populäre Astronomie.*

earth there would be plenty of room inside the sun for the moon to make her monthly journey, since the moon would be little more than half way out to the sun's surface. A spot on the sun having a diameter of 8000 miles, or the size of the earth, would be regarded as fairly small and a telescope would be needed to detect it. Dividing the linear diameter of 865,000 miles by the angular diameter 1920″, it is found that 1″ at the sun corresponds to 450 miles, which is equivalent to 725 kilometers. These are useful quantities to remember, especially when considering the subject of eclipses.

Since the surfaces of spheres are proportional to the squares, and the volumes proportional to the cubes of their radii, it is readily found, by squaring and cubing 109.5, that the surface of the sun is 12,000 times that of the earth while its volume is 1,300,000 times the volume of the earth.

The mass of the sun is 332,000 times the mass of the earth, a relation that can be determined by comparing the distance a body falls towards the earth in a second of time in obedience to the law of gravitation with the distance that the earth falls towards the sun in the same interval of time and also in obedience to the law of gravitation. In one second the earth travels eighteen and a half miles of her annual journey about the sun, but as this is accomplished without friction we feel no sensation from this rapid flight. In going eighteen and a half miles, the earth deviates but one-ninth of an inch from a straight line. As the earth weighs six thousands of millions of millions of millions of tons, which is 6×10^{21} tons, the sun weighs 2×10^{27} tons. This is such a colossal number that it makes little difference in our comprehension of it whether it is the American ton of 2000 pounds, or the English ton of 2240 pounds, or the long ton that we pay for when we buy coal or the short ton that is furnished by the dealer when the coal is placed in the bin.

Knowing the mass and volume of the sun compared with that of the earth, we find the density of the sun is 0.255 times that of the earth, or about 1.4 times the density of water. The attraction of gravity at the surface of the sun is 27.6

times that which it is at the earth's surface, so that a man weighing 150 pounds would weigh over two tons if transported to the sun, and his feet would be so heavy, even if the footing were secure, that he would not have strength sufficient to lift them.

The small density of the sun, being only one-quarter that of the earth, is one of the most significant bits of knowledge connected with the study of the sun. All theories of evolution point to the fact that the sun and earth are made of the same materials. Indeed Rowland was wont to say that if the earth were heated to incandescence it would give a spectrum identical with that of the sun. The low density makes it evident that the sun cannot be a solid like the earth, nor indeed can it be a liquid, and it must therefore be a gas, the terrific heat of the sun being sufficient to vaporize all known terrestrial substances. The condition of immense heat and enormous pressure caused by gravitation on the sun cannot be even distantly approximated in our laboratory experiments. Unquestionably the sun does not obey the laboratory laws to which such perfect gases as oxygen, nitrogen and hydrogen are subjected. For the complete explanation of solar phenomena it is necessary to proceed from known conditions to those impossible to duplicate in the laboratory by the difficult and uncertain methods of extrapolation. The steps of scientific development must accordingly be carefully planned and wisely thought out, or else the path of truth may lead away from the goal of progress rather than towards it.

The spherical portion of the sun that we see is called the *photosphere.* According to Young, *The Sun,* page 109, the "photosphere is a sheet of self-luminous cloud; possibly like the clouds of our own atmosphere, with the exception that the droplets of water which constitute terrestrial clouds are replaced in the sun by drops of molten metal, and that the solar atmosphere in which they float is the flame of a burning fiery furnace, raging with a fury and an intensity beyond all human conception." This notion of the photosphere propounded a third of a century ago has been greatly modified by modern research. By means of convection currents, gases from the interior of the sun are brought to the surface. There set free from the enormous internal pressure and meeting the cooler temperatures of outside space the gases expand. The pent-up energy being suddenly released, there is a rapid fall of temperature, and according to Young's theory, small solid or liquid particles are formed. By gravity these sink back to the solar furnace, there to be changed again to the gaseous form. The rising and falling back again to the solar surface of the " drops of molten metal " cause a continued rain of meteors on the sun's surface. Two decades ago these " drops " played a very important rôle in many solar theories, but particularly in that regarding radiation pressure (Chapter XXI). The temperature of the photosphere " seems to be certainly in excess of 6000° absolute, Centigrade. There are no substances, so far as known, which can exist except as vapors in these conditions. Hence, it seems reasonable to suppose that the sun contains no solids or liquids, unless perhaps in sun-spots, and that its substance, as we see it, and within the layers we see, is altogether gaseous." [1]

In carrying out investigations regarding the sun's surface there are two points of view that astronomers should never forget. The first is that the photosphere can be viewed or photographed only *through* the superposed layers of the solar atmosphere. The photospheric spectrum which must be continuous from red to violet without breaks, can never be obtained. The second point to be remembered may be visualized by analogy with the earth. At an elevation of three and a half miles above sea-level, atmospheric air has its density cut in half. But gravity at the sun is nearly twenty-eight times its value on the surface of the earth. Allowance being made for the hundred-fold diameter of the sun when compared with that of the earth, it is seen that within ten or twelve miles of the sun's surface there would exist one-half of the total material in the sun's various layers of gases were it not for the enormous temperature of the sun. The decrease in pressure upwards from the sun's surface is consequently extremely rapid. The importance of

[1] Abbot, *The Sun*, p. 243, 1911.

this point cannot be over-emphasized, since most investigators seem to forget that the change in pressure must take place at such a very accelerated rate.

The sun may be viewed with a telescope by the use of solar eye-pieces of various kinds; or by projecting the sun on a screen, the solar image being brought to a focus by slightly drawing out the telescopic ocular. If the telescope is of moderate size, and the definition good, the surface of the sun looks like " rough drawing paper, or like curdled milk seen from a little distance." The use of a large telescope and of moments of exquisite seeing that come but rarely reveal an infinite wealth of detail. The best drawings of the sun are by Langley who describes the surface of the sun as that of " snow-flakes sprinkled sparsely over a grayish cloth." Before the application of photography great was the diversity of opinion concerning the ultimate nature of the light-giving particles of the sun. We learned then of " rice grains," of " willow leaves," of " thatch-straw " and of " granules "; and the various camps in favor of one or other designation were about equally divided.

Photographs of the sun may be obtained by the method known to the great host of camera users, that of the focal-plane shutter. This consists essentially of a slit of variable size that can be driven by means of a spring across in front of and close to the photographic plate. By regulating the width of the slit opening and the tension of the spring, exposures may be varied at will. The exposures necessary to obtain good photographs of the sun depend mainly on the size of the telescope, the sensitivity of the photographic plate and the method of development. The modern dry plate of great rapidity does not permit the securing of solar photographs of the greatest detail. With the sun, where there is such an abundance of light, it is unnecessary to make use of the fastest plates which are primarily for the purpose of decreasing the exposures. Better results may be secured by the use of finer grained, slower and more contrasty plates. As a matter of fact, the modern dry plate cannot furnish the exquisite definition secured by the old wet-plate process, and for this reason the superb solar photographs of Janssen at Meudon are unsurpassed even at the present day.

The best conditions for observing the sun are found not more than one per cent of the time spent at the telescope. Under these maximum conditions, the skilled eye can see finer details than can be portrayed on the photographic plate. According to Langley, the "snow-flakes" are in the neighborhood of 50 to 100 miles in diameter, and these in turn are made of flakes similar in form, but of one-fifth the dimensions. These small particles cover but one-fifth of the surface but radiate three-quarters of the total solar light, and hence they must shine with an intensity twenty-fold that of the darker portions of the sun.

The most noticeable feature of photographs showing the whole solar disk is the darkening that is found near the edge of the sun. This darkening is caused by the absorption by the sun's atmosphere, a beam from the limb of the sun passing through a greater layer than one from the center. The sun or moon when rising or setting looks reddish to us on account of the absorption of the blue and violet by our terrestrial atmosphere. In a similar manner, the solar atmosphere absorbs more and more of the violet end of the spectrum as the limb of the sun is approached, and the maximum of radiation is displaced towards the red. A similar shifting of the wave-length maximum is found when comparing the spectrum of a sun-spot with that of the photosphere. Abbot (*The Sun,* page 107) gives measures of the distribution of radiation over the sun's disk from the center outwards to the edge. If the sun could be viewed without the absorptive effects of its own and the earth's atmospheres, the maximum intensity in the spectrum would be shifted by an appreciable amount to the violet, and the sun instead of appearing as yellow in color to the eye would look bluish.

The exquisite photographs of Janssen show the solar granulation in splendid detail, the features being sharp and well-defined. Other parts of the same photographs are quite smudgy in comparison, as if the solar surface were in violent commotion. To these parts Janssen gave the name *réseau photosphérique.* If photographs taken in rapid succession

THE SUN'S SURFACE

" Like snowflakes sprinkled over a grayish cloth."
Photographed by Janssen at Meudon, September 9, 1883.

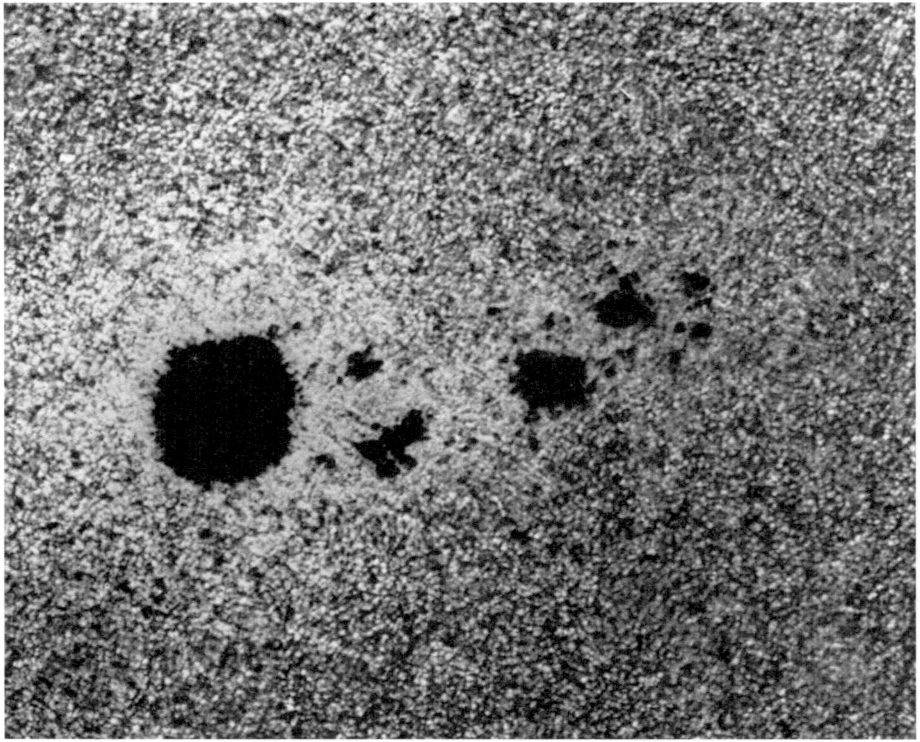

Sun-Spot Group

Photographed by Janssen at Meudon, April 1, 1894.

are examined it is found that the smudgy and ill-defined portions exist at different parts of the solar image. The simplest and most apparent explanation seemed to be that these changes afforded positive evidence of violent commotion on the sun which certainly must exist there on account of the very high temperature. This evident explanation, however, seems not to be the true one. Any motion of the atmosphere of the sun, or of the earth's atmosphere close to or far away from the photographic plate would have the effect of blurring the photographic image. To make a long story short — the general opinion regarding the *réseau photosphérique* is that it is not a solar phenomenon at all, but is caused by the disturbance of the air heated in the telescopic tube by the sun's beams. The portions of the photograph in good definition represent the true granulation of the solar surface. Direct photographs similar to those of Janssen have been made by Hansky of Poulkova and Chevalier of the Zô-Se Observatory in China. Exposures made in rapid succession give the following information regarding the ultimate nature of the photosphere as depicted by photographs: (1), The solar granules have a diameter of 400 to 1200 miles, though at times smaller granules are seen no more than 100 miles in diameter. (2), They are generally circular, or elliptical in shape. (3), They coalesce to form larger particles. (4), These granules are the "clouds" of Young's theory. (5), The life of one of these clouds is very short, the majority of them last for approximately half a minute, and practically none exist longer than a few minutes. (6), The displacements vary widely in direction and in velocity. The movements range from zero to thirty kilometers per second, though occasionally higher speeds are observed. (7), The granules in fact seem to be the summits of a fleecy structure of condensed particles. In fact, they represent[1] on an enormous scale a phenomenon similar in appearance to a storm-tossed and choppy sea when viewed aloft from an airplane.

The most prominent features of the solar surface are the spots. Individual records of these exist as far back as the

[1] Astrophysical Journal, *27*, 12, 1908.

Chinese, but their real history begins with the invention of the telescope; and they were independently discovered by Galileo, Fabricius and Scheiner. There is a great wealth of scientific literature connected with the study of spots — but here it will be possible to give only the salient features, and a brief summary of our present knowledge, which, alas! is far from complete. As in the study of the photosphere, the details of the appearance of spots can be better observed visually than by photography. A normal spot consists of an *umbra,* more or less round, surrounded by a less darkened *penumbra,* the structure of the constituent parts of a spot differing much from each other and from the surface of the photosphere. The roundish shapes of the granular photosphere are changed in appearance to the straw-thatch of the penumbral filaments which exhibit a great wealth of detail, and these in turn transform into the smooth, black, velvet-like appearance of the umbra. The umbra is, however, not uniformly black but is more or less cloudy in appearance when conditions of seeing are at the best. Generally associated with spots are the *faculae,* or bright patches on the sun. These exist at slight elevations above the average surface of the sun, and are best seen when near the edge of the sun where the greater absorption of the sun's atmosphere and the elevation of the faculae make them visible by contrast with the darker surroundings. The umbra of a spot is dark only by contrast with the more dazzling photosphere, yet withal it is not black, for it is more brilliant than the electric arc. During the progress of an eclipse of the sun a spot has been observed by Evershed to be much brighter than the dark limb of the moon occulting it. Langley estimates that the blackest spot gives 500 times as much light as an equal area of the full moon.

Spots vary in size, from a few hundred miles to 50,000 miles in diameter in the case of the very largest spots. Groups of spots may extend across one-sixth of the diameter of the sun, and consequently may be visible to the naked eye when the sun is seen through haze or near the horizon, or when the eyes are protected by smoked glass. Spots have usually a short life, sometimes disappearing in a day or two, sometimes lasting for a month or longer. The longest record is that of the spot seen during the years 1840–41 which persisted for eighteen months. Owing to the violent solar motion the changes in sun-spots are naturally very rapid, the disintegration of spots taking place usually by the formation of a bright " bridge " which may be shot across a spot at a high rate of speed (compared with terrestrial motions), of as much as one thousand miles per hour. The elevation of sun-spots with respect to the general photospheric level is still being actively discussed, even after the lapse of a century and a half since 1769, when Dr. A. Wilson of Glasgow first propounded his well-known theory, that the foreshortening of the penumbral filaments as the spot neared the sun's edge showed that the spots were saucer-like depressions in the sun's general surface. Spots having been seen which confirmed the theory while others seemed to disprove it, the arguments have gone on *pro* and *con*. For a more complete discussion, see Agnes M. Clerke, *Problems in Astrophysics*. In view of the very rapid change in gravitation near the surface of the sun already noted, it is altogether probable that the elevation at which sun-spots exist differs less than fifty kilometers from that of the photosphere. This small allowable difference of altitude is not sufficient to permit the saucer-like sinks needed for Wilson's theory.

The most evident fact concerning the spots is their periodicity, first discovered in 1843 by Schwabe, the average period being 11.13 years. There are marked differences in the length from maximum to maximum, and equally great divergences in the intensity of the various maxima so that it must be said that spots are very irregular in their regularity. The individual periods range between 7.3 and 17.1 years as extremes. But no matter how divergent the period is from the mean, the rise to maximum spottedness always consumes less time than the descent to minimum. On the average the intervals are 4.62 years and 6.51 years, respectively. The sun-spot curve thus resembles the light curve of the average variable star of long period, and also those of the Cepheids. The importance of this fact is here emphasized for the reason that we have learned from the researches of Abbot that the total radiation of the sun varies in amount, and as a consequence the sun must be regarded as a variable star of long period.

Various attempts have been made to examine the sun-spot curve by the methods of harmonic analysis in order to find any secondary periods that may underlie the main period of 11.13 years. The most notable attempts in recent years have been by Schuster,[1] Hirayama,[2] Kimura,[3] Michelson,[4] and Larmor and Yamaga.[5] Kimura and Michelson each examined 160 years of sun-spot records from 1750 to 1910, and although the material for examination was the same, the conclusions reached are greatly at variance. The former decides, "The 11-year period is not so conspicuous as generally considered. Although the most important of all, yet to my surprise there are a great many periodicities lying between 8 and 12 years, most of them being of considerable relative amplitude." Kimura predicts the form of the sun-spot curve up to the year 1950, the predictions giving a maximum of spots at the beginning of the year 1914, with an intensity about equivalent to that of the maximum of 1905–6. According to observations at Mt. Wilson, however, Nicholson[6] finds that the maximum did not take place until August, 1917, over three years later than the predicted time, while the activity was considerably greater than at the preceding maximum. Michelson concludes that "with the exception of the 11-year period and possibly a very long period (of the order of 100 years), the many periods found by previous investigations are illusory." Quite similar deductions are drawn by Larmor and Yamaga who even go so far as to state that when the periodic part is removed the residue of the sun-spot activity is of a fortuitous sporadic character, not amenable to further analysis. Various attempts have likewise been made to explain the sun-spots by means of the attractions of the

1 *Phil. Trans. Roy. Soc.* 206.
2 *Tokyo Sugata, 3,* 9.
3 *M. N. R. A. S., 73,* 543.
4 *Astrophysical Journal, 38,* 268.
5 *Royal Soc. Proc. A., 93,* 493.
6 *Publ. A. S. P., 31,* 223, 1919.

planets, particularly those of the giant of the sun's family. But Jupiter's period is 11.86 years, and even when other periods of other planets are superimposed, no success has followed the attempts to explain sun-spots by planetary influences. The cause of the cycle indeed seems as much unknown today as when a hundred years ago Schwabe first began his systematic observations.

Spots are never found more than 45° from the equator and seldom at the equator. A curious distribution of the spots in latitude manifests itself during the progress of the solar cycle. Approximately at the time of minimum, sun-spots begin to manifest themselves in two zones, more than 30° north and south of the equator. With the lapse of time, the spots are found closer and closer to the equator, the sun-spot maximum taking place with the spots in zones 17° north and south. The disturbance gradually dies out in latitude 8° or 10°, after a lapse of 13 or 14 years from the first outbreak. Before the final flickering out, the new cycle has begun to manifest itself, so that near sun-spot minimum there are found four zones of disturbance, two near the equator, and two farther north and south.

Many serious attempts have been made to connect sun-spots with various phenomena, solar and also terrestrial. The correlation with the solar manifestations mainly rests on secure ground, but with some of the earthly influences, the connection seems rather far-fetched. If it is hotter than the average at a certain locality in the United States, like St. Louis, or if mayhap at the same time, Northern France is having a cold spell, and if coincidently there is a large spot-group on the sun, an astronomer, or usually pseudo-astronomer, is always found who informs the daily press that the sun-spot is the cause of the heat (or cold). The famines in India, the potato crop in Ireland, the price of corn in England, the rain-fall in the Island of Mauritius, the financial panics of Wall Street all have been investigated by statistical methods, and each and all have been found to pass through periods of the same length, and to be connected with the sun-spot period. It has many times been said that "figures never lie." It is quite true that the figures themselves do not tell falsehoods, but many and varied are the interpretations that may be placed on these figures. The president of every big business corporation well knows that if his company shows in two successive years the same approximate amount of earnings, it is very easy for him to declare a substantial dividend, or to charge certain amounts to " improvements," and curtail or even pass a dividend. The whole question depends on whether he wishes to please the fifty-one percent of the majority stock-holders (of whom he is one), or to take advantage of the forty-nine percent (of whom he is not one). It is quite possible, and indeed probable, that the weather and rainfall are connected with variations of solar activity as evidenced by sun-spots, and that other manifestations of meteorological changes are also correlated,—but to prove the connection " is another story." The weather is " not made on the spot," it depends on a vast variety of conditions and it is therefore difficult and well-nigh impossible to single out the solar cause from all of the possible influences that may affect the weather. It might not be out of place to remind all such investigators that a great variety of periods have been found in the sun-spot cycle itself, but that none of the subsidiary periods appear to have any reality, no matter how firmly substantiated by figures they seem to be. Most of the meteorological dependences seem to be equally illusory. There are, however, many terrestrial and many solar phenomena which have a well proven connection with sun-spots. These will be given below in brief form, though some of them will later be expanded more fully.

Records kept at the Greenwich Observatory and extending over nearly a hundred years, show (1), that the diurnal range of the magnetic declination, and (2), that the horizontal force of the magnetism flowing through the earth, follow the sun-spot fluctuations not only in the main 11-year period but even in the small and secondary variations. The parallelism is so intimate that it is at once evident that if the cause can be found of the sun-spot cycle there also will be found the true explanation of the variation of terrestrial magnetism. Although the records are not so complete, (3), aurorae and (4), magnetic storms are more abundant when spots are numerous. The same may be said (5), of faculae, and (6), of prominences. (7) The shape of the corona changes with the sun-spot period; at minimum of spots there are long equatorial extensions, and well defined polar rays. (8) The conclusions of Köppen, Stone, Gould, Nordmann, Newcomb, Abbot and Fowle, Arctowski and Bigelow are that there is a change in the mean temperature of the earth, small in size, but amounting to 0.7° between sun-spot maximum and minimum, the earth being cooler at sun-spot maximum, which, in fact, is quite contrary to the ordinary popular belief.

The contribution of the spectroscope regarding spots and related phenomena will be given in a subsequent chapter.

The state of our knowledge of the angular diameter of the sun is far from satisfactory. The accepted value of this fundamental quantity comes from measurements made many years ago with the heliometer. The discussion by Schurr and Ambronn gave the angular diameter of the sun when at its mean distance from the earth to be equal to 1920″.0 ± 0″.03. The investigations showed that there were slight but unmistakable differences between the polar and equatorial diameters, but in spite of these differences it was assumed that the sun was spherical. In this procedure they followed the example set by the great master, Auwers.

In the present day of refinement in solar research it seems quite unsafe to assume that the sun is necessarily spherical or that the changes taking place in the polar and equatorial diameters are so small that they are beyond the possibility of measurement. The heliometer, as the name signifies, was invented for the purpose of measuring the diameter of the sun, the first heliometer coming from the hands of Fraunhofer (p. 77). It is a very valuable and refined instrument of measurement, and in addition to its use on the sun it has been extensively employed in the determination of stellar parallaxes. The latter research has shown its limitations. When used on the sun, the great heat of the sun causes the same effect which is found in every instrument of precision when exposed to the sun's rays, namely, the instrument is put out of accurate adjustment. The heliometer being specially sensitive to changes in adjustment it is not surprising to find, by reference to the original observations, that the same observer, using the same instrument on two successive days and under good observing conditions, would obtain differences in the measurement of the equatorial diameter of the sun amounting to 1″, 2″, 5″, or even 10″ or more. In addition, every observer has a " personal equation " in that he may constantly measure a quantity smaller or larger than the average observer. Each astronomer who has made visual or photographic measures is familiar with the problem of systematic errors. With the sun it is difficult to know how to allow for irradiation which causes a spreading of the image.

The only practicable method of combining observations inconsistent among themselves is to group all the measures together and take the mean, at times assigning different weights. If the quantity of observations is sufficiently great and the number of observers numerous enough, it is quite safe to assume that the peculiarities of any one individual will have little effect in the final mean. For the determination of the diameter of the sun, Auwers had at his disposal no less than 15,000 observations made by 100 observers working between the years 1851 and 1883. The measures, however, were not all made by the heliometer.

The large systematic errors to which these measures were subject have had a curious effect on future observations, a parallel to which is not found in any other department of astronomical investigation. The result has virtually been to terminate all observational measures of the solar diameter. What would it avail any astronomer if he should measure the diameter of the sun with the heliometer on hundreds of days, spending perhaps many thousands of hours of diligent toil in the research, only to find that his results differed from the generally accepted value? To the astronomical world this difference would probably be regarded as a proof, not that the accepted value was in error or that the diameter of the sun was changing, but rather that the measures of the individual though made with the greatest of refinement were subject to personal equations. At best the astronomer secures little reward for the hard toil devoted to his skilled researches. Moreover, he is a human being and naturally he desires some compensation other than that of advertising to the rest of the scientific world that he is a faulty observer. And this, strange to relate, is the state of affairs in the enlightened days of the twentieth century when hundreds of thousands of dollars are spent each year in solar research! Astronomy appears to be virtually saying that no improvements are possible in the work done nearly half a century ago by the heliometer — and so we shall assume that the sun is spherical and without change.

Are no other methods available? Why not try photography? Surely the great resources of modern astronomy can conquer any difficulties! There are indeed no difficulties of any note connected with the photographic processes, for excellent photographs of the sun are being secured daily. The main difficulty to be overcome is the same one that affects heliometer work, namely, the heat of the sun. This alters the length of the telescope tube and changes the focal length of the object glass so that the exact scale of the photographs is uncertain. These changes, however, do not alter the relative scales of the polar and equatorial diameters. Some indefatigable worker, therefore, has already waiting to his hand some hundreds of thousands of solar negatives to be measured and discussed for the purpose of determining whether or not the sun is spherical.

The only method apparently available of eliminating the effect of the heat of the sun and at the same time applying photography is clearly outlined by Hayn [1] who applied it with great success at the eclipses of April 17, 1912, and August 21, 1914. By means of photographs taken at the time of a solar eclipse, not however, during totality, but during the partial phases, the shape and size of the sun can be determined, the shape and size of the moon also, and in addition, the times of contacts of the limbs of the sun and moon usually secured at eclipses.

A research somewhat similar in character has been carried

[1] *Astronomische Nachrichten, 201,* 185, 1915.

out by Henry Norris Russell [1] who measured photographs taken at Harvard College Observatory for the purpose of determining the position of the moon with respect to the stars. To secure satisfactory photographs, it was necessary to make the exposures on the moon the thousandth part of those required for the stars, and at the same time the telescopic object-glass had to be shielded from the light of the moon so that the photographic plate might not be fogged. The difficulties to be surmounted by Hayn's method during the progress of the eclipse are not as great as those overcome at Harvard in thus photographing the moon. To give information of the highest degree of reliability, it is necessary to know the latitude and longitude and the observed times with great precision. On this account it would be preferable to test Hayn's method at a fixed observatory rather than to attempt it under the temporary conditions of an eclipse expedition. The best locations will be those nearest the path of totality. At the eclipse of September 10, 1923, three great American observatories and the Mexican National Observatory at Tacubaya were conveniently located, the American institutions being Mt. Wilson, Lick and Lowell observatories. With the superb instrumental equipment of the Mt. Wilson Observatory and an eclipse ninety-eight percent total, what magnificent scientific results might not have been obtained if the skies had not been densely cloudy!

At the total eclipse of January 24, 1925, the Van Vleck Observatory with its 20-inch visual refractor was in the path of totality, and consequently a series of photographs were taken between third and fourth contacts. On account of the poor conditions of seeing, everywhere observed at this eclipse, the limbs of both sun and moon were not sharp. Measures [2] of ten Van Vleck photographs and two made with the Yale refractor show the difficulties of making proper allowance for irradiation, or spreading, of the photographic images. It might be possible to allow for uniform irradiation but it is more difficult to take care of differential irradiation due to the fact that the intensity of the solar radiation de-

[1] *Harvard Annals, 72, 76* and *80.*
[2] *Astronomical Journal, 37,* 32, 33, 1926.

creases toward the limb, thus greatly affecting plates taken when the solar cusp is small near the time of totality. Evidently the only practicable method would be to take photographs both before and after totality.

At the brief eclipse of April 28, 1930, three photographs were taken [1] with the 36-inch Lick refractor, in spite of widespread clouds.

The eclipse of October 21, 1930 gave the opportunity of taking sixteen photographs with each of two cameras, one 38-inches, the other 15-feet in focal length. The lenses were fitted with graflex shutters and were stopped down. With each camera, eight plates were taken at one-minute intervals starting shortly after first contact, and another eight plates ending shortly before fourth contact. The plates were used to determine the times of first and fourth contacts. Differences were found in the intensities of the plates caused by variations in transparency and alterations in the rate of the exposing shutter. The plates were measured [2] independently by Marriott and Pitman. The mean of their measures shows that the 15-foot camera gave the time of first contact 1.2 seconds earlier, and the time of fourth contact 1.2 seconds later, than was obtained from the 38-inch camera; an effect unquestionably due to irradiation. Assuming a correction of $+ 1''.50$ to the sun's mean longitude, the photographs showed that the *American Ephemeris* position of the moon needed corrections of $+ 5''.54$ in mean longitude and $— 0''.25$ in latitude. For predicting the eclipse, Washington had used a correction to the moon's longitude of $+ 5''.47$ obtained from occultations. The photographs showed that first contact was 1.5 seconds earlier and fourth contact 0.1 seconds later than the predicted times.

[1] *Publ. A. S. P.*, *42*, 145, 1930.
[2] *Astronomical Journal*, *41*, 129, 1931.

CHAPTER VIII

MODERN ECLIPSES BEFORE 1878

"Tycho sought the truth
From that strange year in boyhood when he heard
The great eclipse foretold; and, on the day
Appointed, at the very minute even,
Beheld the weirdly punctual shadow creep
Across the sun, bewildering all the birds
With thoughts of evening." — NOYES.

ASTRONOMY owes much to the eclipse of the sun visible in Copenhagen on August 21, 1560, and to the fact that a red-headed, freckled-face boy of fourteen, destined to become the greatest and most careful of observational astronomers since the time of Hipparchus, had his keen young imagination fired by watching the "orange ember in the sky wane into smouldering ash." As a consequence, this boy, Tycho Brahe, resolved to devote his life to unravelling the deep mystery of these strange happenings. The romantic incidents of his productive life have been beautifully told by Alfred Noyes. From his own printing press in his observatory of Uranibourg appeared the *Historia Coelestis* in which appears a long list of eclipses beginning with one visible in Rome on March 28, in the year 5 A.D.

At the eclipse of May 3, 1715, Halley referred to that of the year 1140 as the last one previously observed in London. Although visible not far from London, Hind finds from investigation that this eclipse was not seen in the city itself, so that it can be said with certainty that not a single total eclipse of the sun had visited London for 600 years previous to 1715.

The first eclipse of the sun to be carefully observed in the British Colonies of America was that of June 24, 1778, which was watched by the astronomer David Rittenhouse

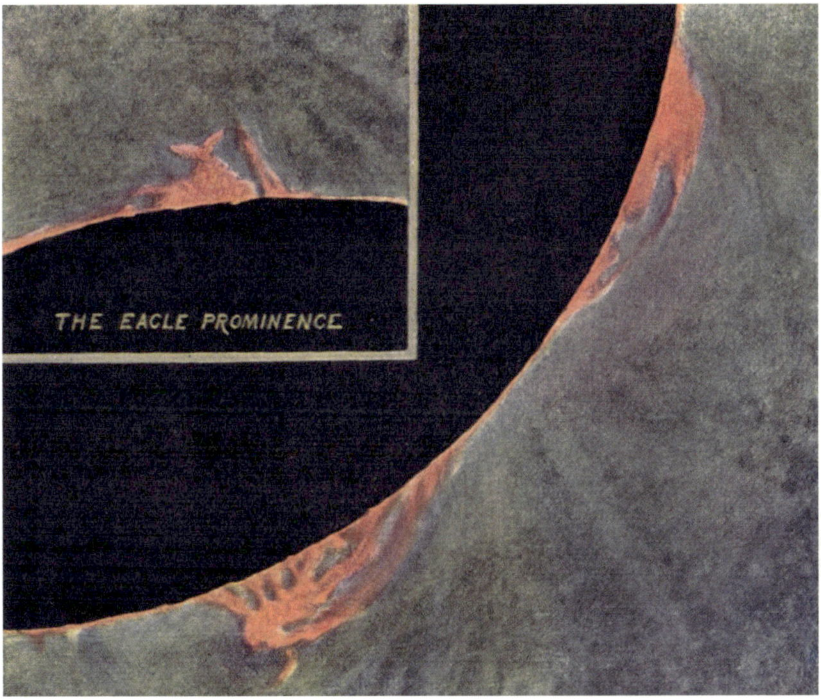

DETAILS OF THE HYDROGEN PROMINENCES, JUNE 8, 1918, INCLUDING THE "EAGLE PROMINENCE"

From the painting by Howard Russell Butler, N. A.

of Philadelphia. The first American expedition was organized and sent out from Harvard College for the eclipse of October 27, 1780. As this took place during the war of the American Revolution, an appeal was made to " the government of the Commonwealth that a vessel might be prepared to convey proper observers to Penobscot-Bay; and that application might be made to the officer who commanded the British garrison there, for leave to take a situation convenient for this purpose.

" Though involved in all the calamities and distresses of a severe war, the government discovered all the attention and readiness to promote the cause of science, which could have been expected in the most peaceable and prosperous times; and passed a resolve, directing the Board of War to fit out the Lincoln galley to convey me to Penobscot, or any other port at the eastward, with such assistants as I should judge necessary.

" Accordingly, I embarked October 9."[1]

Probably on account of an error in the tables, the eclipse was not total where the Harvard party was located. Between the first and second contacts Professor Williams measured the angular length of the moon subtended by the decreasing crescent of the sun. He gives the following description of what appeared shortly before the total phase was expected: " The sun's limb became so small as to appear like a circular thread or rather like a very fine horn. Both the ends lost their acuteness and seemed to break off in the form of small drops or stars some of which were round and others of an oblong figure. They would separate to a small distance, some would appear to run together again and then diminish until the whole disappeared."

Apparently this is a clear description of the so-called " Baily's Beads " observed by Francis Baily at the eclipse of 1836. An excellent description of this phenomenon is given by Agnes Clerke in her *History of Astronomy during the Nineteenth Century,* page 74. Baily gave the correct explanation of the phenomenon he saw as being due to irradiation. This same effect is seen when one holds up his

[1] *Memoirs American Academy of Arts and Sciences, 1,* 84, 1783.

hand to the sunlight. In making the fingers come close together, they appear to touch each other before one feels they are actually in contact. An analogous manifestation is called the " black drop " which caused surprise at the transits of Venus in the years 1761 and 1769, and was the source of great trouble to astronomers at the transits of 1874 and 1882, so widely observed for the purpose of determining the solar parallax. The appearance of " Baily's Beads " is a phenomenon well worth watching and should be attentively looked for just before totality begins and just after it ends. Very excellent observations may be made with a good pair of field glasses or with a small telescope, a large telescope being unnecessary.

Baily was not an astronomer by profession. He was a stock-broker, and fortunately he had been successful in the making of money, with the result that he was able to devote the maturer years of his life to astronomy which he took up as his hobby. His work is but one of the many instances of the great debt of science to the amateur astronomer. One important result of his observations in 1836 was to show professional astronomers that at the time of the total eclipse of the sun there were other phenomena to observe than the mere times of contact of the limbs of the sun and moon.

The eclipse of 1836 witnessed not only the phenomenon of " Baily's Beads " but also an attempt by Forbes to test the physical constitution of the sun's atmosphere by means of the spectroscope. A new era for astronomy had accordingly dawned. An eclipse occurred in Southern Europe on July 8, 1842, and into the narrow track were collected the foremost astronomers from England, France, Germany and Russia. What was observed in 1836 was as nothing compared with the wonders of the eclipse of 1842!

One of the strangest portions of the history of astronomy before the middle of the nineteenth century is the evident lack of interest in, or perhaps one should say, the dearth of accurate observations of the phenomena visible at the time of a total eclipse of the sun. The startling suddenness of the apparition, coming in the early days of civilization without warning, must have brought terror to the hearts of the populace and caused them to fear war or pestilence or the death of a favorite prince. It is but natural that the prehistoric superstition of the dragon swallowing the sun should have spread during the middle ages from the far East to all of the civilized world. One fleeting glance, however, should have revealed, even to the most timorous minded, the pearly-gray light of the corona and brought to view the glow of the rosy-hued prominences. To those of the present generation, the nine hundred and ninety-nine out of a thousand who have never had the good fortune to witness a total solar eclipse, it might not be out of place to point out that a telescope is entirely unnecessary for viewing the beauties of the corona, this being a spectacle that derives its glory from the wide-spread splendor and slight gradations of contrast. A telescope, small or large will of course magnify any particular portion — but to see and enjoy the beauty of the corona as a whole nothing is actually needed but the normal naked eye.

Published references to the corona in the early literature are exceedingly rare. Plutarch and Philostratus give allusions which unmistakably refer to the corona, but apparently the first to take any scientific cognizance of the crown of glory was Kepler who seems to have witnessed the solar eclipse of 1605 in Naples. A hundred years later at the eclipse of 1706, Cassini, who was a practised observer, describes the " crown " of pale light, and he decides that it must be caused by the illumination of zodiacal light; and eleven years thereafter, Halley saw the corona and also prominences, but he was unable to decide whether the corona belonged to the sun or to the moon.

If so little attention was paid to the corona, it is not surprising that even less should be taken of the " red flames," though if one refers to the plate facing page 130 he will see what a brilliant spectacle they afforded in the eclipse of 1918. The first reference to them seems to be at the eclipse of 1706 when they were apparently observed by Stannyan who wrote a description of them to Flamsteed. The first vivid portrayal was by Vassinius of Sweden who observed them in 1733. The Spanish admiral Ulloa observed them while at sea during the eclipse of June 24, 1778, and he furnished a valuable account, with the added explanation that the rosy hues were caused by the sun's light shining through some hole or crevice in the limb of the moon!

The astronomers who witnessed the eclipse of 1842 were entirely unprepared for the phenomena that met their gaze. Baily repaired to Pavia, and made his observations from one of the rooms of the University. One of the professors, out of the goodness of his heart, offered to assist him in any way possible, but Baily informed him that all he wanted was to be " left *alone,* being persuaded that nothing is so injurious to the making of accurate observations, as the intrusion of unnecessary company." Not being content with this gentle hint, the key was taken from the outside of the door and it was securely locked on the inside. Baily's report of the observations made at the eclipse is found in the *Memoirs, R. A. S., 15,* 4, 1846, as follows: " The *beads* were distinctly visible. . . . I was astounded by a tremendous burst of applause from the streets below, and at the *same moment* was electrified at the sight of one of the most brilliant and splendid phenomena that can be imagined. For at that instant the dark body of the moon was *suddenly* surrounded with a *corona,* or kind of bright *glory.* . . . I had indeed anticipated a luminous circle round the moon during the time of total obscurity, but I did not expect, from any of the accounts of previous eclipses that I had read, to witness so magnificent an exhibition as that which took place. . . . The breadth of the corona, measured from the circumference of the moon, appeared to me to be nearly equal to half the moon's diameter. It had the appearance of brilliant rays. Its colour was quite white, not pearl colour, nor yellow, nor red.

" Splendid and astonishing, however, as this remarkable phenomenon really was, and although it could not fail to call forth the admiration and applause of every beholder, yet I must confess that there was at the same time something in its singular and wonderful appearance that was appalling. . . . But the most remarkable circumstance attending this phenomenon was the appearance of *three large protuberances* apparently emanating from the circumference of the moon, but evidently forming a portion of the corona. . . . All of these projections were of the same roseate cast of colour, and very distinct from the brilliant vivid white light that formed the corona. . . . The whole of these three protuberances were visible even to the last moment of total obscuration, at least, I never lost sight of them when looking in that direction; and when the first ray of light was admitted from the sun, they vanished with the corona, altogether, and day-light was instantly restored."

The same appearance was witnessed by Airy, by Arago and others. Arago has an interesting account of the effect of the eclipse on the populace who had gathered in great numbers to watch the phenomenon. "When the sun, reduced to a very narrow filament, began to throw upon the horizon only a very feeble light, a sort of uneasiness seized upon all; every one felt a desire to communicate his impressions to those around him. Hence arose a deep murmur, resembling that sent forth by the distant ocean after a tempest. The hum of voices increased in intensity as the solar crescent grew more slender; at length the crescent disappeared and an absolute silence marked this phase of the eclipse. The phenomenon in its magnificence had triumphed over the petulance of youth, over the levity which certain persons assume as a sign of superiority, over the noisy indifference of which soldiers usually make profession. A profound stillness also reigned in the air, the birds had ceased to sing."

The arrival of totality in Milan was greeted by a great shout, mingled with cries of "Long live the astronomers" who had provided such a beautiful phenomenon to please and interest the populace!

The unexpected nature of prominences and corona seen at the eclipse coupled with the publication in 1843 of Schwabe's discovery of the periodicity of sun-spots caused an unprecedented increase of interest in matters pertaining to the physical constitution of the sun. Various ingenious explanations appeared. While most astronomers believed that the prominences were truly solar in their origin, there were many who thought they were possibly some exhalation in the earth's upper atmosphere, while still others believed that in some manner diffraction round the edge of the moon was responsible for these eclipse envelopes. In passing, we might mention a notion of no less an authority than Halley which was so curious that it should be classed along with William Herschel's belief that the sun might be cool and habitable. Halley[1] thought that the appearances on the eastern and western edges of the sun at a total eclipse might reasonably be expected to be different, for the reason that "the eastern limb of the moon had been exposed to the sun's rays for a fortnight, and as a consequence it would be natural to expect that the *heated lunar atmosphere* might exert some absorbing effect on the solar rays, while on the contrary the western edge of the moon being in darkness and cold for two weeks could exhibit no such absorbing action."

The interest aroused in total eclipses was now so great that astronomers were determined to take advantage of every opportunity, no matter how short the time of totality nor how great distances it was necessary to travel in order to view the eclipses. The eclipse of July 28, 1851, was visible in Norway and Sweden, and English astronomy was well represented in the persons of the astronomer royal Airy, Hind, Dawes, Carrington, Stephenson, Gray, Lassell and Williams. Although Faye[2] still asserted with force that the prominences were merely optical illusions or "mirages produced near the moon's surface," the general consensus of opinion was that the origin of the red flames was to be sought in the sun. To this fire of scarlet hue Airy gave the name of *sierra*.

Any lingering doubts regarding the origin of this sierra were forever dispelled by the observations made at the eclipse of July 18, 1860, visible in America, Spain and Northern Africa. The solution of the problem was accomplished by photography which was applied for the first time at an eclipse with anything like success. As the prominences

[1] *Phil. Trans.*, *29*, 248, 1715.
[2] *Memoirs, R. A. S.*, *21*, 5, 1853.

are red in color to the eye and as the ordinary photographic plate is insensitive to red (plates are usually developed under ruby light) grave doubts were felt whether photographs would be able to portray the red flames. The only thing to do under the circumstances was to "try something and see what happens" (excellent advice for the scientist usually credited to the late Professor H. A. Rowland). Photography had already been applied at the eclipse of 1851 when Busch obtained some feeble impressions of the eclipsed sun by the daguerreotype process. Photography was even attempted in 1842 using iodized paper, but with no results. Warren de la Rue used the heliograph from Kew, enlarging the image before it reached the photographic plate, while Father Secchi employed a six-inch refractor without enlargement. Photographs of both observers were successful. De la Rue was near the Atlantic in Spain while Secchi was on the Mediterranean Coast, six minutes of elapsed time being necessary for the moon's shadow to travel from one station to the other. The conclusions from the 1860 eclipse were: 1. The prominences are rich in actinic power (now known to be due principally to the H and K light of calcium). 2. As the moon passed in front of the sun it progressively covered and uncovered the prominences, thereby demonstrating completely that their origin is strictly solar. 3. During the six minutes of elapsed time, changes were noted in some prominences but no variations in others. 4. The material from which the red flames arise is found around the whole solar globe. This is the *sierra* of Airy, but later, called the *chromosphere* by Lockyer.

The success attending the eclipse of 1860 came almost at the same time with the unraveling of the enigma of the spectrum. After two centuries of slow and painstaking progress, the crucial experiments had been performed by Kirchhoff by means of which the action of the spectroscope was at last understood. As a result, interest in eclipse observations was no longer confined to the astronomer alone, for many investigations were to be undertaken which were not confined by the determination of exact time or position. The birth of physical astronomy, or astrophysics, insured that henceforth eclipse observations would be of quite as much interest to the physicist as to the astronomer, and for the interpretation of these observations, research work in the laboratory was quite as important as that in the observatory. What spectrum would the prominences give? The answer was not difficult. Apparently the prominences were not, as had previously been thought, masses of photospheric material shot up to great heights by some explosive or volcanic action on the sun, since it was evident that the boomerang-shaped protuberance seen at the eclipse of 1851 could hardly have existed under the laws of gravitation. It was top-heavy and could not have hung there above the sun even for the short space of time it was visible if it had been composed of the general material forming the sun's photosphere. And then there was the distinct difference in color noted between the red flames and the body of the sun. Manifestly, since *all* of the gases forming the sun did not take part in the solar outburst, the prominences probably consisted of a few gases only, perhaps one of their chief constituents was the lightest of known gases, hydrogen, whose visible spectrum known from stellar investigations consisted of a strong line in the red, another in the blue, and a series of others coming closer together as the violet end of the spectrum was approached. The red line of hydrogen seemed to give a color not differing materially from the red of the prominences. If therefore the prominences were actually outbursts of hydrogen gas heated to great temperatures in the solar furnace, the eclipse spectrum would be vastly different from the ordinary solar spectrum. The band of light would not be continuous from the red to the violet end, nor would any dark Fraunhofer lines be visible. Since the prominences were probably gaseous, their spectrum, as known from the laws of Kirchhoff, must consist of bright lines on a dark background, an emission spectrum. It appeared therefore that the prominence spectrum would consist of a few bright lines only, the red and blue lines of hydrogen, and the series towards the more refrangible end, more difficult to see on account of the fact that the human eye is not sensitive to violet light.

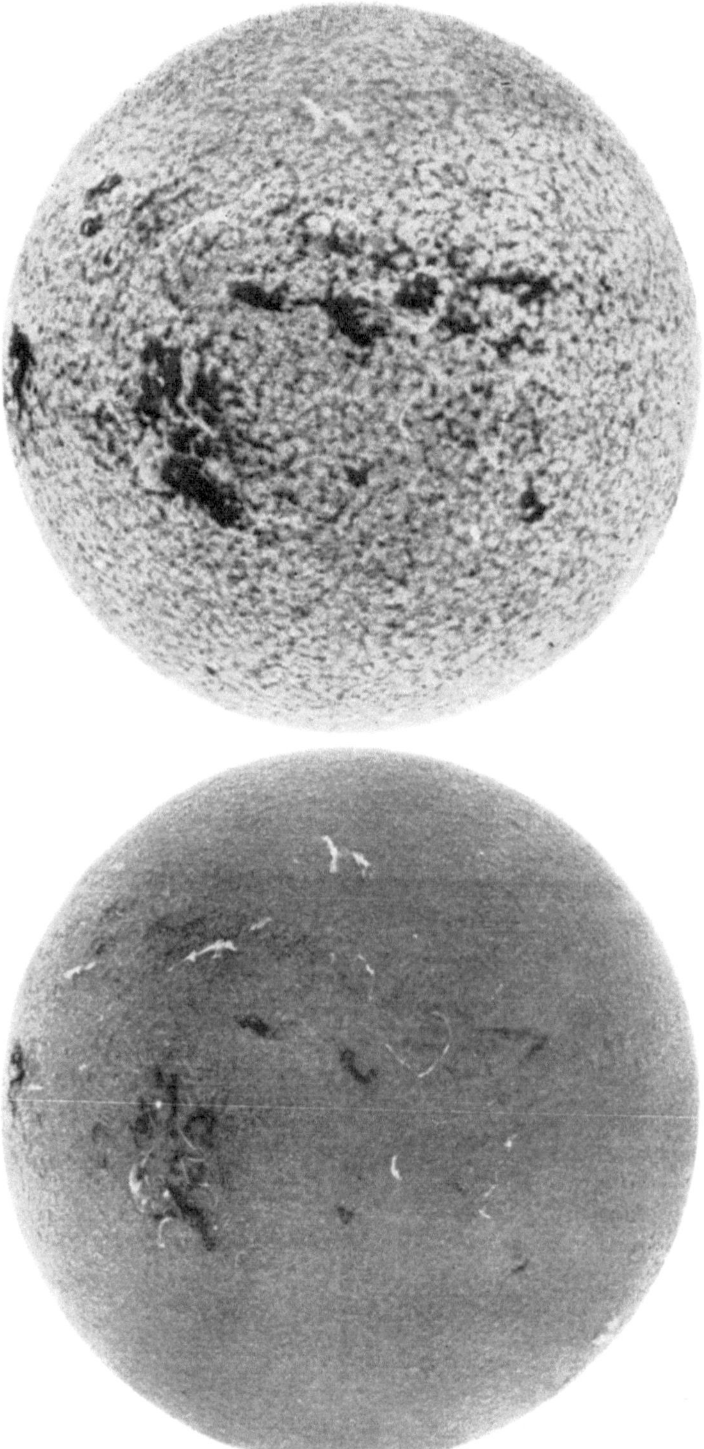

OUTER LAYERS OF THE CHROMOSPHERE

Photographed by Deslandres by spectroheliograph at Meudon, France, on June 7, 1919. *Above:* K3 of calcium. *Below:* Hα of hydrogen. Note the differences in appearance of the two photographs and the fine detail that appears in the hydrogen photograph.

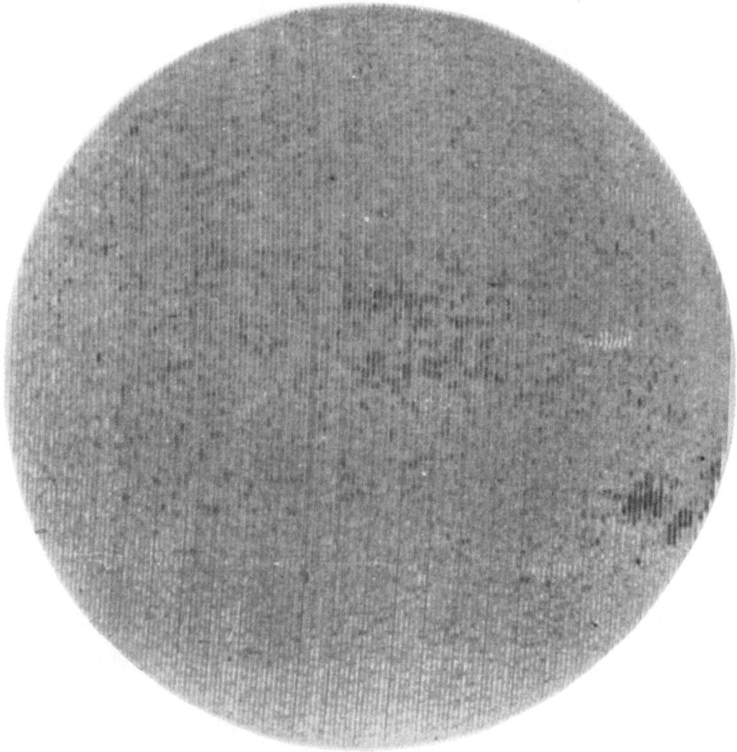

TEST OF RADIAL MOTIONS IN THE SUN
Photographed by Deslandres at Meudon with the K1 ray of calcium, April 11, 1910.

But there was no eclipse on which to " try and see what happened " until August 18, 1868, so it was necessary for the physicists and astronomers to possess their souls with patience; but alas! the eclipse was visible only in far-off India, the Malay peninsula and Siam. The distances were great, but the problems were important, and accordingly several expeditions, two British, two French, one German and one Spanish were found in the eclipse track. The greatest success attended the observation of Janssen. The slit of his spectroscope directed to the edge of the sun revealed the spectrum of the prominences consisting, as had been thought, of a series of bright emission lines, most prominent among which were three lines, the red and the blue lines clearly belonging to hydrogen, but one of almost equal brilliancy in the yellow. The color of the yellow line seemed to match the D-lines of sodium, ever-present in laboratory experiments, but why should the gas sodium, comparatively heavy, be found in the prominences?

The brilliancy of the prominence lines was so remarkable that Janssen determined to seek them again after the eclipse was over. If they were solar in origin they must be found on the sun every day, varying it is true in shape and dimensions. The only reason why the prominences cannot be seen any day is the same reason why the stars are not visible in daylight, — the glare of the earth's atmosphere, especially when close to the sun, being very great. Abolish the atmosphere where the observer worked, and the stars and prominences would at once become visible; and since the moon has little or no appreciable atmosphere, the prominences would be seen each day without an eclipse to the " man in the moon " if such a person existed. But how get rid of, or diminish, the glare of the earth's atmosphere to such an extent that the prominence luminosity will be more intensive than the light of the earth's atmosphere? This feat can be accomplished by the spectroscope. The emission lines, C and F of hydrogen, found in prominences, are monochromatic in character, that is, these lines betake approximately the nature of mathematical lines and show no appreciable width. As a matter of fact, it is impossible to diminish their intensity by increasing the dispersion of the spectroscope, whether this increase is accomplished by the addition of extra prisms or by the employment of a grating. Increased dispersion, however, spreads out these lines in the spectrum to greater distances apart, but their intensities are not thereby diminished. On the other hand, the continuous background of the solar spectrum can be weakened at will by merely increasing the amount of the dispersion, a fact which is at once evident since the light passing through the slit is by an increase of prisms spread over a greater area. Reflected sunlight, whether from the moon's surface, from the planets, from a silvered mirror or from particles of dust in the earth's atmosphere, gives the solar spectrum. Consequently, the prominences may be made visible by the spectroscope by the simple process of increasing the dispersion to such an extent that they can shine by contrast with the weakened atmospheric glare. As a result of these ideas, Janssen looked for and found the prominences after the eclipse was a thing of the past. The same or similar ideas had occurred to other workers with the spectroscope, notably to Huggins and to Lockyer in England. Without having been present at the eclipse, the latter tried for the prominences and found them for the first time on October 20, 1868. Lockyer sent a record of his observations to the French Academy, and without having heard of the Englishman's results, Janssen sent on to Paris the report of the work he had done at the eclipse and afterwards. By a strange coincidence, the papers from both investigators were read at the same sitting of the Academy, in honor of which event a medal was struck bearing the likeness of both Lockyer and Janssen.

Thus in the moment when the chemical nature of the prominences was discovered by the spectroscope, these objects ceased to be phenomena confined to eclipses only. As a happy and most fortunate result, a two-fold benefit thus accrued to the astronomer: freed from the necessity of observing prominences during the all too brief moments of a total eclipse he could devote his energies to other investigations. Not only could prominences be observed without an eclipse, but by the same methods, researches could be carried out on the solar envelope from which prominences arose, the chromosphere. Frequent violent eruptions on the sun carried the solar flames up to great distances and with enormous velocities, and these phenomena were observed visually by Young, Lockyer, Tacchini and a host of other investigators, many important researches [1] being carried out. The invention of the spectro-heliograph in 1893, independently and almost simultaneously, by Hale and Deslandres permitted an attack on these problems by the help of photography.

The great triumph of the spectroscope in 1868 gave evidence to the solar astronomer of the important problems awaiting solution, and it appeared almost certain that each observation carried out with care would be a valuable discovery. What was the corona? Did it shine by its own light, or by reflected sun's light? Was there any connection between the prominences and the corona? How far out did the corona extend, and were there any changes in its form that could be detected? Was the explanation of the dark Fraunhofer lines of the solar spectrum the true one, and was it possible that eclipses could help in the problem?

The eclipse of August 7, 1869, crossed America diagonally from Alaska to North Carolina. The United States government made a large appropriation to help defray the expenses, and as a result of this and also on account of the favorable time of year, the eclipse track in the United States was almost one continuous observatory, so thickly were the astronomers scattered. To give a list of those who saw the eclipse would be practically equivalent to making a record of every astronomer of any importance who was then living in the United States. Those too "were the good old days" before railroad executives were harassed by government regulations and Interstate Commerce Commissions and when free transportation for passengers and goods could be furnished. The following is copied from the *Report on Observations of the Total*

[1] See Young, *The Sun.*

Eclipse of the Sun, August 7, 1869. On page 3, J. H. C. Coffin, U. S. N., Superintendent of the Nautical Almanac, reports, " Col. Thomas A. Scott, vice-president of the Pennsylvania Central Railroad, furnished a special car, and with the cordial coöperation of Mr. Robert Harris, general superintendent of the Chicago, Burlington and Quincy, and Mr. C. E. Perkins of the Burlington and Missouri Railroad, provided free transportation for these parties, with all their instruments and apparatus to and from the places of observation. Free passes to ten or twelve others over the same routes were also granted." In the same report, page 115, Professor Morton estimates that " an expense of $1500 was spared the government appropriation."

Fortunately, clear skies greeted the observing parties. Little of the important work accomplished will be noted in detail here. Spectroscopically, the most valuable discovery was that the spectrum of the corona was continuous but was traversed by a single green ray. This green line was detected independently by both Harkness and Young, the latter identifying its position as coinciding with the line numbered 1474 on Kirchhoff's scale. But since this line 1474 is due to iron, it was surprising and perplexing in the highest degree to find it present in the corona and reaching such great heights above the sun's surface. In spite of the apparent coincidence, it was evident that the substance causing the green line was not iron. To it the name *coronium* was given, — and today after more than a half century of active research we know little more of coronium than when it was first discovered. At this same eclipse, Professor E. C. Pickering employed a portrait lens, thus recognizing its value in the portrayal of the heavens. As he was located at Mt. Pleasant, a little farther west than the rest of the expeditions, to him belongs the honor of securing the first successful photograph of the corona in America. Other photographs were secured by Winlock and others on a larger scale and with good definition which showed the corona and prominences.

The United States Congress appropriated the sum of $29,000 for observations at the eclipse of December 22,

1870, visible in Spain, Northern Africa, Sicily, Greece and Turkey. America was represented in the eclipse track by the following: Pierce, Newcomb, Harkness, Hall, Eastman, Winlock, Young, Langley, Pickering, Peters, Watson, Clark, Ernst, Willard, Ross, Gannet, and General Abbott. The British government granted £2000 and a ship. A great variety of observations were projected and only partially carried out on account of clouds that prevailed almost everywhere, observations photographic, spectroscopic, photometric and polariscopic. Lockyer met shipwreck and clouds, but was rewarded in the end by a glimpse of the corona for one brief second and a half! Janssen made good his exit from beleaguered Paris in a balloon, but though he succeeded in escaping from the bullets of the Germans, he was forced to capitulate to obscuration by the clouds.

The most conspicuous success awaited the efforts of Professor C. A. Young of Princeton who had foretold, and whose eye was the first to see the "*flash spectrum*." According to the theory of Kirchhoff (p. 97) the spectrum of the photosphere would be continuous from red to violet without any dark lines were it not for the overlying solar atmosphere. Here in the so-called *reversing layer*, the gases are at a lower temperature than in the photosphere, and on account of these cooler conditions the photospheric light is absorbed by the reversing layer, and the resultant spectrum is that of dark lines on a bright background. As already explained, if the photosphere could be removed, then the gases forming the reversing layer, since they are at a high temperature, would give a series of bright lines on a dark background, which is technically called a "reversal" of the Fraunhofer spectrum. To describe the appearance in 1870, one cannot do better than to quote from the words of the discoverer:[1] "The observation is possible only under peculiar circumstances. At a total eclipse of the sun, at the moment when the advancing moon has just covered the sun's disk, the solar atmosphere of course projects somewhat at the point where the last ray of sunlight has disappeared. If the spectro-

[1] Young, *The Sun*, p. 83.

scope be then adjusted with its slit tangent to the sun's image at the point of contact, a most beautiful phenomenon is seen. As the moon advances, making narrower and narrower the remaining sickle of the solar disk, the dark lines of the spectrum for the most part remain sensibly unchanged, though becoming somewhat more intense. A few, however, begin to fade out, and some even turn palely bright a minute or two before totality begins. But the moment the sun is hidden, through the whole length of the spectrum, in the red, the green, the violet, the bright lines flash out by hundreds and thousands, almost startlingly; as suddenly as stars from a bursting rockethead, and as evanescent, for the whole thing is over in two or three seconds. The layer seems to be only something under a thousand miles in thickness, and the moon's motion covers it very quickly.

"The phenomenon, though looked for at the first eclipses after solar spectroscopy began to be a science, was missed in 1868 and 1869, as the requisite adjustments are delicate, and was first actually observed only in 1870."

The same phenomenon was witnessed by Pye, a member of Young's party. The bright lines were so numerous that the impression was gained that every one of the thousands of Fraunhofer lines was reversed from dark to bright, while "the phenomenon was so sudden, so unexpected, and so wonderfully beautiful as to force an involuntary exclamation."[1] Professor Young called this sudden transition the *flash spectrum*. The same phenomenon was witnessed at the eclipse of December 12 of the following year by Maclear, Herschel, Lockyer and Fyers, the eclipse being total in India, Ceylon and Northern Australia (where clouds interfered). Again at the annular eclipse of June 6, 1872, seen by Pogson in India, and at the total eclipse of April 16, 1874, witnessed by Stone in South Africa, the flash spectrum was observed. This is one of the most interestingly beautiful and important of the phenomena connected with total eclipses. During the past half century the flash spectrum has been carefully observed at each eclipse. More

[1] *Memoirs, R. A. S., 41,* 435.

recently it has been photographed in exquisite detail, and has also been photographed at Mt. Wilson without an eclipse. (See Chapter XV.)

Apparently, it had been almost definitely settled that the corona is a truly solar appendage, for in addition to the spectroscopic evidence in the matter, the photographs in 1871 showed the same details in the coronal streamers from observing stations widely separated, an effect which could not possibly take place if the corona was terrestrial in origin. Such then was the state of scientific information regarding the corona as it existed *before* the eclipse of 1878 which was visible in America. Little was known with exactitude, there was very much of perplexity and uncertainty and doubt. The unsatisfactory nature of the whole subject may perhaps be envisaged in one sentence copied from a book by one of the most assiduous of spectroscopic investigators of the time, Norman Lockyer. " Now the whole phenomenon of the corona may be defined in two words, *cool prominences.*"

If the knowledge of the corona was disappointing and perplexing, the decisions of the spectroscope regarding the reversing layer were equally uncertain. Kirchhoff's theory demanded that there should be a layer of vapors relatively cool surrounding the photosphere. Where was this layer, close to or far away from the sun? Was the layer thin, or one more extensive? At the eclipse of 1870, Young had discovered a flash spectrum apparently of thousands of lines suddenly turned from dark to bright. This phenomenon was visible for only two or three seconds at the beginning and ending of totality, and in consequence, the reversing layer must be very shallow, approximately 600 miles in thickness. But were *all* of the Fraunhofer lines turned from dark to bright? Since the appearance was visible during the few hurried and excited seconds of an eclipse, it was manifest that no exact comparison with Fraunhofer lines was possible, nor would be possible until the time should arrive when the flash spectrum could be recorded by photography. Moreover some spectral lines were seen to turn bright half a minute or more before the beginning of totality, so that it was evident that not all of the reversing layer was confined to the 600-mile limit. Still other difficulties presented themselves. If the absorption did take place in a layer of the solar atmosphere, then the light of the photosphere passing tangentially through this layer at the sun's limb should experience an absorption greater in amount than that from the sun's center. In consequence, the spectra of the center and limb of the sun were compared in great detail, with the resulting discovery that any differences in the relative intensities of the lines were of such minor character that they could readily be explained by the slight darkening of the limb compared with that of the center. Kirchhoff believed that the spectral lines of any element like sodium were *characteristic,* or in other words, the same element gave always the same series of lines no matter how the element was vaporized. But when the flame spectrum was compared with that of the higher temperature of the electric arc, and the still greater temperature of the electric spark, great differences in the relative intensities of the lines were at once observed. Lockyer was the first to call attention to, and emphasize the importance of these changes. Take the element calcium, for instance. With a Bunsen flame and with small dispersion, the chief line visible is in the red end of the spectrum. At the temperature of the arc, the strongest line is in the blue, at 4227A. The red line is also visible and also two lines in the violet, the H and K lines of the solar spectrum. Increase the temperature of excitation by using the electric spark, and the two violet lines greatly increase their intensities and become much stronger than the blue line, while the red line practically disappears. Similar changes of intensity were found by Lockyer in magnesium, lithium, iron and other elements examined, and these conclusions have been abundantly verified by all observers since his time. Still further difficulties presented themselves. When the wave-lengths for the various metals were determined, it was discovered that different elements had spectral lines with apparently identical wave-lengths. It was found that some of these common lines were due to impurities in the metals examined, but when these lines were omitted from consideration there were still many lines evidently common to two or more elements. It was consequently manifest that the identification of lines in the solar spectrum was not the simple operation Kirchhoff supposed it to be, and that "the more observations were accumulated the more the spectroscopic difficulties increased." [1] Although some of the theories and conclusions of Lockyer were incorrect, nevertheless spectroscopists are under a great debt to him for the splendid series of researches carried out both in the laboratory and in the observatory.

To solve some of these difficulties, Lockyer busily investigated the spectra of many elements by his well-known method of *long and short lines*. If an electric arc is arranged horizontally and its image is projected on the vertical slit of the spectroscope, it is seen at once that the lengths of the lines in the spectrum vary considerably. Since the core of the vapor between the two carbon poles must be much hotter than that on the outside edges, it was evident to Lockyer that the short lines were high temperature lines that become visible only at the hottest point, that of the core, while the long lines were those which could exist at different temperatures, at the great heat of the center of the arc and at the lesser temperatures of the outside edges. This method of long and short lines thus appeared to give a ready and convenient means of separating the lines of highest temperature, which were comparatively few in number, from the balance of the spectrum.

By arranging the spectra of stars in an orderly series, beginning with the white stars, and passing through yellow stars to those red in color, it was noted by Lockyer that there was a steady increase in the total number of lines in the spectra, there being few lines in the white stars other than the hydrogen series, while the red stars possess an enormous number of spectral lines. Thus in the progression in the reverse direction, from red stars back to white, the spectra show fewer and fewer lines and become simpler

[1] Lockyer, *The Chemistry of the Sun*, p. 176.

in character. The same progression from red to white stars saw an increase in intensity in the H and K lines due to calcium. And as Lockyer was the first to recognize the importance of temperature in altering the character of spectra, so to him likewise belongs the great honor of recognizing that changes in stellar spectra show that the white stars are hotter than the yellow and these in turn are hotter than the red. (Present-day researches show that as the stars increase in temperature in the stellar series, the H and K lines do not steadily increase in intensity but reach a maximum strength, then begin to fade away and are entirely wanting in the hottest stars, the blue-white helium stars of spectral type B and stars of type O.)

Momentous in the highest degree therefore were the discoveries of Lockyer showing the importance of temperature in the interpretation of spectra, — but his explanations of the underlying causes have not borne the weight of time. His was the *dissociation theory* that the chemical atoms were continually broken up into elements less complex in structure, which exhibited simpler and simpler forms of spectra as higher and still higher temperatures were reached. As the white stars are those of the highest temperature, and as they show practically nothing but the spectrum of hydrogen which is the element of smallest atomic weight, it must represent the simplest element with the simplest spectrum. (The most recent researches show that though there may be dissociation of the chemical atoms this takes place in a manner vastly different from the dissociation demanded by Lockyer's theory).

The sun being but a typical yellow star, this theory can be put to the test in attempting to explain the phenomena of the solar atmosphere as revealed by eclipses. Lockyer assumed "that in the reversing layers of the sun and stars various degrees of *chemical dissociation* are at work, which dissociation prevents the coming together of the atoms which, at the temperature of the earth and at all artificial temperatures attained here, compose the metals, the metalloids and the compounds."[1] In conse-

[1] *The Chemistry of the Sun*, 201.

quence, there were grave doubts expressed by Lockyer whether the chemical elements known from laboratory experiments could at all exist at the great heat of the sun except in the cooler parts of its atmosphere. It was imagined by him that this atmosphere consisted of successive layers " like the skins of an onion," the layers next the sun obviously being the hottest. Hence in the interior layers could exist only " those constituents of the elementary bodies which can resist the greater heat of these regions." The spectrum of these inside layers, if such a spectrum could be obtained apart from that of the rest of the sun, would therefore not be a reversal of the Fraunhofer spectrum. According to Lockyer's hypothesis, the *whole* of the solar atmosphere is effective in the production of absorption. Young's observation of the flash spectrum demanded a shallow reversing layer of a few hundred miles, but Lockyer's theory refused to admit the existence of a reversing layer separate from the superincumbent strata. In other words, there could be no division possible into reversing layer, chromosphere and corona; these were but different manifestations of the solar atmosphere, the corona being regarded as the outermost and cooler parts of an atmosphere having a composite existence and obeying the laws of gravitation. In consequence of this theory, although Lockyer had himself witnessed the flash spectrum at the eclipse of 1871, he at first was forced to doubt, and then actually to deny the existence of the shallow reversing layer. The flashing out of lines that he observed " has been called the reversing layer; but I do not now (1881) believe that it is the reversing layer for a moment, for, when it comes to be examined, we shall probably find that scarcely any of the Fraunhofer lines owe their origin to it, and we shall have a spectrum which is not a counterpart of the solar spectrum." (*Loc. cit.* p. 360.) The solution of the problem could not be effected visually during the brief seconds available at a total eclipse for observation of the flash spectrum. The interpretation by one astronomer of what was observed should be entitled to as much weight as the opinion of another. There was no hope of a solution of the problem until the time should arrive when photography could come to the rescue by furnishing a permanent record of the flash spectrum which could be compared line by line with the Fraunhofer spectrum in order to see whether the one spectrum is the exact reversal of the other.

While these observations, epoch-making in their importance, were being made on prominences and reversing layer, the corona was not forgotten. Strange as it may now seem, there were still many astronomers of repute who believed that the origin of the coronal light should be sought, not in solar but in lunar or terrestrial causes. There were even two theories based on the moon, one that the corona was due to the diffraction of solar rays which pass near the moon's edge; the other, that the phenomenon was due to reflection of solar rays from the irregularities of the moon's surface. Another curious theory which found great favor at the time was that the corona was due simply to glare in the earth's atmosphere. As a result of this hypothesis the corona would necessarily be a phenomenon entirely local in its structural character, details appearing differently to observers at separate localities. If due simply to atmospheric glare, the coronal details should be found projected also on the dark moon.

There were now available four different methods for attacking the corona: first, visual observations by the naked eye, supplemented by the telescope; second, photography; third, polariscopic observations; and fourth, the spectroscope. The polariscope had already shown that part, at least, of the coronal light was reflected sunlight. This conclusion was corroborated by the discovery by Janssen at the eclipse of 1871 of dark Fraunhofer lines in the coronal spectrum, chief among which the D line of sodium was recognized. The green emission line discovered at the eclipse of 1869 was observed again in 1870 and 1871, Tennant in the latter year discovering that this ray was quite as conspicuous in a rift in the coronal light as in the adjacent streamers. The corona appeared thus to be shining from the luminosity of some unknown gas, which

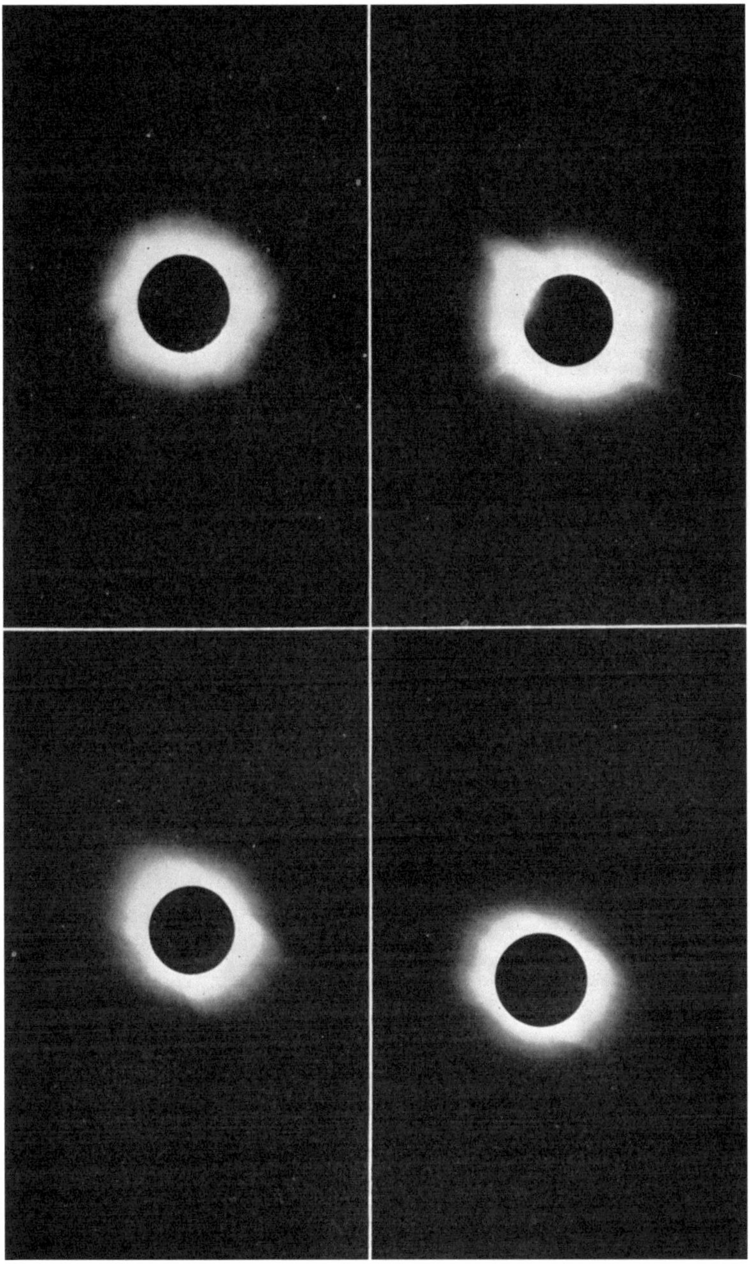

1893
Chile

1898
India

1900
Georgia
U.S.A.

1901
Sumatra

THE CORONA PHOTOGRAPHED BY LICK OBSERVATORY PARTIES IN DIFFERENT QUARTERS OF THE WORLD

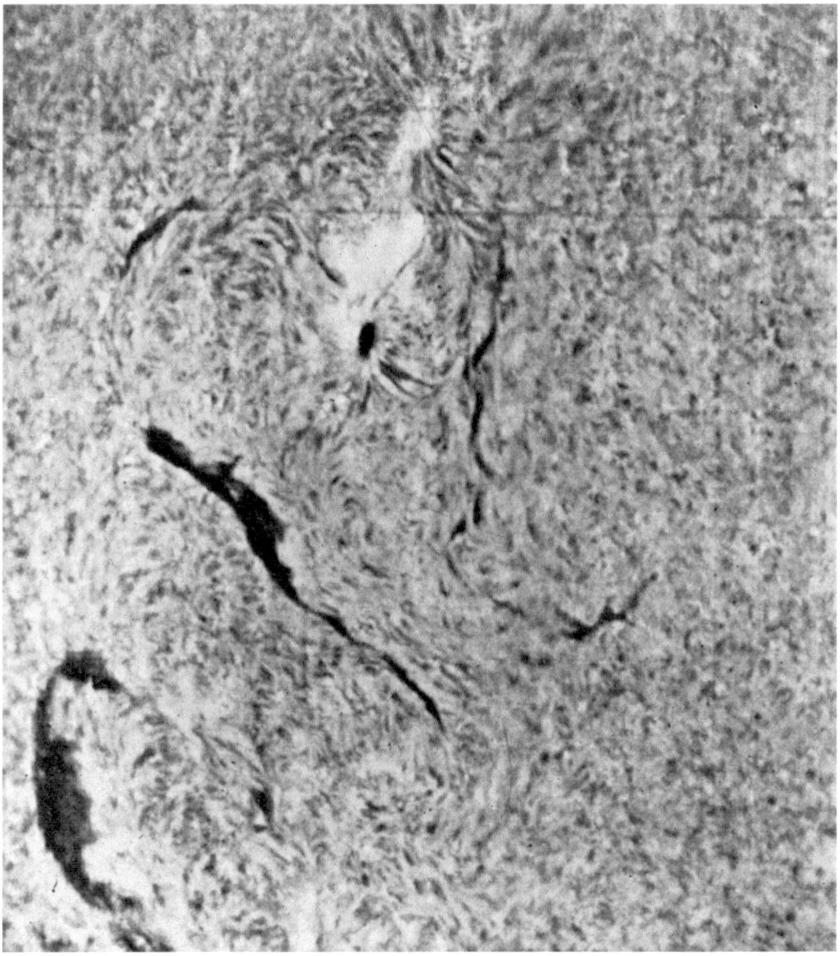

PROMINENCES SEEN IN PROJECTION AGAINST THE SUN'S DISK
Photograph from Mt. Wilson Observatory.

strange to say shone as strongly in the dark regions of the corona as in the bright streamers.

In 1871 for the first time, and due to a suggestion by Young, a slitless spectroscope was tried. With the use of such an instrument at mid-totality, the emission lines appeared as *rings* of light, from the extent of which one could ascertain the height of the various gases forming the corona. By the help of photography in 1871, Lockyer showed that hydrogen extended uniformly about the sun to the enormous height of 200,000 miles. Could the solar atmosphere possibly extend to this colossal distance? Let us reason by analogy with what we know of the earth. Its atmosphere, as known from observations of meteors, extends about one hundred miles above sea-level. The sun is more than one hundred times the diameter of the earth and with the same relative distribution the solar atmosphere might conceivably extend 10,000 miles from the solar surface. But an atmosphere possesses mass and is obedient to the law of gravitation, and since gravity on the sun is twenty-seven times that on the earth, we would expect that a true atmosphere on the sun could not extend to elevations of more than 4,000 miles. But the hydrogen observed was at distances of 200,000 miles, with the green ring reaching the still greater extent of 300,000 miles, whereas the coronal streamers were seen to stretch out several millions of miles from the sun's surface. Apparently, the spectroscope had not solved many of the difficulties of the coronal structure, but rather had succeeded only in complicating matters, for to add to the difficulties already great, it was now necessary to explain how it was possible that the luminous gases hydrogen and coronium could extend to the very great distances revealed by the spectroscope. No terrestrial origin could be found for the green coronal line, nor did Young's discovery in 1876, that Kirchhoff's "1474" was a double line help solve the problem. The perplexities were indeed very great. A faint ray of hope appeared in an unexpected quarter. In 1866, shortly after the great November meteor shower, Schiaparelli proved that the Perseid meteors moved in the same path as Tuttle's comet of 1862, while the Leonids and the Temple comet had identical orbits. This double coincidence between meteor and cometary orbits was corroborated in 1872 when it was found that the Andromedes, or Bielids as they are now called, had the same path about the sun as the lost Biela comet. The importance of meteors in any cosmical process was thus realized, and it was but natural that attempts should be made to solve the coronal puzzle by means of the meteoric hypothesis, — but as we shall see later, with little success. Newton and Cleveland Abbe in America and Lockyer in England pinned most faith to the meteoric explanations.

CHAPTER IX

NINETEENTH CENTURY ECLIPSES AFTER 1878

THE total eclipse of July 28, 1878, was observed in the United States from Wyoming to Texas. On account of the great interest in and the vast importance of the problems to be solved, the eclipse was widely observed from some twelve stations by about one hundred astronomers.

Ever since the discovery of the sun-spot period in 1843 and the finding that the earth's magnetism possessed the same period, astronomers had been on the alert to ascertain whether other solar phenomena moved in cycles parallel to the sun-spot curve. It had already been found that both prominences and faculae were more numerous when spots were great in number, and naturally the question was asked whether the coronal streamers did not originate with greater energy when spots were at a maximum. Here was an opportunity to test any conclusions, for Wolf's sun-spot number for July 1878 was 0.1, representing a minimum of spots, while the number for December 1870 was 135.4, a time of maximum of spots.

In the state of Colorado the moon's shadow path crossed the Rocky Mountains. Pike's Peak (altitude 14,400 feet) was occupied by Professor Langley and party. At their station before the day of the eclipse there was hail, rain, sleet, snow, fog and every form of bad weather which continued for a week, and to add to the discomforts, the horrors of mountain sickness had to be overcome. But on July 28, the weather conditions along the eclipse track were of the very best.

The first change noted in the corona of 1878 was the enormous decrease in total lustre, when compared with the coronas of 1870 and 1871, Harkness estimating the luminosity to be only one-seventh of the corona of 1870, while Lockyer regarded 1878 to be one-tenth of the brightness of the 1871 eclipse. The decrease in brilliancy was accompanied by a remarkable and unexpected change in shape. To Janssen in 1871, the dark moon looked like the center of a giant dahlia, the corona being nearly circular in outline. In 1878, the streamers along the sun's axis were much shorter in length but much more pronounced in character, these polar rays resembling more than anything else the lines of force around a magnet. But the most astounding phenomenon was the enormous extent of the coronal streamers along the sun's equator. Langley in the pure, rare air of Pike's Peak followed these streamers to six diameters of the moon on one side, but on the other side where he had been more intently watching, to the colossal length of twelve diameters, or more than ten millions of miles! These equatorial extensions were confirmed by Simon Newcomb in Separation, Wyoming, by Cleveland Abbe farther down the slope of Pike's Peak, and by almost every astronomer who witnessed the eclipse. The perplexities surrounding the corona were accordingly multiplied many-fold, for how could a solar atmosphere obeying gravity exist at the huge distance of ten million miles from the sun's surface? Young and Abbe saw long faint beams shining along the sun's axis.

Remarkable as were the visual phenomena manifested, their testimony was no whit stranger than the revelations by means of the spectroscope. The hydrogen and green coronium emission lines were visible, but with such vastly diminished intensity compared with the eclipse of 1871 that most observers completely missed seeing them. They were, however, visible to Young, Eastman and some others. If the emission lines were weak, the Fraunhofer lines of the corona were comparatively strong, showing that the reflected light near the sun's limb was relatively stronger than in 1871, a fact confirmed by observations with the polariscope.

The eclipse of 1878 showed the long equatorial wings of the corona, strong polar brushes, faint incandescent light of coronium and hydrogen, and light reflected strongly from material particles near the sun's limb. Were each of these four special features unalterably connected with the condition of the sun-spot minimum, or did they happen merely by chance? Time alone could furnish the answer. Great progress was made at this eclipse in the photography of the corona, particularly by the use of portrait lenses which were successful in portraying a mass of detail in the inner and brighter corona, but failed to show the outer streamers. Photographs of these faint extensions must needs wait until some date in the future when plates of greater sensitivity could be produced.

Another important observation at the eclipse of 1878 was the discovery (?) of two bright star-like objects by two American astronomers, Swift and Watson. The objects could not be identified with any of the fixed stars, and it was therefore necessarily assumed that they were small planets moving about the sun inside of the orbit of Mercury. The reputations of these two astronomers for careful observing were so great that it cost the science of astronomy a quarter of a century of eclipse observations before it was finally decided that no intra-Mercurial planets exist which are as large or as bright as the objects supposed to have been seen.

The next eclipse to be observed was that of May 17, 1882, the forerunner in the Saros of the eclipse of May 28, 1900. The 1882 eclipse was seen in Egypt with a brief duration of totality amounting to seventy-four seconds. This eclipse is memorable on account of the bright comet that was seen and photographed near the sun, the comet not being observed either before or after the eclipse. The photographic plates had now become more rapid, the dry plate having been invented, and accordingly the astronomers had to their hands better facilities for attacking the corona with camera and spectroscope. Also for the first time we hear of the prismatic camera, which is a slitless spectroscope, with a photographic plate to take the place of the observing eye-piece; and this instrument, particularly in the hands of Lockyer, was to play an important rôle in eclipse spectroscopy. Eleven years having elapsed since 1871, the year 1882 was one of maximum sun-spots, there being no less than twenty-three separate spots on the face of the sun on the day before the eclipse. The form of the corona in no way resembled that of the minimum of 1878 but bore a striking resemblance to the crown of glory of 1871, the shape being more nearly rectangular, or even star-like, and the long equatorial extensions and strong polar brushes being entirely lacking. The spectroscope also revealed vast differences from the eclipse of 1878. The corona as a whole was more brilliant than that of the preceding eclipse, and the emission lines of the spectrum were obtained both by a slit spectroscope and by the prismatic camera. With the former instrument, Schuster photographed about thirty lines in the spectrum of the corona. Many new spectral lines were visible in the red and violet to Tacchini and Thollon respectively. These lines were seen and photographed during the progress of totality, and not near the beginning or end of the total phase. Apparently the lines did not seem to belong to the flash spectrum and must have their origin in the true corona. But for the first time a suspicion seems to have been aroused that the lines might after all be due to prominences and chromosphere and not to the true corona, for Schuster observed the H and K lines of calcium to appear bright even across the face of the dark moon where no light at all was supposed to exist! Evidently the chromospheric light was reflected by some atmosphere somewhere, either in the higher reaches of the sun's atmosphere directly in line with the center of the dark moon, or in the atmosphere of the earth. The first condition could hardly be possible and that left no contingency other than the second. Schuster's observation was not the first to reveal bright lines on the dark face of the moon because as early as 1870 Young had perceived bright hydrogen lines.

If the lines of emission were stronger in 1882 than in 1878, it was not so with the dark Fraunhofer lines. The spectroscopes revealed the continuous spectrum in the brighter inner corona, but farther from the sun the dark lines due to reflected photospheric light were observed both visually and in the photographs, but these lines were not so strong as in 1878. Sun-spot maximum appeared therefore to correspond to a star-like corona, with no polar brushes, strong coronium and other bright coronal lines, but with the Fraunhofer lines intrinsically weaker than at sun-spot minimum.

The direct photographs of the corona by Schuster were in better definition and showed more details than those of previous eclipses. In fact, the impressions made on the plates were so strong that Dr. Huggins obtained the idea that it might even be possible to photograph the corona without an eclipse. For observing the prominences without waiting for an eclipse, the spectroscope had already been utilized to get rid of the glare of the sunlight in our own atmosphere. It is not possible to make use of the spectroscope in the same manner for obtaining coronal photographs, for the simple reason that the bright-line radiations of coronium are not sufficiently strong in character to enable the coronium light to shine in contrast with the enfeebled solar glare. Many attempts to photograph the corona in full sunlight were made by a variety of different methods lasting over a number of years, the astronomers being urged on by great hopes since the first trial photographs seemed to predict success. The details of the early work of Huggins, and the later researches will be deferred until a subsequent chapter.

One-third of the way round the globe eastwards from the Dutch East Indies, where the eclipse of 1901, May 18, was observed, brings one to the middle of the Pacific Ocean. The eclipse track of May 6, 1883, lay almost entirely across the water, but fortunately, in its path there was a small coral reef, only seven miles long, and unknown ten years previously. The importance of the discoveries of 1882 and the fact that the eclipse of the year 1883 was of the very long duration of more than five minutes, attracted to Caroline Island, astronomers from America, England, France, Austria and Italy. The great risks taken by eclipse expeditions in the tropics of being overtaken by clouds was shown in this eclipse, but fortunately a clear spell between two periods of clouds was experienced. The general features of the corona greatly resembled that of the year before, the sun continuing to show many spots. Owing to the long exposures possible, excellent photographs of the corona were secured, and for the first time in the history of eclipses, greater extensions of streamers were photographed than were visible to the eye.

The most important observations were unquestionably by means of the spectroscope. Up to this time, all observations during totality had shown a continuous spectrum of the corona close to the sun, and farther out faint Fraunhofer lines, with the bright green line of coronium crowning the whole. According to the dissociation theory of Lockyer, neither continuous spectrum nor dark lines could exist there, and "if these statements regarding the corona were strictly accurate my hypothesis was worthless."[1] Hence, a careful search was made by Janssen for Fraunhofer lines. They were found by him in great numbers, thus confirming the observation of 1882. To make assurance doubly sure, the dark spectrum lines in the corona were successfully photographed. As a result of these observations, Janssen concluded[2] that "the basis of the coronal spectrum was formed by the complete Fraunhofer spectrum, and that, therefore, there exists in the corona, and above all in certain localities of it, an enormous amount of reflected light; and since we know that the coronal atmosphere is very rare, it follows that these regions must abound in cosmic matter in the state of solid corpuscles, in order to explain the abundance of reflected sunlight."

Spectroscopic observations of great interest were made on the corona by Hastings. He used a 60° prism attached to a six-inch telescope, there being two totally reflecting prisms placed outside the slit so that the spectrum of two opposite sides of the sun could be brought together and examined by comparison. The observations were confined to the green coronium line. At the beginning of totality, this line was 12′ in length and very bright on the eastern

[1] Lockyer, *Chemistry of the Sun*, p. 365.
[2] *Comptes Rendus*, *92*, No. 10.

limb, while on the western limb it was only 4' in length and comparatively faint. As the eclipse advanced, the inequality vanished, at mid-totality conditions were equal, while at the end of totality the lines on the western limb were the longer and brighter. Such a great change could not be explained by assuming that the moon in its motion progressively covered and uncovered the bright coronal radiations, and accordingly, Hastings attempted to explain his observations on the assumption that the outer corona has no real existence but that its appearance is caused by diffraction round the edge of the moon. On this hypothesis, the true corona is confined to a very narrow ring around the sun, the light from this inner ring of material substance being widened by diffraction to form the outer corona which thus takes upon itself all of the appearances of reality. To the astronomers who had seen the great extensions of 1878, it was hard to believe that diffraction of light could adequately explain the detail of the coronal streamers at the great distance of twelve diameters from the sun's limb, but it was equally difficult to understand how luminescence could exist in a solar atmosphere at the colossal distance of ten million miles from the sun's surface. If the coronal light were reflected, it could not be seen unless reflected from material particles, and if the light were intrinsic, how could it have any existence in an atmosphere so infinitesimally rare? The answer given by Hastings denied the solar origin of the corona, and seemed to be a step backward. Apparently there was no way out of the quandary, but to wait for future eclipses.

In the attempt to secure information regarding the flash spectrum three distinct improvements in the line of attack were inaugurated in 1882: 1. Eye observations were no longer to be trusted exclusively, the spectrum must be photographed. 2. A grating was used, before, during and after totality. 3. A moving plate was utilized with an integrating prism spectroscope to secure photographs before, during and after totality. The grating showed little in addition to the H and K lines seen near the limb throughout totality. The photographs attempting to secure the reversing layer succeeded only in imprinting the hydrogen lines and comparatively few of the brighter lines, in fact the total number of lines were certainly too deficient to guarantee the confirmation of Young's belief that the whole Fraunhofer spectrum was reversed in the flash spectrum.

The total eclipse of August 29, 1886, was visible in the West Indies. The energies of many of the observers were devoted to testing the method of Huggins of photographing the corona without an eclipse. For this purpose, fifteen separate photographs were taken on the day before the eclipse and a series of twenty during the partial phases, these photographs to be compared with plates obtained during totality. The conclusions were quite definite, for not a single one of the coronal details was found on the plates taken outside of totality, and it seemed therefore necessary to decide that it was impossible to photograph the corona, except within the limits of a total eclipse, at least under the conditions of hazy sky and low sun that had prevailed.

Tacchini made a careful comparison of the prominences observed spectroscopically before and after totality with those seen directly during the total phase, and he concluded that all the prominences showed themselves larger and taller during an eclipse, the upper portions being white in color when the prominences exceeded 1′ of arc in height. The differences of apparent height may find a ready explanation in the effect of contrast with the background, inside and outside of totality, but the matter of the color of the prominences could not be so readily settled. For many years "white" prominences found a conspicuous place in spectroscopic literature. Tacchini observed the flash spectrum visually. Turner attempted to observe changes in the coronal streamers resembling currents, but obtained no results of value.

One of the expeditions at this eclipse experienced a series of accidents due to the unavoidable necessity of employing volunteer observers as assistants. Some of the incidents might have been amusing if an opportunity had been afforded on the morrow for making another attempt, — but at the rare event of a total eclipse such untoward happenings are not laughing matters, but become almost tragedies. Some of these mishaps were: failure to get the sun's image on the photographic plate in the most important instrument, the breaking of the polar axis just before totality, in the next important instrument, the failure of an assistant to make exposures, the standing of two native policemen in front of the photometer during totality, the seizure of the plates by well-meaning but ill-advised customs officials, thereby causing a delay in the development of the plates until the following May by which time the plates had greatly deteriorated. Eclipse observers generally are not anxious to repeat such a chapter of accidents.

The eclipse of the following year, August 19, 1887, was one of widespread disappointment, for the projects so carefully prepared ended only in failure to secure results, not through any fault in the plans themselves but on account of the astronomer's enemy, clouds, which prevailed almost everywhere. Fine weather, due to holes in the cloudy sky, prevailed at several of the stations in Russia and Japan, however, and some photographs and observations of value were secured.

The year 1889 brought two eclipses, both extensively observed. Here was inaugurated the splendid series of expeditions sent out from the Lick Observatory. The path of the eclipse of January 1 crossed Nevada and California, and the photographers near the line of totality were so well organized for the work by Mr. Charles Burckhalter that an excellent series of photographs of the corona resulted. The best photograph secured at this eclipse, and in fact the very best obtained at any eclipse to this date, was secured by Barnard. The equipment was very meager. The largest lens employed was $3\frac{1}{2}$ inches aperture, stopped down to $1\frac{3}{4}$ inches, and of 49 inches focus. Barnard's success depended on an accurate adjustment of the instrument but more specially on the skill and care with which the plates were developed. Barnard was a professional photographer before he was an astronomer, and he was thoroughly familiar with the best methods of developing a plate in order to bring forth all of the latent detail. The eclipse of December 22 of the same year was successfully photographed at Cayenne in the West Indies by the Lick party consisting of Burnham and Schaeberle. The photographs showed that changes had occurred in the corona since the eclipse of the beginning of the year. The earlier eclipse took place near sun-spot minimum and exhibited the equatorial extensions of the corona. The eclipse of December 22 is memorable from the death of Father Perry a few days after the eclipse, a martyr to the cause of science. This brave man, though greatly weakened, took part in the eclipse work, and having found as soon as totality passed that everything had passed off well, he called for three hearty British cheers, — which unfortunately he could not himself lead.

The greatest success attended the observations of April 16, 1893, largely through the use of apparatus much more powerful than had ever been employed before at an eclipse. The most conspicuous advance came to the party from the Lick Observatory in Chile who used a camera of five inches aperture and forty feet focus for securing photographs of the corona. Schaeberle decided to point the objective directly at the sun and to mount it on one fixed pier and the movable photographic plate on another, both piers to be wholly free from contact with the great tube extending between lens and plate. The slide carrying the photographic plate was the only moving part, and its motion was so regulated by means of inclined planes as to give it the same velocity and direction as the sun's focal image during the eclipse. The details of erecting this instrument, known as the Schaeberle mounting, are found in the *Contributions from the Lick Observatory, No. 4.* A careful focus was secured and beautiful photographs were obtained showing the prominences and inner corona with a definition which left little to be desired.

Optical power, up to then unprecedented, was employed in the prismatic camera designed by Lockyer and used by Fowler in West Africa. The camera had a focal length of 7 feet 6 inches with a prism of 45°, giving a dispersion of about two inches from F to K. Fowler secured photographs which were supplemented by a series obtained by Shackleton in Brazil, with the result that the positions of 164 chromospheric lines were measured between F and K. Deslandres, at the same eclipse, attempted to measure the rate of rotation of the corona by observing the relative displacement of the spectra of two regions of the corona at opposite sides of the sun placed in juxtaposition. A grating spectroscope was used, and the conclusion reached was that the corona partakes of the general rotation of the sun. Unfortunately, there is no justification for this deduction by Deslandres since the measures made by him were on the H and K lines which belong to the chromosphere and not to the corona. One of the most important results of this eclipse was that it became possible for the first time in eclipse spectroscopy to separate clearly the spectrum of the corona from that of the chromosphere, and it was henceforth no longer assumed that a spectral line visible during totality belonged of necessity to the corona.

The eclipse of 1896, taking place on August 9, was observed by a large number of expeditions. An English party consisting of Christie, Turner and Hills went to Japan where also was one from the Lick Observatory and another American expedition headed by Todd, and also two Japanese parties.

Lockyer went in H. M. S. *Volage* to Norway where a large party of seventy-five, including officers and sailors of the ship, took care of a large and varied program. In one department of the work, for instance, in the sketching of the corona, a competition was started on board ship by thirty-five volunteer sketchers, an artificial corona being exposed to view for 105 seconds, the time of duration of totality. Sixteen who showed the greatest proficiency were selected for sketching of the corona on eclipse day. But alas, for " the best laid schemes o' mice and men," — the clouds prevailed almost everywhere except where there was a small English party consisting of Stone and Shackleton. The latter was successful in timing his observations with the prismatic camera so well that the long desired photograph of the flash spectrum was at last secured. Although the focus was not of the very finest, still there were shown a total of 464 lines in the spectrum between F and K. This photograph, taken by one of Lockyer's assistants, sounded the death-knell of the dissociation theory, — but Lockyer still refused to be convinced that the flash spectrum was a reversal of the Fraunhofer spectrum. His argument was a very simple one, which was, that between F and K, 5694 Fraunhofer lines were tabulated by Rowland, while in the eclipse spectra of 1893 there were but 164 lines and in 1896 but 464 lines, consequently showing but three and eight percent, respectively, of the Fraunhofer lines reversed in the flash spectrum. Lockyer however failed to draw attention to the fact that Rowland's atlas was secured with a much greater dispersion than that used at the eclipse and with vastly superior definition. Lockyer's conclusion,[1] the result of the spectra at these two eclipses, was that "the chromosphere is a region of high temperature in which there is a corresponding simplification of spectrum as compared with the cooler region in which the Fraunhofer absorption is produced." The manner of settling the question, the way of advancement for future eclipses was clearly indicated: the flash spectrum must again be photographed and with increased dispersion, and great care should be exercised to see that the exposures were made at the correct times, with as good focus and definition as possible.

Such photographs were secured at the eclipse of January 22, 1898, visible in India, where such excellent conditions of weather were experienced that then partially compensated for the ill luck of the previous eclipse. The largest expedition in point of numbers was that under the direction of Sir Norman Lockyer located at Viziadrug on the West Coast, the astronomers being assisted by the officers and men of H. M. S. *Melpomene*. The program was an extensive one, embracing visual and photographic observations of the corona, the most important problem being a spectroscopic attack on the chromosphere with two large prismatic cameras of six and nine inches aperture. By these two instruments about sixty photographs were secured, the exposure times

[1] *Recent and Coming Eclipses*, p. 111.

varying from 1 to 59 seconds. These included two series of ten snap-shots at the beginning and another ten at the end of totality and a number of exposures of different lengths during totality. Christie and Turner, representing the British Joint Permanent Eclipse Committee, were at Sahdol. Copeland was at Goglee, Newall and Hills were at Pulgaon, Campbell of the Lick-Crocker expedition was at Jeur, while at Talni was located Evershed and also Mr. and Mrs. Maunder.

The most important problem was that of the flash spectrum, and fortunately, successful photographs were secured by Fowler and Dr. Lockyer, by Campbell, by Hills, by Newall and by Evershed. A discussion of the spectra by Sir N. Lockyer again confirmed him in the opinion he had held since 1873, that many strong chromospheric lines were not represented among the Fraunhofer lines, while many of the dark lines found under ordinary conditions in the solar spectrum did not appear as bright lines in the flash spectrum. He therefore concluded that the flash did not represent the spectrum of the reversing layer. It is true that the hydrogen series and the helium lines of the flash spectrum are not found in the Fraunhofer spectrum, and also that there are great differences in intensity between the two spectra, and in a sense therefore one spectrum is not the exact reversal of the other; but none the less, it is impossible to reach any conclusion other than that practically every strong dark line in the solar spectrum is present as a bright line in the flash spectrum. The matter will be treated more fully in Chapter XVI. The excellent photographs obtained by Fowler and Lockyer permitted a discovery of the greatest value to future scientific development, namely, that "enhanced" lines, or those which are relatively stronger in the spectrum of the spark than in the arc, are specially prominent in the spectrum of the chromosphere.

Exquisite photographs of the corona with the 40-foot camera were secured by Campbell, while Mrs. Maunder with a Dallmeyer lens of only one and a half inches aperture photographed the faint extensions of the corona running out to nearly six diameters from the moon's limb. The corona of 1898 presented a mixed aspect, a combination of the polar brushes observable at sun-spot minimum being combined with the quadrilateral shape of sun-spot maximum.

One contribution of great importance was the measurement of the wave-length of the green coronium line. For nearly thirty years since its discovery, it had been assumed that the coronium line was identical in position with the chromospheric line at 5316.8 A. Lockyer, Fowler, Evershed and Campbell independently found the coronium line to be farther to the violet, at 5303 A. That the value of this wave-length was now found for the first time, in spite of observations made at several eclipses, will show more clearly than words can express that the eclipse spectra prior to 1898 were poor in definition and small in dispersion.

Newall used a spectrograph with two slits with which he hoped to secure photographs to test the rotation of the corona. Unfortunately, the slits were placed 8′ from the sun's limb and the coronal light was too feeble to impress any traces on the plate. Newall observed the corona with a polariscope while Turner attempted to achieve similar observations by photography.

A new epoch of accuracy in photographing the chromospheric spectrum having been begun in 1898, it was but natural that every effort should be made to continue the success of this work in 1900, in order to secure, if possible, still greater definition with larger dispersion. The eclipse track of May 28, 1900, lay over the southeastern part of the United States, and after crossing the Atlantic Ocean, passed over Portugal, Spain and Algeria. On account of its easy accessibility to American and European astronomers the eclipse was witnessed by a greater number of observers than ever before in the history of eclipses. Fortunately good weather was experienced almost everywhere. The program was a wide and varied one, and it will be possible here to mention only a few special lines of work and record the names of comparatively few of the many observers.

Photographs of the corona were taken in numbers to the hundreds, or even to the thousands, by small, medium,

The Corona of "Minimum Type"
Photographed May 28, 1900 by E. E. Barnard and G. W. Ritchey.

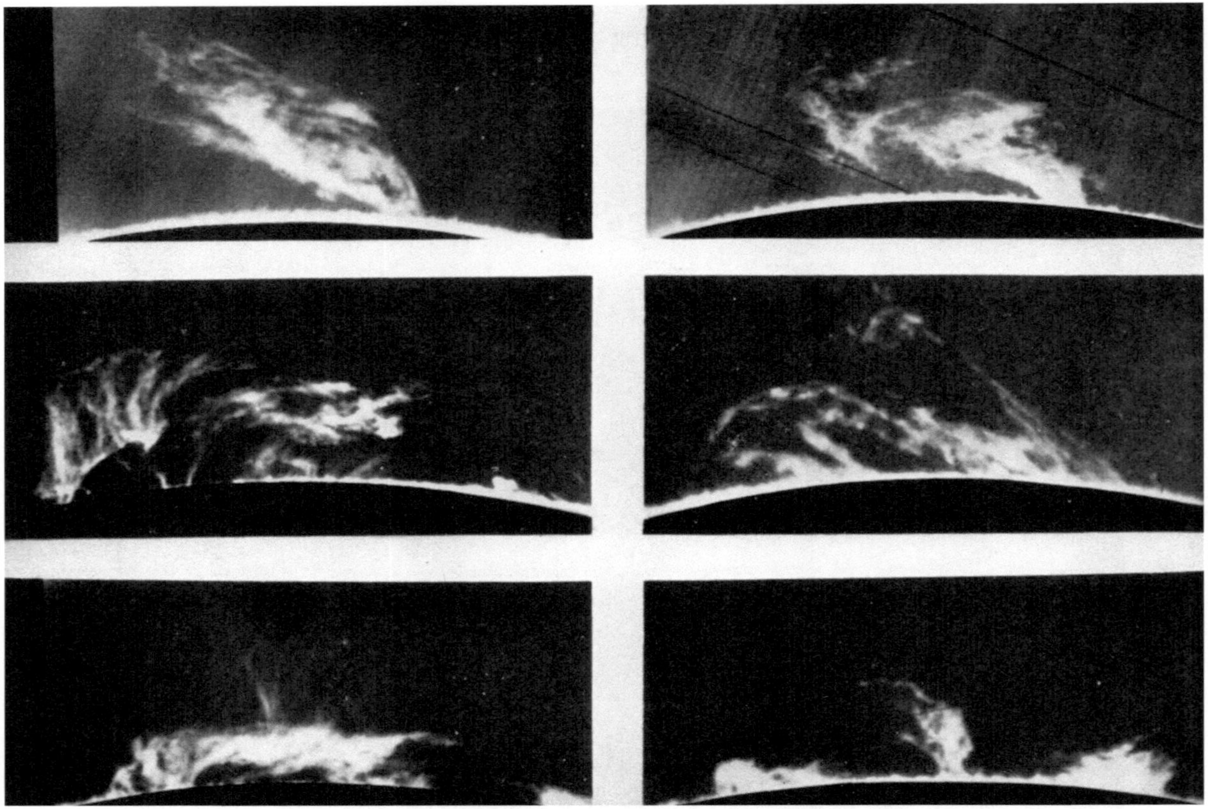

Prominences Photographed on the Sun without an Eclipse

Photographed by Slocum with 40-inch Yerkes refractor.
The rapid changes appear to indicate horizontal currents.

large and huge sized cameras, the greatest focal length being that employed by the party from the Smithsonian Institution who utilized a lens of twelve inches aperture and 135 feet focal length. The photographs on this large scale were in good definition and they showed a great wealth of detail in the inner corona. Excellent photographs were secured at Wadesboro, N. C., by Professor Barnard and Mr. Ritchey with a horizontal camera of 61½ feet focus. A reproduction of the exposure of thirty seconds is given on page 164. This eclipse having taken place near spot-minimum shows strong polar brushes and long equatorial streamers.

Both in America and in Europe, an extended attack was made on the flash spectrum by slit spectroscopes, by prismatic cameras and by gratings, plane and concave, used both with and without a slit. The United States Naval Observatory had three stations in the field, at Barnesville, Ga., at Griffin, Ga., and at Pinehurst, N. C. The three concave gratings used by Ames, Crew, and Humphreys were each employed with a slit. The photographs showed nothing. With plane grating and quartz lens, Huff secured a well-exposed spectrum, — but the focus was not of the very best. Better results were secured by Frost who used a concave grating, objectively without slit.

In Europe, Sir Norman Lockyer, located at Santa Pola in Spain, carried out an extended series of observations similar in scope to those made at Viziadrug in 1898. The nine-inch prism was combined with a camera 20 feet in focal length in order to secure a great linear dispersion in the resulting photographs for the purpose of measuring the heights of the various layers forming the sun's chromosphere. Successful photographs of the flash spectrum were secured by Fowler and Dr. Lockyer, and also by Dyson at Ovar, and Evershed at Mazapan in Algeria. The latter station was selected so that it might be as near as possible to the edge of the band of totality so that the photographs of the chromosphere might be obtained in high solar latitudes. Unfortunately, through an error in the *Nautical Almanac,* Evershed found himself just outside, instead of barely inside, the path of totality. The series of photographs obtained, however, were of fine definition and were specially valuable in affording a means of comparison with photographs of the flash spectrum which have usually been taken near the solar equator. This comparison shows that the spectrum of the chromosphere is the same at the sun's polar regions as at low latitudes, and it appears fairly certain that the spectrum of the sun's limb is as constant in character as the ordinary Fraunhofer spectrum. In this connection it should be borne in mind that Evershed's photographs showed the flash spectrum where the moon was practically at grazing incidence with the sun, and consequently the layer of the chromosphere photographed must have been very close to the edge of the photosphere.

Fortunately, the Algiers Observatory was in the shadow track, and here through the kindness of the director, M. Trépeid, Mr. Wesley was given an opportunity to observe the corona visually with an equatorial coudé of 0.3 meters aperture. Mr. Wesley's long experience in the study of the corona had well fitted him for the task of finding out whether more detail could be observed visually than is portrayed on the best photographs. The conclusion reached was very definite that the photograph, if on a sufficiently large scale, exhibits all of the coronal features that can be seen with the eye.

Successful polariscopic observations were made by Turner, Newall, and others. The inner corona showed marked polarization, a result which was difficult to reconcile with the absence of Fraunhofer lines in the spectrum of the corona.

Interesting observations were made by Abbot with the bolometer in measuring the radiation of the corona at different distances from the sun's limb. Results of great value were secured, but they showed the necessity of confirming them by observations at succeeding eclipses.

CHAPTER X

PERSONAL EXPERIENCES IN 1900 AND 1901

THE first total eclipse witnessed by the author was that of May 28, 1900. The moon's shadow path in America traversed the southeastern states from New Orleans to Norfolk; it then crossed the Atlantic Ocean, passed over Spain and the Mediterranean and left the earth's surface at sunset in Northern Africa. Good weather conditions had been predicted on both sides of the Atlantic and so it was unnecessary for the European astronomers to travel to America or for the Americans to cross the Atlantic eastward in order to insure a promise of successful operations. A few astronomers, however, did make the sea journey. As an illustration of how little the average European knows about America — or rather it should be said, *did* know of the United States in the year 1900 — it can be said on good authority that before crossing the Atlantic to witness the eclipse, the head of one of the expeditions appealed to the United States government for a guard of soldiers to protect the lives of the party from the wild natives (sic) of North Carolina. On arriving in New York, however, the fears were effectually dispelled when the party found themselves aboard a luxurious Pennsylvania railroad train and discovered that they themselves, their baggage and their instruments were routed through to their destination, — and entirely free of charge.

In view of the accessibility of the eclipse track in the United States and on account of the fact that it was a very favorable season of the year for a trip to the South, it was but natural that such great American institutions as the Lick, Yerkes and Naval Observatories should send well-organized expeditions, and that individual astronomers should gather in great numbers to witness the fascinating phenomenon. Most of the American astronomers were seeing their first eclipse and it might well have been thought that they would suffer from " nerves " at the exciting moments of totality; but there were many seasoned veterans in the ranks, men like Young, Langley and Campbell, who had witnessed eclipses in foreign lands.

Perhaps the most important part of the program for 1900 was the spectroscopic work. In 1896, after many trials, the first good photograph of the flash spectrum was obtained by Shackleton. In India in 1898, as a result of better definition, the spectrum photographs exhibited a wealth of detail that added greatly to our knowledge of solar physics. Evidently the procedure for the next eclipse should be the attempt to secure spectra with increased dispersion, so that wave-lengths should be determined with greater accuracy, thus permitting more reliable information regarding the sources of the spectral lines and rendering more positive our knowledge concerning motions of rotation of the solar envelopes. America is the home of the diffraction grating, plane and concave, brought to such perfection by the refined labors of Professor Henry A. Rowland of Johns Hopkins. Among the more powerful instruments brought into service were three concave gratings, each used with slit, and two plane gratings each used objectively without slits. Each of these five gratings were employed in connection with a quartz lens. Ordinary photographic plates were used, and as these are but little sensitive to green, yellow and red light, it was decided to concentrate on the blue and violet parts of the spectrum; hence the employment of quartz, rather than glass, lenses. The concave gratings were mounted in the ordinary Rowland manner in the attempt to secure sharp definition and large dispersion — but the dispersion was too great for the light available and no lines were found on the photographs. Huff obtained well-exposed plates with a plane grating, but the focus was good for a short region only near wave-length 4000 A.

The experience gained in 1900 was put to use in the plans for the eclipse of the following year. On May 18, 1901, the moon's shadow touched the earth's surface at sunrise on the

THE PASIG RIVER AT MANILA

KRAKATOA BY ITS ERUPTION IN 1883 CAUSED A LOSS OF 40,000 LIVES

A Cog-railroad and Tree Ferns in Sumatra

The Water Buffalo Takes His Daily Siesta

east coast of South Africa. After passing over the Island of Mauritius, the shadow traversed the Indian Ocean at cannon-ball speed and touched land shortly after noon on the west coast of the Island of Sumatra in the Dutch East Indies. After crossing over northern Borneo, Celebes and New Guinea, the shadow left the earth's surface at sunset far south of the Philippines. This eclipse was specially important on account of the very long duration of totality, six minutes. The location that afforded the best living conditions and promised the greatest success from favorable weather was the west coast of Sumatra, but such a site would carry the American astronomers as far away from home as they could possibly get — half way round the world.

In view of this great duration of totality and the importance of the observations to be attempted, the United States government, through Congress, appropriated money to equip and send out an expedition of thirteen members, two from the Smithsonian Institution, six from the U. S. Naval Observatory, and five guests of the latter, all of whom had had eclipse experience in 1900, the author being one of the five guests. In order to arrive in the East Indies in plenty of time before the day of the eclipse, May 18, the expedition left San Francisco on February 16 on board the U. S. Army Transport *Sheridan, en route* to Manila. On the journey westward three delightful days were spent in Honolulu, viewing the points of interest of that far-off "Paradise of the Pacific." At that time the Hawaiian Islands were cut off from communication with the outside world, except in so far as news was brought by steamer, for there was no cable to Honolulu — and it was before the days of wireless.

On March 26th, the U. S. S. *General Alava* left Manila to carry the expedition the remaining 2200 miles to the west coast of Sumatra where it was intended to locate for the purpose of the eclipse observations. After coasting along the Philippines and the north shore of Borneo, the equator was crossed on March 31st. Father Neptune came on board, and the reception to him was right royally given by the man-o'-war's men. April 2nd found us in the straits of Sunda which separate Java from Sumatra, and the ship passed within half a mile of the island volcano Krakatoa whose eruption in 1883 resulted in the loss of 40,000 lives. The huge tidal wave following the explosion, which inundated the whole surrounding country, left a Dutch man-of-war high and dry a mile and a quarter inland, and 76 feet above high water mark; and this wave was felt even in the English Channel, 11,000 miles away. The air wave sent out by the eruption was traced by barometers through seven complete circuits of the earth. Fine particles of dust were shot up to enormous distances in the atmosphere, causing the brilliant sunsets noticed for many months in 1883.

After coasting along the west of Sumatra, we entered the pretty little land-locked harbor of Emma Haven, probably the first American government ship that had ever entered port there, and were then at the end of our sea voyage of over 10,000 miles.

After official calls on the governor of that peaceful Dutch colony, we proceeded to Padang, the capital, just four miles distant, and obtained our first view of the island which was to be our home for the next two months. First impressions were formed at the Oranje Hotel, a typical hostelry of the East. The building, half hidden by cocoanut and tropical palms, is of one story with high thatched roof. At the front is the wide open verandah — the parlor and reception room of the hotel — and back of this, running through the middle of the building, is the dining room. On either side are the spacious bed rooms, each with its wide verandah — exceedingly inviting places in the hot afternoons. The bed is the chief curiosity to the American, for it is a four-poster, with finely-figured mosquito net and of the huge dimensions of 7 x 8 feet. Over the mattress is spread a daintily embroidered sheet, but there are no bed-covers — and indeed, under the equator, few blankets are needed. Of pillows, there are several, the ordinary bed having four and two " Dutch wives " — the name given to a long bolster used to keep the occupant of the bed cool.

A day in the East is begun by awaking about six o'clock. Coffee is immediately brought by the *jungus* — for no white man does any manual labor, and each has his native valet —and then, clad in pajamas, one risks the bath. Such a thing as a bath tub is unknown, the bath being taken in a cemented room with a cistern of water at the back, over the top of which is placed a wire screen to keep out inquisitive foreigners. Dipping water through a rectangular opening in the screen by means of a bucket, and throwing it over oneself, constitutes the bath—and a very excellent one it proved to be. After leisurely dressing (for no one is ever in a hurry), a light breakfast is partaken of, consisting usually of bread and a few cold meats. The men come to breakfast dressed for the business of the morning, in their white suits, or often in their pajamas, consisting of trousers of cotton stuff of the most marvelous colors and patterns, and a coat or *kibaya* of white, made after the Chinese style without collar. But how tell of the dress of the ladies? For they appear at breakfast in costumes in which an American girl might be ashamed to be caught in her boudoir: a native skirt or *sarong* reaching to the ankles made of picturesque cotton stuff, and a lace-edged linen jacket. This is the dress of the native women who wear neither shoes nor stockings. The Dutch women have adopted their costume *in toto,* except that they usually add a pair of gold embroidered slippers. After breakfast, the serious business of the day is taken up until it is time for the midday meal, or *rijsttafel.* This, as the name signifies, is largely made up of rice—and it is a most astonishing meal. Into a large soup plate is put a liberal supply of splendidly cooked rice, and on top of this, chicken, meat, potatoes, and portions of fifteen or twenty curried dishes. It makes an astounding looking mess, but after a little experience in selecting the proper proportions and combinations of the curry and spices, the whole makes a very palatable dish. This is only the first course, to be followed by potatoes fried in cocoanut oil, and excellent chicken, with a third course of fruit. The next two hours of the day are universally given up to the siesta. About four o'clock, the East again awakes, and after a cup of tea or chocolate, takes another bath and dresses for the afternoon. The Dutchman seeks his club and plays a game of billiards or cards. Games requiring much exercise do not find favor in his eyes, but he has a taste for horse flesh, breeds most wonderful little ponies, and is always riding or driving. The ladies now appear for the first time during the day in European costume; and the hours before late dinner are given up to social functions. If these are of a ceremonious nature, the men must wear a black tail coat, and at least carry a hat, even if it is not worn.

A nice little informal gathering takes place before dinner each evening on the hotel verandah, where the men drink their *pitje* of gin and bitters, furnished by the hotel to all comers. Dinner at 8:30 or 9:00 has nothing remarkable about it, being very similar to our own. As soon as dinner is over, negligee is resumed, and the day is at an end.

Within a week there arrived in Padang two other parties besides our own from the United States, and scientists from England, Holland, France, Russia, Japan, and even India, about eighty astronomers altogether from all parts of the world, all bent on learning something about the sun.

The Dutch were very kind and generous in their treatment of the foreign scientists, and did everything in their power to make the two months' stay in their island as easy in becoming acquainted with strange conditions and as enjoyable, as possible. Passes were furnished to each astronomer for passage over the "Staatsspoorweg op Sumatra," the railroad owned by the government which runs from Padang into the interior about 100 miles, one spur reaching Fort de Kock, the other Sawah Loento. These passes included not only the transportation of our persons, but also of freight and baggage, which is not carried free in the island. Labor was largely furnished without cost, machine shops were put at the disposal of the astronomers, and their slightest wish was readily met and generally anticipated. In short, these Hollanders were perfect models of hospitality.

In a few days after arrival, the astronomers had separated to the different locations determined on as bases of operations. The English divided into two parties, one going to Sawah Loento, where were located Mr. and Mrs. Newall, the other to an island off the coast of Sumatra where the party under Dyson could have the assistance of the officers and men of H. M. S. *Pigmy,* detailed there to help in the observations. The Dutch located not far from this island but on the mainland, and the astronomers had the help of the gunboat *Sumatra.* Perrine of the Lick Observatory of California decided on Padang as likely to be most favorable for his researches.

The Naval Observatory divided their party into three sections, and decided to locate along the line of the railroad. The engineering principles under which this road was constructed were marvels of simplicity. If a small hill was met in laying out the line, the rails were laid over it, if the hill proved too steep for this, the tracks went around it. As the road starts from sea level and reaches an elevation of 4,000 feet in less than 100 miles, following a valley between mountains of 9,000 feet, there are many steep grades and sharp curves; in fact, for half the distance it becomes a cog road. But it is one of the most picturesque routes imaginable, running now through the green rice fields, now past fields ready for the sickle — for there are no seasons, and sowing, reaping, and threshing may be under way in the same field — now through a grove of cocoanut and banana palms, and again through a dense tropical jungle, where a dozen chattering monkeys scared up by the train go swinging away from limb to limb. On our first trip over the road we had just come to a bridge and were going down a steep grade when the Malay brakeman in some way lost his headgear. While we sat there for half a second wondering what he would do — for we had never seen a native without his *topi* — he solved the difficulty by jumping off the train, running back after his cap, and after a short run of 50 yards, catching the train again, which all this time was moving at its usual speed. At one of the stations where we got off for a few minutes to look around, we were immediately the center of a large crowd of most curious natives. They offered us bananas — *pisang* as they call them — and oranges, mostly green in color, little pieces of sugar-cane on slivers of wood, and many strange and wonderful looking messes. They eyed us with much wonder, and attempted a few words in Malay, which we however understood about as well as they did our English. In a few days however, they became acquainted with our mission, and spoke of us by the Dutch word "Zoneclips," while we, on the other hand, learned to make ourselves understood in Malay.

It was necessary to select the three Naval Observatory eclipse stations on the line of the railroad, on account of the difficulty of transporting heavy instruments and supplies into the interior. At Fort de Kock, near the edge of the moon's shadow path, three members under the direction of Professor W. S. Eichelberger, U. S. N., investigated the corona and the atmosphere of the sun with photographic telescope and spectroscope. At Solok, the main station of the party was located under the direction of Professor A. N. Skinner, U. S. N. Here was Professor Barnard with a photographic telescope of 61½-foot focus, which would give an image of the sun about seven inches in diameter. Barnard planned to give several short exposures, together with a single exposure of two and a half minutes, on a plate 40 inches square, hoping thereby to obtain some exquisite details of the corona. At Solok were also located several spectroscopes under the supervision of Jewell who was engaged in various interesting objects of research.

Abbot of the Smithsonian Institution was investigating along two separate lines; first, with four telescopes mounted so as to photograph a large region in duplicate in the vicinity of the sun, he was attempting to discover new members of the solar system revolving inside of the orbit of the planet Mercury (if there be any such objects). His second task was a plan for measuring the heat of the corona and of the dark side of the moon turned towards us with the delicate bolometer, which is in reality nothing but a very sensitive thermometer capable of detecting the heat of an ordinary candle at the distance of five miles and of measuring differences of temperature to the one-millionth of a degree.

The author was in charge of the U. S. Naval Observatory party at Sawah Loento at the terminus of the government railroad. The *sawah,* or irrigated rice-fields, is one of the prettiest sights imaginable. These fields are usually on the side of a hill, for the west coast of Sumatra, where the astronomers were located, is extremely rough and rugged. The sides of the hills have been terraced in order that the water may flow from one level to the one next lower, the enormous amount of manual labor required to make these terraces being for the most part performed by the women. The rice grows under water, but artificial irrigation is not needed, for it rains almost every day, not merely a shower, but with an average fall of half an inch for every day of the year. A view of these rice terraces with their vivid tropical greens — and such brilliant color is never seen in the temperate zones — is certainly a magnificent sight. All trace of the *sawah,* however, has departed from the village of Sawah Loento, and its place has been taken by the Ombilian Coal mines, belonging to and operated by the Dutch government, which are remarkable for having a vein of rich coal forty feet in thickness.

The elevation of the railway station at Sawah Loento, according to the Dutch maps, is 262 meters above sea level, or 859 feet. The height of the eclipse station above this was measured by means of an aneroid barometer as 400 feet, which therefore made the elevation of the eclipse camp about 1,260 feet above sea level. A short distance away was the expedition from the Massachusetts Institute of Technology consisting of Burton, Hosmer, Harrison Smith and Matthes.

The solar problems to be investigated by the Naval Observatory party at Sawah Loento were along two separate lines: to photograph the corona with a camera 104 inches long, with lens six inches in diameter, and secondly, to photograph the spectrum of the chromosphere with a plane grating and quartz lens, used objectively without slit. To mount the instruments, piers of brick were constructed. Tents were used as coverings for the instruments, with an extra one for a storehouse, four tents altogether being set up. As the Boston party had reached Sawah Loento about ten days before us, they had learned to some extent the best way in which to proceed. The benefit of their experience was freely imparted, and much valuable information cheerfully given. To Meinheer van Lessen, the chief engineer of the coal mines, many thanks were due for the very generous way in which he looked after the supplying of all building material. The bricks, made near the coal mines, were transported to the eclipse location in three stages; first by rail to the residence of the controller; second, by the slow-moving "kreta kerbau," drawn by the sturdy water-buffalo; and third, for the remainder of the distance, about a third of a mile, by coolies. The slowness of the last part of the operation can perhaps be imagined when it is stated that in a basket slung on a bamboo pole on the shoulders of two coolies, five ordinary sized bricks would be carried. This was indeed the minimum load, but the maximum was never more than ten. It was very interesting to see six coolies, using three bamboo poles, carrying a barrel of cement. The coolies provided were convicts. The way in which the Dutch East Indies treat their prisoners is one of the interesting features in the management of the Malay. If a native of Sumatra commits a crime, he is sent to one of the other East India islands; and similarly natives of the other islands are sent to Sumatra to serve out the terms of their punishment. Consequently, in a penal settlement in Sumatra are natives of Java, Borneo, Celebes and of some of the smaller islands, but no Sumatrans. The reason for this strange separation is that there is great enmity between the different races, and so if a Javanese prisoner tries to escape he is immediately apprehended by the first Sumatran who meets him and sent back. And thus it is that there are large penal settlements with no surrounding walls and very few guards.

Such a settlement was at Sawah Loento. The convicts, to the number of 3,000, were employed in the coal mines. There was no guard over them except the " mandur " or policeman, one of themselves raised to a position of authority and responsibility, and answerable for the conduct of the coolies under him. Notwithstanding these conditions, one could go around with perfect safety and with less fear of molestation than in New York City. In fact, the pistols we had brought, with which to defend ourselves from cannibals, were soon packed carefully away in the bottom of our trunks.

A Sumatran Beauty

Market Scene in Sumatra

Sumatran Mother and Half-caste Daughters

East Indians in Picturesque Garbs

In order to make the convicts more satisfied with their lot, the Dutch authorities paid them the amount of seven cents Dutch money, 2.8 cents American, per day. There seemed to be a great amount of sickness among the convicts, possibly due to their confinement in the hot country, and several hundred of them were in hospitals or were put on sick duty without pay. These coolies were provided for us by the controller free of charge, and we were able to have about as many as we needed. They were very slow and not over fond of work, but nevertheless they could carry bricks about as fast as our Malay " tukang " could lay them. He used to squat at his work, and it took five days for him to lay 2,200 bricks. But the piers were finally built, the tents raised, and the instruments gradually mounted.

The day of the eclipse dawned clear, and our hopes were that these favorable conditions would remain until after totality, which occurred shortly after noon. First contact was observed in a perfectly clear sky, but soon after this, clouds began to gather, and a quarter of an hour before second contact the sky was completely overcast.

The disappearing crescent of the sun was observed with a binocular before one barrel of which was arranged a small plane grating in such a way that with one eye the spectrum could be seen, and with the other eye the sun itself. With this, shortly before the time of second contact, bright lines were seen for a few seconds at F and H and in several places in the green and yellow, but these disappeared almost at the instant of being seen, the sun being completely hidden by clouds, and the flash passed without our being able to see it.

Toward the middle of totality, conditions became a trifle better, so that it was possible to see, through clouds, the corona extending for about half a diameter from the sun, and with the small spectroscope to trace the form of the coronium line quite distinctly. During no time of the 5^{m} 41^{s} of totality was an unclouded view of the corona obtained, but nevertheless, the clouds were so thin at the end of totality that the second flash was beautifully seen.

One hour after the total phase the clouds cleared away and a perfect sky remained for the rest of the day. Alas!

that the eclipse did not occur at one o'clock instead of twelve!

The pre-arranged program was carried out, however, in its entirety as if it had been clear.

Powerful spectrographs were used also at the two other Naval Observatory stations. At the main station near Solok, Jewell had a concave grating, but as the result of the experience in 1900, the instrument was used without a slit. Dinwiddie had a prismatic camera under his care, a single large prism of 60° angle, kindly loaned by the Smithsonian Institution. Near the edge of the moon's shadow-path at Fort de Kock, Humphreys had a huge concave grating of nine inches aperture, with lines three inches in length and a radius of curvature of thirty feet. This was the largest grating ever ruled on Rowland's dividing engine and it was constructed specially for the purpose of this eclipse. It was not a perfect specimen of a grating for the diamond point had broken down during the course of the ruling. This station was purposely located near the edge of the shadow in order to photograph the low-lying layers of the chromosphere. Weather statistics did not promise favorable skies at Fort de Kock — but strange to relate, it was the only locality in Sumatra where astronomers were stationed that clouds were not experienced. The main station at Solok was far from being fortunate, and the astronomers there succeeded in getting — almost nothing, so dense were the clouds throughout the whole of totality.

From Jewell's many years at Johns Hopkins University, working as assistant to Rowland, he had had vast experience in the development of spectroscopic plates, and hence the task of developing the precious negatives secured was handed over to him. While this was being done, and while the instruments were being dismounted and boxed and crated for their long journey home, we had some days to look around and observe the manners and customs of the people. The natives are all of them frightfully lazy — or shall we say only tired? But it is the same disease that afflicts the natives all through the East. The Filipinos are troubled in the same manner, a similarity not to be wondered at, for Sumatrans and Filipinos both belong to the same great parent family. It was probably for this reason that the natives of the Dutch East Indies were so interesting.

The Malay is naturally a man of easy-going, indolent character, who never gives open expression to a sense of astonishment or fear, and is probably little affected by these feelings. When alone, he is gloomy and taciturn, never either singing or talking to himself. The upper classes are exceedingly courteous, yet this outward refinement, strange to say, co-exists with the most pitiless cruelty and contempt of human life, traits which belong to the dark side of their character; and herein lies the explanation of the many diametrically opposed judgments which have been given us of the native of the East. There is another trait we must not forget, and that is an insatiable love of gambling which no laws seem to be able to suppress.

These are the characteristics of the Malay, but they describe the Filipino just as accurately as they do the native of Sumatra. We were almost two months in Sumatra and in that time were thrown constantly with the natives who, as servants and coolies, were used continually. It became necessary to learn Malay and speak it; this however proved not to be a very difficult task as the language contains neither declensions nor conjugations.

The horned-roofed houses were very picturesque, and near Fort de Kock one of these was seen being modelled in gold and silver filigree work (at which the natives are very skillful), for a present from the colony to the Queen of Holland, to whom they are very loyal. The natives have a pretty legend, the *Menangkarbau,* about the origin of the horn-shaped roofs. This relates that there was once eternal enmity between the Javanese and Sumatrans, first one side conquering and then the other, and it seemed likely that the bloody conflict would continue without anything ever being decided. Finally the sages of the two peoples hit upon a plan to the effect that each nation should choose a champion from the animal world, that these should meet in mortal combat, and that whichever animal was victor, that nation was to be considered the conqueror for all time. The Javanese chose a vigorous young tiger, under the impression that their champion surely could not be beaten, while the forefathers of the Sumatran natives picked a strong two-year-old karibou bull. Fancy, if we can, the two tribes in half circles meeting to see the fray, the shouts of rejoicing on one side and the woes of the other when the karibou succeeds in using its horns to such advantage that the tiger is killed. To perpetuate this victory, the Sumatrans forever after build their roofs to show the karibou horns. Usually the roof has one pair of horns, but it may have two or three pairs, the extra pairs, we are told, signifying that a daughter is married in the house. For the family in Sumatra is maternal, with the mother as head of the house, so when a daughter is married she brings her husband to live in her mother's home. Strange to relate, this form of family government coexists with Mohammedanism, where a man may have three or four wives. In such cases, however, we were never given to understand just how a man divided his time, and how he could live peaceably with so many mothers-in-law. Women do most of the work, till the soil, gather the crops, thresh the rice and carry all the burdens, while the man evidently considers himself a superior being and succeeds in doing very little work. No wonder then he has three or four wives to work for him!

The Dutch have kept the original tribal relation and each tribe still has its chief, who is however, little more than a figurehead. Each man is required to perform one hundred days labor each year for the government, and as a great amount of this is put on the roads, excellent carriage roads appear even in almost inaccessible parts of the island. The principle under which the Dutch run their colonies has, by some writers, been described as merely for the purpose of making money out of them, and hence it has been the policy to keep the natives from becoming educated or Christianized, to keep out European immigrants and capitalists and to preserve the whole trade as a monopoly of the home government. And so we hear of the " culture system," but the products under the " system " have gradually been reduced, until now in Sumatra the chief ones are coffee and tobacco — and who has not heard of " Sumatra wrappers "? Natives are obliged to plant a certain number of coffee trees each year, and sell the product at a fixed rate to the government. The Dutch have, however, been exceedingly successful in keeping the Malays docile and contented, and it might be well for the Americans to study the results of these three hundred years experiment on Malays of exactly the same characteristics as the Malays in the Philippines, the Filipinos.

After traveling half way round the world in search of knowledge, to have cloudy weather during the precious six minutes of the eclipse was indeed heart-rending, and the majority of the astronomers were pretty blue as a result. But when the plates came to be developed — and what a boon the photographic plate has been to the astronomer — it was found that the clouds had not interfered quite as much as expected. And so, down on the coast, Dyson and the Dutch parties, with the assistance of their warships, succeeded in getting, with telescope and spectroscope, some really excellent photographs of the corona and " flash." The corona could not be traced a very great distance from the sun, but many exquisite details were seen in the inner corona. At Padang, Perrine of the Lick Observatory had the same kind of cloudy weather as experienced by nearly all the other scientists, yet the results of his work were very satisfactory in spite of the clouds. The photographs of the corona taken with the 40-foot photoheliograph, like those of the English and Dutch parties, showed splendid detail in the inner corona, but they were particularly interesting from an appearance in the northeasterly portion of the corona, as if an explosion had taken place. Perrine later found that this remarkable disturbance was immediately above the prominent and only sun-spot visible during eleven days, thus showing an intimate connection between sun-spots and disturbances in the sun.

The three parties at Sawah Loento suffered likewise from the clouds. The main work of Burton and his party from the Massachusetts Institute of Technology was investigating the magnetic disturbances while the eclipse was in progress; this work, however, could be carried out as well in cloudy weather as in clear. The Boston party found the disturbance not so great as at the eclipse of 1900. Newall of Cambridge, England, had a very complete spectroscopic program to carry out, including many interesting researches, among which was an attempt to measure the velocity of rotation of the corona and to see whether or not this halo shines by its own inherent light.

The few days previous to departure were spent at Padang where we had the pleasure of meeting many of the visiting astronomers whose accommodation taxed to the utmost the capacity of the leading hotels, the Oranje and the Atjeh. On the evening of May 27, the United States consular agent gave a dinner and farewell ball in honor of the American astronomers and naval officers. Many prominent people were present, including Governor Joekes, also the officers of H. M. S. *Pygmy*. After the ball was over, the author slept in one of the beds of the Oranje Hotel with Professor Barnard, Dr. Abbot and one other. Our ship, the U. S. S. *General Alava,* had waited in Emma Haven to carry us back to Manila and we sailed at 10 A.M. May 28 on our homeward voyage. The eclipse of 1901 was a thing of the past! On June 8 we entered Manila Bay through the north channel, the Boca Chica, and anchored off Cavite. After a fortnight's wait in Manila, we sailed aboard the U. S. Army Transport *Indiana,* and reached San Francisco on July 16, five months from the day we had departed.

What scientific results had been obtained from the long journey and the many months spent away from regular duties? When the photographic plates were studied in detail it was found that much of great value had been secured in spite of the clouds, and as a result the Sumatra eclipse had contributed some very important information regarding the problems of solar physics.

CHAPTER XI

THE SPANISH ECLIPSE OF 1905

THE next eclipse to be widely observed was that of August 30, 1905. On this day the eclipse began in Manitoba and, after crossing through Northern Canada, it left Labrador about 8 A.M. on its trip across the Atlantic. Shortly after noon the shadow cut into Spain, then on through the Mediterranean, Northern Africa and Egypt, leaving the earth's surface at sunset on the coast of the Indian Ocean.

Spain was chosen by the majority of astronomers, both because the duration of totality was longer, and because the promise of good weather conditions was better; and here in a path one hundred and twenty miles in width running diagonally across the peninsula, hundreds of astronomers, American and European, were gathered.

The party sent out by the United States government was under the general direction of Rear-Admiral Colby M. Chester, U. S. N., superintendent of the Naval Observatory. Three men-of-war were furnished by the Navy Department for the purposes of the expedition, the U. S. S. *Minneapolis,* U. S. S. *Dixie* and U. S. S. *Caesar,* the first named being the flagship of the squadron.

The three vessels left separately from the United States about the end of June and met in Gibraltar about the middle of July. "Gib" is one of the most interesting places in the world, especially when entering on a naval vessel. It was a glorious sight, as we steamed in at dawn on board the *Minneapolis,* to behold the wonderful rock and, sheltered at its base, the Mediterranean squadron of the British navy, consisting of eight battleships and eight first-class cruisers, under the greatest of English admirals of the time, Lord

Charles Beresford. The morning of our arrival was spent in firing and acknowledging thunderous salutes, and in making official calls. To carry out properly these acts of courtesy between the American and British nations, it was necessary to fire no less than one hundred and fifty-two rounds of ammunition. On the morning of our second day in Gibraltar, the British squadron sailed, and it gave us an idea of the quality of the greatest navy in the world to see the splendid, seamanlike manner in which the big ships got under way without confusion, and one by one in perfect order departed from the crowded harbor.

After leaving Gibraltar and entering the blue waters of the Mediterranean, the *Minneapolis* steamed along the coast of Spain for about four hundred miles and anchored in the harbor of Valencia, the first American man-of-war to visit a Spanish port since the Spanish-American War.

It had been decided to divide the Naval Observatory expedition into three, sending two parties to Spain and one to Africa. In Spain the parties were located, one at the edge of the path of totality at Puerto Coeli, and the other near the central line at Daroca.

Daroca is in the heart of old Spain, about forty miles from Saragossa, and as a railroad had been there only four years it was a *terra incognita* for modern tourists — for which we were duly thankful. Our six weeks' stay there was a happy commingling of hard work — and there was plenty of work to do — with pleasant experiences in getting acquainted with Spanish life and people. The site for the town is indeed a peculiar one, in a valley so surrounded by hills that each heavy rain storm used to flood the city, until about 1600, a tunnel was constructed through one of the hills to carry away the waters. The tops of these hills are crowned with walls and forts, most of them constructed by the Moors a thousand years ago, some of them by the Catholic Spanish since that time. There is one tower of special interest, and still in good state of preservation, which is said to have been built by the Romans before Saguntum was founded, and it is, therefore, more than two thousand years old. (The railroad from Valencia passes through Saguntum

Filipino Homes Built on Stilts to Get Out of the Wet

Horn-shaped Roofs and Handsome Wood-carving in Sumatra

The Roman Fort (2200 years old) and Moorish Fortifications at Daroca, Spain

An Old Aqueduct at Daroca, Still in Use. Looking South to the Valley

where Hannibal and the Romans had their memorable fight in B.C. 238.)

The Spaniards received us with open arms and did everything in their power to assist in our work and to make our stay in their midst as pleasant as possible. As no one in the place could speak English, it was necessary to make ourselves understood in their language.

To help in the erection of the temporary observatory, six sailors were sent in from the *Minneapolis,* and all hands, astronomers and sailors, worked each day from early morning till late at night, building piers, erecting telescopes with houses to shelter them, mounting spectroscopes, and fixing up a meteorological observatory. After the carpenters and machinists had finished their work of construction, it was necessary for the scientists to focus and adjust, to see that everything was in good working order, and to make trial photographs. A few days before the eclipse, the party was increased in size to thirty-five, officers and sailors having come up from the ship for the purpose of assisting in the observations. Frequent drills were held in order to familiarize each one with his part and thus to be sure that everything would go right and that no precious seconds would be wasted at the time of eclipse.

The location of the eclipse camp was half a mile south of the town, in the midst of a beautiful, fertile valley. From there, while we worked, we could catch glimpses of scenery typical of Spain. The first feature to attract one's attention is the extremely barren aspect of the country, which is in sharp contrast with the garden-like appearance of England. The hills of Spain were in early times densely wooded, but now are almost entirely devoid of trees and look from a distance as if there were not a particle of vegetation on them. Moreover, the rainfall is so slight that agricultural pursuits must rely upon irrigation, and thus it is only the valleys that are green and cultivated. In such a valley along the shores of the little river was our eclipse camp located. The greenest field was decided upon as the site of the observatory, and upon application to its owner for permission we found that he was quite satisfied to allow his plot of ground to be used, but thought some compensation should be made for the valuable crop of grass that might possibly be raised during the summer. On receipt of one hundred pesetas, he forthwith proceeded to take a fatherly interest in all of our doings, and explained scientific matters to every one as if he had been chief of the expedition. His field became the center of interest in the community, and people came from all sides to look upon the strange doings. As a prominent trait of the Spanish peasant seems to be a great and overpowering curiosity, we had pleny of onlookers; and when the mayor and a few of the most prominent citizens were invited to look at the moon through our five-inch telescope, we were rather surprised — to put it mildly — to find over one hundred people turn up, when only a half score had been invited. Their curiosity took the form only of making each and every one in the town intensely interested in what was going on, and to show that interest they turned out in force each afternoon to see how matters were progressing. It might be asked, what their attitude was towards these Americans who had so lately beaten them in the small war. Before the expeditions reached Spain, it was feared that perhaps there might be some friction on that account, but these fears were not realized. As a matter of fact, the only person we met who seemed to have any feeling at all in the matter was a former soldier in the Spanish army. He had seen service in the Philippines, had been captured and thrust into prison by the Filipinos, had been rescued by the Americans, and as a result he had only the kindest feelings towards everything belonging to the United States. As for the rest of the people, they seemed to have forgotten all about it, or else they did not know there had been a war, for it must not be forgotten that only about one quarter of the people in Spain can read and write.

After this my third total eclipse, I can confidently say that observations at such a time consist of much hard work and many nerve-racking experiences. The astronomer is never on hand sufficiently long beforehand to take things quietly and easily, he must work under conditions to which he is totally unused, and over his head hangs the knowledge that everything must be completed by a certain day and a certain hour, for the eclipse cannot be postponed, and there is no second trial in case of failure. In addition to working hard all day as carpenter and instrument maker, the astronomer must stay up half the night adjusting his instruments on stars, so that during the last few days before the eclipse very few hours of sleep each night are obtained. However, in spite of the many difficulties that were continually cropping up, the mounting and adjusting of the instruments was practically completed by August 25, when our observing party was swelled in numbers by the officers and men from the *Minneapolis*. From then until eclipse day the time was spent in putting the finishing touches on the work of adjustment, and in having frequent drills in order to insure that everything would go without a hitch.

What was the promise of weather for the important day? We had been closely scrutinizing the weather each day to see what conditions we were to expect, and were much pleased to find that the sky was usually clear just after noon, the hour when the eclipse was to occur. August 29, however, had been cloudy all day so that on eclipse day we had to go to camp early to test our final adjustments, go through drills once more and to be sure that all the apparatus worked smoothly. The skies were clear and our hopes for success were high. Outside the roped-off enclosure, the whole town of Daroca was assembled, for it was naturally thought by the people that nowhere could the eclipse be seen so well as where the astronomers were located.

At 11:52 A.M. a little indentation was seen on the western limb of the sun, and the eclipse had begun. The skies were clear with the exception of a cloud here and there, and our most ardent wish was that the clouds would leave the sun clear for the next couple of hours. For the first hour that the moon was creeping over the sun there was nothing of very great moment to notice, but for the next twenty minutes till 1:12, when the sun was blotted out, we were each of us filled with expectancy, for matters began to take on a weird and unnatural appearance. The little blotches of light under the trees, instead of being familiar circles, were little crescents, exact counterparts of the sun itself. The darkness began to make itself really felt, and without looking at the sun one would know that something out of the ordinary was happening, for the gloom did not in the least resemble that of sunset. A hush fell upon the crowd of assembled and talkative Spaniards when, ten minutes before totality, a big cloud drifted over the sun. Would this cloud move away? Or were we going to be disappointed? It hung there for a space of time that seemed to be an age, while in reality it was only five minutes. It was a big scare, but when that passed, with a shout from us all, there wasn't another cloud anywhere to bother us. Seventeen seconds before the calculated time, with the last disappearing ray of sunlight, the corona broke forth into view. What a magnificent sight it was shining out with its pale, pearly light for a couple of diameters round the edge of the sun, with its streamers and brushes of delicate light! True to prediction, the corona was almost square in shape, and was not at all alike in appearance to the other coronas the writer had seen in 1900 and 1901, with their long fish-tail extensions along the sun's equator and short-curved streamers near the sun's poles. In the upper left hand quadrant, huge red flames sixty thousand miles high could be seen with the naked eye, and these with a closer view with the telescope resolved themselves into a forest-like structure. Close to the sun the corona was very bright, in fact so brilliant that the eye was not readily able to take in all the details of the faint streamers. As a pictorial effect, without the long equatorial extensions, this corona was much inferior to the two last ones seen. Still it was a magnificent sight, and we were more than thankful for having clear skies for making our observations.

When totality first started we were each and all of us much too busy to take very much notice of our immediate surroundings or even of the corona itself. We could not help becoming aware that our Spanish onlookers outside the ropes were appreciating the show in the skies provided for them without expense. From the noise made each one seemed to be telling his neighbor at the top of his voice just how it happened and what there was worth seeing, and this in spite of the fact that the mayor of Daroca had generously provided half a dozen members of the civil guard to preserve order and keep quiet. For the first half minute the din was so great that it was impossible to hear the seconds counted, or to know exactly when to begin and end the exposures of the photographs. When the Spaniards had quieted down, after their first outburst, all that was heard in the eclipse camp was the steady count of the observer calling out the seconds as they passed, the quiet words of the observers giving commands to their assistants and the click, click of the various pieces of apparatus as exposures were made and plate holders removed. Everything passed off without a hitch, and with the first reappearance of the sun our work was over and we could take a long breath.

We had been favored with clear skies. How many others were equally fortunate? It did not take us long to find out, for the Spanish government had installed right in our camp a telegraph office, and for fifteen days no less than three operators were at our service to send and receive our messages; and for this not a single cent of money was asked or expected. It was found that fifty miles to the west of us, at Alhama, where were the observers from the Lick Observatory under Professor Campbell, there were thin clouds, while one hundred miles to the east along the Mediterranean coast, the Englishmen were even more unfortunate in having the clouds denser. In the northeastern part of Spain at Burgos, more astronomers were located than at any one place, and here too was King Alfonso of Spain. Five minutes before totality it was pouring rain, but as if by a miracle a little blue patch of sky appeared, and the eclipse was seen under perfect conditions. The weather along the eclipse track was: in Labrador, cloudy (no observations made); in Spain, cloudy and clear; in the islands of the Mediterranean, cloudy; on the coast of Africa, slightly cloudy; but farther inland and along the rest of the eclipse track the skies were perfect. All three parties of the Naval Observatory were fortunate in having their work unhindered by a single cloud.

My own work was entirely spectroscopic. The photographic plates were developed within the walls of the college of Daroca, and in the long hours necessary for this work I was greatly encouraged and assisted by my good friend the rector of the college, Padre Felix Alvirez. Daily intercourse with this reverend father endeared him to me very much; and Señors Lorente, Soria and Padre Felix made my stay in Daroca one of the most interesting spots of my whole life by the kindness with which they bore my imperfect Spanish, by the interesting bits of history they told of Daroca, and by the deep insight each gave into the courtesy of a Spanish gentleman's heart. The developed photographs of the flash spectrum showed exquisite definition over a wide range in wave-lengths. Evidently I had stored up for myself the material for months and years of intensive study, the results of which will be told in Chapter XVI.

CHAPTER XII

THE AMERICAN ECLIPSE OF 1918

THIRTEEN years after the splendid observing conditions of 1905 in Spain I witnessed my fourth total eclipse of the sun, again as a member of the expedition representing the United States Naval Observatory. On June 8, 1918, the shadow of the moon touched the earth's surface on the Pacific Ocean, far south of Japan. Owing to the revolution of the moon about the earth and to the rotation of the earth on its axis, the shadow crossed the Pacific Ocean at a speed of over a thousand miles per hour. It was well after noon before the shadow reached the American continent, and the eclipse began in the state of Washington. Here the width of the shadow was only sixty miles, so that only those fortunate enough to be within this narrow track were able to see the eclipse in its totality. The eclipse passed southeasterly through Washington, Oregon, Idaho, Wyoming and Colorado in succession. In Colorado, the shadow had dwindled to forty miles in width. After passing through some of the central states, the shadow left the United States at Florida and left the earth's surface in the Atlantic, off the coast of the Bahama Islands.

The eclipse was seen almost exclusively from the United States, and so it will be known as the American Eclipse of 1918. Since more than half the civilized world was in the grip of the tremendous war, it was necessary for American astronomers in the year 1916 and early in 1917 to make their plans to insure that this eclipse should be well observed. Before our own country had become involved in the war, Congress had been asked for, and had made, a special appropriation to defray the expense of equipment and travel for the party from the U. S. Naval Observatory.

In order to help the astronomers of the country to make

as intelligent a choice of an eclipse site as possible, the Naval Observatory, in 1917, had prepared a large scale map of the United States showing, among other things, railroad lines, contour lines and the location of towns within the eclipse track. The city of Baker in eastern Oregon seemed to be the ideal spot for the government party. The question of clear skies was the all-important one for the proper location of an eclipse party, but, fortunately, the U. S. Weather Bureau had a regularly equipped station at Baker, and a record of many years' continuous observations seemed to be the ideal method of securing the desired information regarding the probabilities of good conditions on the day of June 8. As the Weather Bureau promised an absence of clouds and rain with an abundance of clear skies, Baker was chosen with the great hope that it would live up to its good reputation in the matter of weather.

We were in Baker exactly six weeks before eclipse day, and the time was none too long. The apparatus was sent forward by through freight, and though we greatly feared delays, it arrived safely the second day after our own arrival. To assist in the work of erecting the apparatus, the superintendent of the Naval Observatory had requested the services of five sailors from the U. S. Naval Station at Bremerton, Washington, who were in charge of a chief petty officer. The sailors were carpenters and machinists who assisted the astronomers in splendid style so that ten days before the eclipse, when the balance of the party began to arrive, the apparatus was all erected and partially adjusted, and there remained only the perfecting of the adjustments in order to be ready for the all-important day of the eclipse.

For direct photography of the corona, the largest camera was one of sixty-five feet focal length arranged horizontally, the light from the eclipsed sun being reflected by a coelostat mirror. The spectroscopic work of the Naval Observatory party called for the use of three concave gratings, each used objectively without slit. The largest instrument was a twenty-one-foot Rowland grating of six inches aperture and 15,000 lines per inch. This had a spectrum specially bright in the first order on one side, the grating being kindly loaned by Professor J. S. Ames. Photographic films two by twenty-four inches were used, and it was planned to work from the extreme ultra-violet as far to the red as the length of the films would permit. The second concave grating was of ten feet radius and 15,000 lines per inch, the grating belonging to the Naval Observatory. This was used in the first order and gave the same dispersion as the instrument employed in Daroca, Spain, in 1905. By the use of special emulsions kindly prepared by Dr. C. E. K. Mees of the Eastman Kodak Company, an attempt was made to photograph farther to the red end than in 1905. The third one was a very short focus grating of two meters radius and of six inches aperture belonging to the Astrophysical Observatory of the Smithsonian Institution. This grating gave very brilliant spectra but with little dispersion. The films for this spectrograph were stained with dicyanine.

Fortunately for the work of preparation, and true to the prediction of the U. S. Weather Bureau, no rain fell during the entire stay of the astronomical party in Baker. According to the "oldest inhabitant," the season was unusually dry even for eastern Oregon. By some mysterious force unknown to the astronomers, the eclipse seemed to exert some potent influence over the weather. At any rate, it was asserted by many of the rural papers that no rain could be expected until the eclipse was over. But if an absence of rain was experienced there was no lack of clouds, nor were the clear skies we had been led to expect afforded us. As the time for the eclipse drew nearer, the continued appearance of clouds began to cause anxiety among us. Would they interfere with the eclipse, and at the last moment make all the weeks of careful preparation of no account? If this had indeed happened, it would not have been the first event of the kind. Unfortunately for the astronomer, his work is always at the mercy of the clouds and the weather. But to have the whole work fail through clouds at the time of the few precious minutes of the total eclipse — that is indeed the keenest sort of disappointment! Some astronomers seem to be always unlucky and always experience cloudy weather on their eclipse expeditions, while on the other hand others are always lucky, and sometimes after all hope is abandoned, a rift will appear in the clouds and the eclipse at totality be seen in all its glory. Would we at Baker be lucky or unlucky, would the clouds interfere or not? Nearly all the days spent in Baker, according to the classification of the U. S. Weather Bureau, were actually clear. A " clear " day is not necessarily cloudless from morning till night, but rather one when the " sky averages three-tenths or less obscured, from sunrise to sunset." Clouds, however, gathered almost every day shortly after noon, and this condition was usually accompanied by very high winds that at times rose to the strength of a mild gale. The eclipse was to occur during the middle of the afternoon, and at this time of day the skies were generally overcast. These same conditions prevailed over the whole of the western United States along the path where the astronomers were located. It was well to be an optimist under such conditions of sky, for the pessimist became more and more wretched as the day of the eclipse drew near and his law of averages showed him the almost certain chance of thinly clouded sky during the total eclipse.

Fortunately, so far in my eclipse experiences I had been among the lucky astronomers. In 1900, at my first eclipse, the weather was ideal — not a single cloud in the whole sky. In 1901, I was a member of a rather large party which traveled half way round the world, of which only four of a total of thirteen saw the eclipse, the other nine witnessing the eclipse eclipsed by clouds. I was one of the fortunate four. Again, in 1905, there were many clouds which spoiled the researches of many parties. At Daroca, in Spain, a few minutes before totality a dense cloud covered the sun, but it cleared away before the all-important time and the total phase was seen through a brilliantly clear sky. Three lucky chances out of three made a fine average. The hope was that June 8 would make it four out of four!

By May 30, the whole party had assembled in Baker. A full week was given up to the final adjustments, and to the drills that were to play such an essential part in the work on eclipse day. During the partial phases of the eclipse, very few observations of importance were to be made; all observations of value came during the period of totality which lasted for one hundred and twelve brief seconds. If a slide of a plate holder should stick in place so that it could not be removed, or a lens were not uncapped at the proper time so as to let in the light, the whole work of an instrument might come to naught. On each day of the week preceding June 8, drills were gone through several times, in the morning and again in the afternoon. These drills were so well carried out that on eclipse day each and every one performed excellently the task allotted to him with the result that everything passed off without a single hitch.

As the days in June progressed towards the eighth, there was an air of excitement as each astronomer grew more keyed up to the task before him. Would Saturday be clear? But more especially, would the two minutes from 4:04 to 4:06 P.M. be clear? The skies were anxiously watched during the last days, but alas! almost every day at eclipse hour they were overcast. The optimist reasoned that if it were cloudy all the days before June 8, then on eclipse day perfect weather would surely be forthcoming; while the pessimist on the other hand argued that so many cloudy days meant still one more of the same character, so there would be no use trying to do anything.

Saturday, June 8, dawned with the sky overcast with thin, filmy clouds. The sun was well visible through these clouds, however, and it was possible to examine again the focus that had been obtained with the spectroscopes and with a touch here and a touch there to decide that everything was in perfect condition. During the morning the drills were again practised, and these seemed to promise success. The weather during the six weeks had not held up the work, and everything that thought and work could do seemed now to have been accomplished. The astronomers who had been on the ground for the whole six weeks of preparation had the pleasant consciousness that all of their allotted tasks had been completed, that every little detail had been thought of and that perfect success would certainly crown their efforts if only the clouds would clear away. But during the course of the morning, the clouds grew thicker instead of thinner, and it did indeed seem as if there was little chance of a clear sky.

The first contact was to take place at 2:30 P.M. Shortly after noon, the city of Baker took upon itself the aspect of a holiday. Though the day was Saturday, all stores were closed from three until five in the afternoon so that everyone should have a chance to see the phenomenon. Naturally everyone in Baker wished to go to the eclipse site at the Fair Grounds, to watch the astronomers at work. At the eclipse in Spain, this had been permitted with the result that the whole town had assembled, each inhabitant jostling his neighbor to get as close as possible, and each apparently talking at the top of his lungs, with the result that such a din arose when the eclipse became total that it was impossible to hear the seconds counted off to give warning to the astronomers when to change their plate holders.

In order that this might not happen again, the residents of Baker were told that the gates of the Fair Grounds would be closed, and that absolutely no one would be admitted within the enclosure, and the mayor of the city sent a guard of Boy Scouts to see that these orders were obeyed. Most of the townspeople repaired to the hills to the southeast of the city from which there could be obtained a fine view of the valley and the Elkhorn range, and they were directed to look especially for the shadow of the moon which would come across the landscape at the speed of thirty miles a minute or 1800 miles per hour.

No appreciable improvement in the skies was observed from noon to the time of first contact. Through a thin patch in the clouds, Mr. Hammond, using the five-inch visual telescope, observed the beginning of the eclipse and made a record of it. The clouds if anything became thicker after this so that at three o'clock it was impossible to see even where the sun was. Little thin rifts appeared at times, so that it was possible to see the moon encroaching on the face of the sun. At three thirty, a patch of brilliantly blue sky was seen off to the northwest and as the precious minutes dragged along it became evident that the clouds were moving in such a way that it was quite possible that the blue patch

The "Heliosaurus" Prominence at the Eclipse of 1918 (Barnard)
The disk shows the size of the earth.

American Eclipse of 1918

Lick Observatory photograph with 40-foot camera showing " Eagle " and other prominences.

would reach the sun in time for totality. Fifteen minutes before the total phase the clouds were so dense that had totality occurred then, the scientific results would have been nothing; but the blue sky was coming nearer and it might arrive in time.

Without looking at the sky, one realized that something unusual was happening. The light of the sun became so feeble that even the birds felt the unnatural aspect of things and sang their songs as if they were going to rest. The cocks in the nearby farm crowed. The wind, which was ordinarily blowing at this hour, was quiet. All nature was hushed. Even the seasoned astronomers who had seen two or three eclipses before felt the thrill of the unusual spectacle. And still the question was, would the clouds clear away in time?

At five minutes before totality the warning signal was given by Chief Petty Officer Welsch of the U. S. Navy who was to watch the chronometer and count the seconds. This signal summoned each man to his post. One last look was given to the apparatus to see that everything was in place, the plate holders were adjusted, — and then we waited. " Two minutes " before was called out, and then " one minute," still again " thirty seconds " before the expected time of totality. The clouds by this time had thinned considerably, the patch of blue sky was only a short distance away. The plan had been that after the signal of " thirty seconds," there should be nothing said until the word " Go " told that the total eclipse had begun. I was to watch for this with a pair of binoculars, before one glass of which a direct vision spectroscope had been arranged. This was the plan followed in Spain with complete success. But due to the thin clouds at the beginning, it was impossible to see the spectrum lines with the spectroscope, and the signal " Go " was actually given by Mr. Hammond who was using the five-inch telescope. No sounds disturbed the work of the party except the call of the seconds as the time passed and the brief words of command and shift of plate holders as each member of the party did his allotted task. Ten seconds after totality commenced, the clouds, thin at the beginning, had still further thinned, and at mid-totality the conditions were even further improved. What a gorgeous spectacle then met the eye! The sun was now in a very thin wisp of cloud with blue sky on either side. Although the cloud would undoubtedly detract from the scientific results, still it greatly enhanced the pictorial effect. The corona could be seen stretching for a short distance from the sun's edge, but most remarkable of all were three great tongues of flame, one immediately at the top of the sun, one on the left hand edge, and still a larger one on the right edge of the sun. These shone with a brilliant scarlet light, and made the eclipse of 1918 memorable as the eclipse of color. As the end of totality approached the thin clouds became still thinner, — and two minutes after the eclipse was over the sun had reached the blue patch of sky. If the eclipse had occurred only two minutes later, or if the party had been only half a mile to the northwest, the sky conditions would have been perfect! If, as I have already said, the eclipse had taken place fifteen minutes earlier, the scientific results would have been nothing at all. The optimists had won out.

We had indeed been fortunate. But farther west at Goldendale, Washington, where the Lick-Crocker party was located, a change of weather had happened which amounted almost to a miracle. The account by Professor Campbell runs as follows: "The total phase of the eclipse occurred at 2:57, local mean time. By great good luck a small rift in the clouds formed mostly at the right place and right time. The clouds uncovered the sun and its immediate surroundings less than a minute before totality became complete, and the clouds again covered the sun less than one minute after the total phase had passed. The small clear area was very blue and the atmosphere was tranquil." [1]

The developed photographs exhibit the painstaking care of the astronomers in procuring the precise focus, with the result that all of the photographs show exquisite definition. The thin clouds did not interfere at all with the details of the prominences or flames surrounding the sun. Those taken with the sixty-five-foot camera exhibit the prominences in splendid detail on a scale where the sun is more

[1] *Lick Observatory Bulletin, 10*, 2, 1918.

than seven inches in diameter. The longer exposures for procuring the extensions of the corona were not quite so successful, since the thin, fleecy clouds cut down the fainter streams of coronal light. The smaller cameras showed the same results as the larger ones — splendid detail in the inner corona, but the corona not of very great extent. All the photographs unite in showing many polar rays, and they also exhibit some plumed arches of great beauty. The corona appeared to be of the sun-spot maximum type, but with more polar streamers than were expected.

The spectroscopes procured photographs of exquisite definition, but these photographs suffered greatly owing to clouds which cut down the amount of exposure that at best is none too great.

What was perhaps the most interesting piece of scientific work accomplished at the 1918 eclipse owes its conception to Mr. Edward D. Adams, of New York, who has shown his great interest in science by the founding of the Ernest Kempton Adams fellowship which is awarded each year by Columbia University for researches in the domain of pure science. Upon becoming a member of the United States Naval Observatory party, Mr. Adams took upon himself the responsibility of trying, by some method, by photography, by a drawing, or by a painting, to procure a reproduction which would show the beauties of the corona, and which should be true not only as to form but more especially as to color. Unfortunately for science, it is impossible to obtain a satisfactory representation of the corona and the sun's surroundings by photography. The corona is very brilliant near the edge of the sun, but the intensity fades very rapidly. The eye can take cognizance of the details in spite of the great changes in brilliance, but not so the photographic plate. To obtain the faint extensions of the corona which are readily visible to the naked eye, a comparatively long exposure is necessary. This long exposure causes so much overexposure in the brighter inner regions of the corona that all detail there is lost by being burnt out. Short exposures give us the inner corona in exquisite detail, but the outer corona is then lost through shortness of exposure. Many attempts have been made to cut down the relative exposure by means of mechanical devices — but none of these have been entirely successful. Heretofore, the only success in representing the corona has been obtained by taking photographs with different times of exposure and with different cameras in order to procure photographs with detail both in the inner and brighter parts of the corona, and in the fainter outlying portions. After the eclipse is over, a composite drawing is usually made from the examination of different photographs. This method has given several satisfactory drawings, but they still have left much to be desired. However perfect they may have been as drawings, they took no note of color. Mr. Adams took upon himself the task of finding the right man to draw and paint the corona. Color photography could not help out in procuring the right color, and there was left only the possibility of finding an artist who would have the true scientific spirit, and who could combine an accurate sense of form with a refined perception of color. Mr. Adams was successful in finding Mr. Howard Russell Butler, a portrait painter of note, who has developed a shorthand method of noting both form and color.

During the eclipse, Mr. Butler sat on a lofty perch overlooking the eclipse instruments, and from which he could obtain a fine view of the sun. The task he had taken to himself was no small one. As a portrait painter he usually asked for ten or twelve sittings of two hours each: now he was asked to render his subject in 112 seconds. And moreover this was the first corona he had ever seen!

The methods followed by Mr. Butler in painting the corona have been described by him in *Natural History, 19,* 244–271, 1919, and reprinted as Vol. II, part 6, of the *Publications of the Leander McCormick Observatory.* An abbreviated summary of his description is herewith given:

" The method of working finally adopted may be called a shorthand method. It was to have a sheet of white cardboard on the easel with a series of concentric circles and radii drawn upon it in advance. One of these circles was to have the same diameter as the photographs of the moon to be taken in the sixty-five-foot camera, namely, seven and three-eighths inches. There was to be an inner circle of half this diameter and outer circles whose diameters were respectively one and one half, two, and two and one half times that of the inner circle. I expected to use the seven and three-eighths inch diameter, and did actually use it, but I was thus prepared, in case of an unexpectedly extended corona, to reduce the scale to one half and get everything on the cardboard. In front and beneath my cardboard was a finished sample picture of a corona, painted in advance as I *expected* it would appear, and my plan was to indicate by initials at points on my cardboard the variations of color from this picture; thus *b* was to mean a variation toward blue from the sample picture, and *y* more toward yellow. I wrote out the procedure as follows and tacked it alongside the easel. Practice enabled me to allot a certain number of seconds to each item.

Procedure	*Seconds*
Note value and color of sky	10
Draw value line on moon	10
Note colors of moon	10
Draw outline of corona	20
Use Zeiss binoculars	20
Record positions of prominences	10
Note color and value of prominences	10
Note colors and values of corona, etc.	20
	110

" Then my plan was to paint a first picture from this resulting memorandum, while the impression was vivid, and as soon as there was sufficient light to proceed by.

" While disappointed in not seeing the corona in a cloudless sky, the thin veil had its advantage from the artist's standpoint. It added mystery and the effect was picturesque. The brilliant corona burned through the thin veil as if it were not there. Probably only the outside edges of the corona were affected.

" On the tenth, the photograph negatives were shown to me. Those of the sixty-five-foot camera were seven and three-eighths inches in diameter, the others considerably smaller. I now saw, in minute detail, the two prominences which I had recorded and the mighty cyclone which had been increasingly revealed as the eclipse neared its end, because of the direction of the moon's motion. There were many other minor prominences.

" I now made careful drawings of these prominences from the negatives and of the variations in shading of the surrounding corona. Many arches were found springing over the prominences, and a few rifts of dark channels radiating from the limb but never coming very close to it. The negatives showed very clearly the hairy polar rays, not always radial in direction, and the beginning of a wing springing from the upper right-hand limb of the sun.

" Three paintings were made, the first immediately after the eclipse, the second on the succeeding day and the third after all data had been secured. This final painting is the one reproduced on page 60." The original size of the painting is 49 x 33½ inches.

One of Mr. Butler's paintings has been presented to the American Museum of Natural History in New York by Mr. Edward D. Adams. This canvas is mounted in the Astronomical Room which is kept darkened but with the corona painting illuminated by indirect lighting. Those who have been privileged to see this painting have pronounced it a thing of rare beauty. The astronomers who saw the 1918 eclipse and who have seen the picture look upon it as a marvel of perfection, true both as to form and color, a great work of art which has the added advantage of being scientifically accurate. In the same room of the great Museum are paintings of the 1923 and 1925 coronas, also from the skilled hands of Mr. Butler, the three coronas forming a triptych with interesting contrasts. A reproduction of the 1923 painting is found facing page 204 and that of 1925 as the frontispiece illustration.

The results from the expeditions of 1919 and 1922 which were devoted mainly to tests of the Einstein theory of relativity will be found in Chapter XXIII.

CHAPTER XIII

ECLIPSES SINCE 1923

WHAT a splendid opportunity seemed to await the American astronomer at the eclipse of 1923! On September 10, the moon's shadow touched the earth's surface at sunrise in the Pacific off the coast of Japan. The shadow traversed the ocean at a speed well over a thousand miles per hour and appeared off the coast of southern California somewhat after noon. After crossing Lower California, Mexico and Yucatan, the eclipse ended at sunset in the Caribbean Sea north of British Guiana.

It seemed very fortunate that the total eclipse track was to pass over a portion of California where one naturally expects during the summer months superb conditions of weather. Everyone, indeed, has heard of the boasted climate of southern California, the " land of sunshine and flowers." In order to supplement the regular observations of the U. S. Weather Bureau, the Eclipse Committee of the American Astronomical Society for four years previous to 1923 had had special observations made of the wind and weather during the first two weeks of September and at the hour of the eclipse. As the result of all the information available it seemed that one of the best spots for an eclipse expedition was San Diego, the only large city in the United States inside the path of totality. Within the memory of the " oldest inhabitant " there had not been a single cloudy day on the tenth of September. To make matters almost ideal, the total eclipse came at one o'clock when the danger from sea-fog was reduced to a minimum. A conservative estimate placed the chances of perfect conditions at least ninety per cent. For the first time in the history of science, the astronomer was able to insure his expedition from ill-luck of any kind, from clouds, from a sudden gust of wind that would shake the instruments or from any lack of adjustment that would endanger the final perfection of the photographs. Although eclipse insurance was offered to the expedition from the Leander McCormick Observatory, and at a very small premium, it was not accepted, for it seemed poor business to add to the expenses by insuring against the clouds that weather statistics showed to be so improbable.

The author had made arrangements to observe his fifth total eclipse of the sun as a member of the party from the United States Naval Observatory in the event that the government institution should send its own expedition. A special eclipse appropriation not having been passed by Congress and a Naval Observatory expedition thus being impossible, the necessary spectroscopic apparatus was loaned to the Leander McCormick Observatory. To add to his kindness, the superintendent of the Naval Observatory personally made arrangements with the Secretary of the Navy and the Chief of the Bureau of Equipment for the observers and equipment to be carried by naval vessel from Hampton Roads on the Atlantic to San Diego.

On arrival in San Diego it was found that the U. S. Army was not to be outdone in generosity by the Navy. Before leaving the East, permission had been secured from the Chief of Coast Artillery for the University of Virginia party to locate within the military reservation of Fort Rosecrans. To put it mildly, the writer was amazed to have the commanding officer of Fort Rosecrans extend an invitation, on behalf of the government, for the Virginia party to take up residence in officers' quarters in the Army post. Accordingly, six weeks before the eclipse, the McCormick Observatory party found themselves in a furnished house of ten rooms and two baths, the former home of a major. Meals were sent in from the company kitchen — and Rosecrans boasted of the " best cook in the Army "!

This army post is located on Point Loma, a peninsula lying between the Pacific Ocean to the west, and San Diego Bay on the east. On the top of the ridge which runs out to the old Spanish lighthouse is a famous drive, and from the Point itself there is afforded one of the finest views in the world overlooking San Diego Bay and the city, North Island (with Rockwell Flying Field and Naval Air Station), Coronado Beach and Tent City; to the south lay the Coronado Islands, belonging to Mexico; to the east and southeast the mountains; to the west the blue water of the Pacific.

San Diego has every reason to be proud of her city, located as it is in one of the finest spots on the globe and blessed with an agreeable climate that might well be the envy of any city in the world. We from Virginia, a state which has never been backward in painting a halo around everything connected with the Commonwealth, were much interested in the spirit of civic pride and aggressiveness, which at times may have even bordered upon boastfulness. Our six weeks' stay was filled with pleasant memories and we only wish that each and every total eclipse might be observed under the congenial surroundings of Fort Rosecrans.

Many are the problems that may be profitably attacked at the time of a total eclipse. One of the most important at the 1923 eclipse was the confirmation of the bending of the rays of light from a star as these rays pass close to the edge of the sun, as predicted by Einstein. We had no equipment for this research and we preferred to let others tackle this problem while we devoted ourselves to following up the line of attack carried out at four previous eclipses. It was therefore decided to have two eclipse stations, each equipped with a powerful spectrograph. One of these was located on Point Loma, the other at Lakeside, twenty-five miles inland, and near the edge of the shadow cast by the moon. The essential part of each spectrograph was a concave grating ruled by Rowland of Johns Hopkins. Each of the gratings used was made by ruling on a spherical concave mirror of speculum metal 15,000 lines to each inch.

At North Island was the powerful battle squadron of the U. S. Naval Aircraft forces. Here was a chance to employ photography from the air on any of the problems that could be solved by this method. It is manifestly impossible to use any but comparatively small cameras from an airplane and to give but very brief exposures. On account of the short exposures permitted no spectroscopic work could be attempted from the air, no investigation of the Einstein effect and no photography of the corona that demanded large focal scale. Airplane photographs could not compete with those taken from a fixed installation on terra firma. In the event of clouds and the possibility of soaring above them in a machine, airplane photographs might be taken, but there would be little of scientific value in photographing the corona on such a small scale. There seemed only one direction in which photography from the air could assist the astronomer, and that was in the attempt to find the position of the moon in the sky with greater accuracy.

The program finally adopted consisted in attempting, from five separate stations along the northern edge and one at the southern edge on the shore of Mexico, to photograph from the air the ground underneath intersected by the edge of the shadow of the moon. For this purpose it was necessary to use the best of mapping cameras known to the photographers and to choose special sites to photograph where the terrain would offer as great contrast as possible between a point just inside and one just outside the moon's shadow. Incidentally it must be admitted that so little was known of the amount of light to be expected a few yards outside of the moon's shadow that there was some doubt as to the final success of the investigation. But there was nothing to do but to " try and see what happened." To supplement this work from the air, there were two parties of sailors, each of seventy-five, located at the expected edge of the moon's shadow. The sailors were put at intervals of ten feet at right angles to the edge of the shadow, and each was instructed to note whether he could see the corona or not. If he could see the corona during totality, he was presumably inside the penumbra of the moon's shadow.

In addition to the pilots with mapping cameras, there were others with movie cameras stationed over Point Loma to photograph the coming of the eclipse, each provided with a navy chronometer-watch to detect the accurate time, while another plane went aloft equipped with self-recording apparatus to register temperature, humidity, etc., during the total phase.

During the erection of our apparatus it was interesting to watch the gradual installation of the gigantic equipment to be used on eclipse day near the tip of Point Loma by the astronomers from the Mt. Wilson Observatory. This great observatory, the best equipped and most famous in the whole world, is located on Mt. Wilson near Pasadena. Being therefore only about 140 miles from the eclipse site, it was possible to transport all of their instruments by motor truck from observatory shops to eclipse camp. The problem of transporting the heavy apparatus was a very simple one for the Mt. Wilson staff, and one with which they are very familiar.

To continue the magnificent work begun by Michelson in measuring the angular diameter of the star Betelgeuse, it was found to be necessary to separate the plane mirrors of the interferometer to the great distance of fifty feet. As this meant too large a span to be attached to the 100-inch reflector, a separate mounting was devised. This consisted in a bridge-like truss attached to a heavy polar axis driven by clock mechanism. The clock and central section of the mounting having been completed, the director of the Mt. Wilson Observatory decided to utilize this interferometer structure on which to mount cameras, spectroscopes, photometers, etc., for the eclipse program. Here was a startling innovation in eclipse work, to put all of the instruments on one mounting, "the eggs all in one basket"—but the scientific world has learned to have confidence in the judgment of the Mt. Wilson astronomers. Their longest camera was thirty feet in focal length, used to photograph the Einstein effect. Another camera half this length was for the same purpose. There were cameras to portray the beauties of the corona in various scales and in different colors of light, spectroscopes to ascertain the constitution of the corona, instruments to photograph and to measure visually the intensity of the light of the corona at various angles from the edge of the sun. The instruments (see page 208) numbering fifteen made probably the most complete equipment that had ever been assembled for photographing a solar eclipse. The personnel included about thirty members of the Mt. Wilson staff while an auxiliary party of twenty computers and friends of the staff were prepared to watch and measure shadow bands, etc.

To the astronomers gathered for the eclipse the weather seemed " unusual " (we understand this is the first word that a California baby learns at its mother's knee). At any rate the cloudless skies that we had been led to expect at 1:00 P.M. were not always forthcoming. During the first two weeks of our stay at Point Loma, high fog at noon was the rule rather than the exception. Still it was a long time to the eclipse, and conditions would undoubtedly improve —and they did. The next two weeks gave perfect skies, an absence of wind and altogether ideal conditions. Would this last, or would another cloudy spell come? We were optimists and believed implicitly what our friends the Californians told us.

Saturday, September 8, was a cloudy day and the eclipse would have been under poor conditions. Sunday was even worse. What would Monday bring? Lieutenant Wyatt in charge of the weather observations at North Island made daily predictions of weather conditions to serve as a guide for airplane flights. He telephoned me twice on Sunday. His report was, " There has been excessively hot weather in Imperial Valley for the past week. A secondary low over Northern California makes the prospect of clear weather tomorrow very doubtful. If the back of the hot wave is broken there may be a change. This will be signalled by a brisk wind from the west. Unless this wind springs up in the night, I am afraid for you tomorrow."

At eight-thirty eclipse morning we were off on our trip to Lakeside. As we got inland from the Pacific, out of reach of the high fog, and where ordinarily we got out of the clouds, the conditions improved but little. At ten o'clock at Lakeside we found high cirrus and the conditions looked hopeless. But there were three good hours until the time of totality and much to be done in the final preparations! By twelve o'clock we heard the airplane aloft which was to observe over Lakeside and the pilot waved to us in friendly greeting. The clouds were not very heavy but there were no clear patches anywhere. We kept a stiff upper lip and

Mt. Wilson Instruments Being Assembled for the Eclipse of 1923

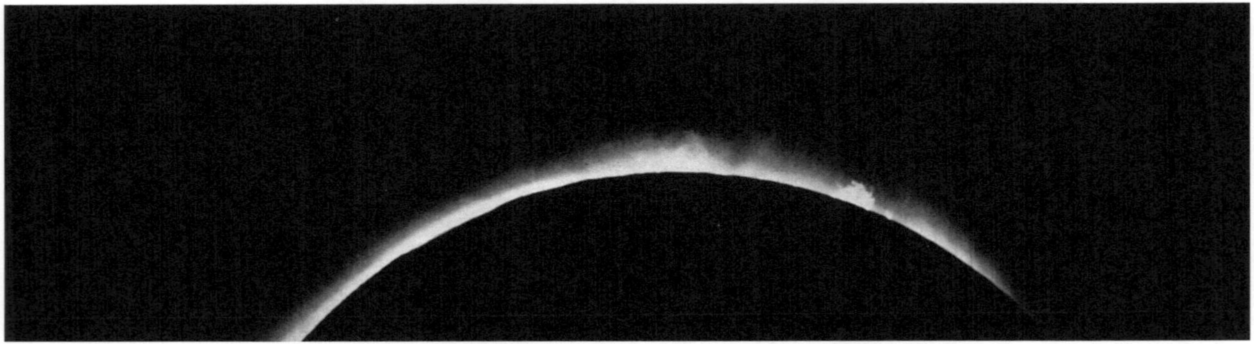

Disturbed Region in the 1923 Corona
Sproul Observatory photograph with 65-foot camera.

refused to believe that after so much bragging California was going to get a black eye.

Still the clouds were not too thick to prevent us watching the diminishing crescent of the sun as totality approached. Nature was hushed but the cocks were crowing lustily as if night were falling. At the very second when expected from the revised times sent out from the *American Ephemeris,* the sun was blotted out and a faint trace of corona appeared. But what a bitter disappointment! We carried through our program, exposing eight plates in the seventy seconds of totality, knowing full well that the developed plates would show not the slightest trace of light.

Well, there was no use crying over it (although one was tempted to). We had done our best and the fault was not ours. If misery likes company, it was evident that the clouds were general and not local. As quickly as a high-powered car could take us to San Diego we telephoned to Point Loma only to find as we expected, that conditions there were even worse and practically not a trace had been seen of the corona. Radio soon told us that the large assemblage of seventy astronomers on Catalina Island, including the Yerkes [1] and Harvard parties, Plaskett, Stebbins, Fox, Wilson and others, had suffered a like fate. Even the usual luck of the Lick party had deserted them, for conditions where Dr. Campbell and the Lick expedition were located at Ensenada [2] were about as bad as could possibly be.

And this was the record for " sunny California "! Not a single expedition greeted with good conditions, and the whole scientific work a dismal failure! There was nothing left to do but pack up and go home — and then begin to get ready for the next eclipse.

The weather conditions in Mexico did not give promise of the superb conditions that had been expected in southern California, so it was all the more gratifying to find the good luck experienced by the astronomers in old Mexico. Four separate expeditions were wholly or partially successful. The Mexican government financed two expeditions, one from the

[1] *Popular Astronomy, 32,* 205, 1924.
[2] *Publications A. S. P., 35,* 275, 1923.

Mexican National Observatory of Tacubaya under the direction of Gallo, the other [1] consisting of Schorr, Ludendorff and Kolhschütter from Germany. The Steward Observatory of Arizona had an expedition [2] in the province of Sonora and the Sproul party [3] was located at Yerbanis. Director J. A. Miller had the capable assistance of Heber D. Curtis, director of the Allegheny Observatory.

The Sproul expedition had a varied program of work, but mention will be made here of two items only, photographs with Einstein cameras of fifteen feet focus and photographs on a larger scale of sixty-five feet focus, the cameras in each case being pointed directly at the sun. The photographs for the Einstein effect, one of which is reproduced facing page 212, showed the corona in splendid detail. The star images however were not of the best definition, and as the photographs were made through thin haze, it was decided by Miller that he would not measure the plates to obtain the deflections of the star images.

After the ill luck experienced by the American astronomers in 1923 where conditions beforehand had seemed so promising, it seemed almost foolhardy to prepare for an eclipse which was to take place at nine o'clock on a winter's morning on the Atlantic coast in the United States with the sun at best only eighteen degrees above the horizon. Would snow cover up the astronomers as had happened on January 14, 1907, the preceding eclipse in this cycle of the Saros? The most optimistic placed the chances of good weather about fifty per cent. With such poor prospects it was natural that none but American astronomers would plan to observe the eclipse. In fact, the chances of success seemed so remote that the Lick Observatory, with its splendid record of carefully planned observations secured at many eclipses, decided against sending an expedition across the continent. After attempting a very large program at the 1923 eclipse, the Mount Wilson Observatory contented itself in 1925 with attacking about one-fifth of the problems planned for the previous eclipse.

[1] *Sitz. Preuss. Akad.*, 83, 1925.
[2] *Publications A. S. P.*, *36*, 170, 1924.
[3] *Astroph. Jour.*, *61*, 73, 1925; *Sproul Obs. Publ.*, No. 7, 1925.

What a surprise the weather again had in store for us on January 24, 1925! The day before the eclipse was one of gorgeously perfect blue skies. Would the morrow provide equally good skies? The clear skies continued throughout the night but it clouded over completely at six in the morning — and totality occurred three hours later. The largest group of astronomers was assembled at Middletown, Connecticut, at the Van Vleck Observatory, and here were parties representing the Mount Wilson Observatory, Leander McCormick Observatory, Harvard University, Universities of Wisconsin and Illinois, United States Bureau of Standards, and many others. What a dejected crowd we were at eight o'clock when we had gathered at the Van Vleck Observatory to observe first contact, the beginning of the eclipse. There was nothing but clouds everywhere! A quarter of an hour later a ray of hope appeared; there was a blue streak of sky low down in the northwest — and the clouds were coming from that quarter. Would it clear off in time? Luck was with us. Five minutes before totality a cloud, very thin and very fleecy, hung over the sun. It was not thick enough to do much damage and it was moving slowly. We hoped it, too, would go. When the timers called out "two minutes," the cloud was almost gone. Now it was beginning to get quite dark, a weird and unnatural pall coming over the landscape. The observers outside noticed shadow bands flickering over the snow. At one minute before totality, with the thin crescent of the sun growing very small, the atmospheric conditions seemed perfect, the thin cloud had gone!

How fortunate we were that the weather defied all the laws of averages on January 24, 1925! With clouds hovering everywhere over New York and New England, the surprising fact was that clear skies greeted most of the astronomical expeditions. It was cloudy throughout Michigan and Ontario, cloudy in Buffalo, but clear at Ithaca, Poughkeepsie, Middletown, New Haven, Nantucket and New York. It is estimated that ten million people in New York State and New England were given an opportunity of witnessing the gorgeous spectacle. The great newspapers of New York City asserted that no single event in the past decade has aroused such widespread enthusiasm. As totality lasted for the brief time of two minutes or less, and as the scientific investigations were nearly all crowded within the period of the total phase of the eclipse, it is probable that no single event in the history of man has had so many words, per minute of the duration of the event, written about it as has the 1925 eclipse. As a spectacle this eclipse suffered from taking place so early in the morning. If the darkening had come on during the middle of the day with the sun high up in the sky, the psychological effect would have been greater. The shape of the corona corresponded closely with that expected from the condition of the sun near spot minimum. To the right of the vertical, however, there was a long pointed shaft of light stretching up more than a degree. The remarkable feature was the total lack of rosy color visible to the naked eye, no prominences being readily visible, and as a result the corona lacked color as if, possibly, to reflect the feelings of the observers who everywhere worked with the thermometer below the zero mark of the Fahrenheit scale.

Several unusual features concerning this eclipse are worthy of note. The eclipse track crossed over the Van Vleck Observatory with its visual refractor of 20 inches aperture. Here was an opportunity of photographing the corona with a large telescope and with yellow light. Miller of the Sproul Observatory used the photographic lens of 63-feet focal length, mounted as in 1918 and 1923, pointed directly at the sun. Naturally there were hosts of cameras of smaller focal length. On account of the high possibility of clouds, the United States Navy employed the dirigible *Los Angeles* which flew out to sea and carried a battery of cameras with which to photograph the corona. Even in a gigantic dirigible the platform carrying the cameras is not very steady and it is very difficult to keep the cameras pointed at the sun. Naturally it was possible to utilize the cameras only of short focal length and to give brief exposures only. If it had been known beforehand that the skies were to be clear at eclipse time, the *Los Angeles* would not have taken part in the eclipse program. Its photographs showed the general features of the corona but they had little of scientific value com-

CORONA OF SEPTEMBER 10, 1923
Sproul Observatory photograph with 15-foot camera.

THE CORONA, JANUARY 24, 1925
Photographed by Slocum at Middletown, Conn.

pared with those taken by cameras with more stable foundations and greater focal lengths. Many airplanes were utilized for the same mission.

Many attempts were made to measure the intensity of the corona, among which should be noted the work of Nicholson and Pettit, by Stebbins and King, and by Coblentz and Stetson, all three parties gathered in Middletown; and also by Parkhurst, located farther west in Ithaca. Curtis at New Haven, Anderson and Mitchell, both in Middletown, photographed the flash spectrum, the first named paying special attention to the red end and reaching to wave-length 8800 A on the photographs. The results obtained by these various observers will be duly recorded on the appropriate pages later in this book.

The developed photographs all show the effect of the low altitude of the sun and the consequent poor seeing. As eclipses are of such rare occurrence it is fortunate that coronal details do not require the finest conditions of seeing for their precise portrayal. The details of the coronal streamers are not inherently sharp and clear-cut in their nature, and consequently a little blurring of detail caused by poor atmospheric conditions has little deleterious effect on the photographs. The case is, however, very different with the prominences. By nature these phenomena have more definite outlines. The best photographs with which to test the quality of the seeing at the 1925 eclipse were unquestionably those taken with the 20-inch refractor of the Van Vleck Observatory. This telescope is regularly employed in the determination of stellar parallaxes by photography. The focus consequently has been thoroughly well determined. The plates taken by this telescope (which has the largest aperture ever used to photograph an eclipse) were a great disappointment, due mainly to the lack of sharp detail shown in the inner corona. The fault did not lie in the telescope or in any lack of careful adjustment, but was to be found in causes beyond the control of the observer, namely, the poor definition resulting from bad seeing. The same qualities of poor seeing are found in the spectral images of these prominences in the flash spectrum taken without slit.

At the eclipse of 1923, the preparations made to photograph from airplanes, so as to fix with greater exactness the place of the moon in the sky, came to naught because of clouds. In 1925, on account of the southern edge of the moon's shadow passing over New York City, an excellent opportunity was at hand to secure many observers to determine the exact edge of the shadow.

Calculations based on the *American Ephemeris* had foretold that the edge of the moon's path would cut Riverside Drive (which runs north and south along the Hudson River) somewhere between 83d Street and 110th Street, with a total uncertainty of approximately one mile. To make certain that the astronomers were not mistaken, observers were located at each intersection of city blocks all the way from 72d Street to 135th Street, usually on the tops of apartment houses, so that a better view might be obtained. Sixty-nine men were employed and each was furnished with a piece of darkened glass and was instructed to look at the sun at the time of totality in order to see whether the corona was visible or whether there was a thin edge of the sun left shining. All of the observers were instructed to report to a central office immediately after the eclipse. Only one of the total of sixty-nine was in doubt as to what he saw, and the sixty-eight gave a clear-cut verdict. The observer at 240 Riverside Drive had seen the total eclipse while a man located at 230 Riverside Drive had seen a small sliver of the sun exposed, indicating that it was a partial eclipse. The distance between these two men was about two hundred and twenty-five feet, which included the width of 96th Street. The edge of the shadow of totality was, therefore, pinned down to within two hundred and twenty-five feet on the west edge of Manhattan Island. Other observers along the East River were successful in making similar observations with the result that the moon's path across New York City is accurately known. It is indeed surprising to find such unanimity of opinion among untrained observers who were all witnessing their first eclipse. Apparently it must be very easy to make up one's mind as to whether the corona is or is not visible. Similar attempts have been made at former eclipses but have always met with failure. None of the observers saw the edge of the moon's shadow as it lay upon the ground in spite of the excellent opportunity afforded on account of the ground being completely covered with snow. Evidently the edge of the shadow is not sharply defined but the light tapers off gradually.

The keen interest of astronomers in total eclipses of the sun again manifested itself in the Dutch East Indies on January 14, 1926. No less than nine expeditions from Europe and America were gathered for the purpose of working ardently and enthusiastically for the brief period of four minutes. Three expeditions, numbering twenty people, traveled from the United States halfway round the globe in the hope of having clear skies during the few precious minutes of totality.

In recounting the history of eclipses, repeated attention has been called to the many failures of eclipse expeditions due to the fickle weather. This has been especially true in the tropics. Where the writer was located in Sumatra for the 1901 eclipse the annual rainfall was the goodly amount of 186 inches or an average of half an inch per day.

Unfortunately, at the 1926 eclipse the weather lived up to its poor reputation. Clouds were the rule rather than the exception. The British party at Benkoelen on the west coast of Sumatra had the best luck. Their program [1] was almost exclusively spectroscopic. Excellent spectra were obtained which will be described in a subsequent chapter.

Although the party from Swarthmore [2] was located not far from the British in Benkoelen, they did not fare so well with the weather. Successful photographs were secured with the large camera of 63-feet focal length pointed directly at the sun, with twin cameras of 15-feet focus for testing the Einstein deflection, and with smaller cameras. Curtis again used the short focus concave grating for the flash spectrum but on account of the haze and the deterioration of the plates stained with dicyanine no spectra were obtained.

The Harvard expedition, consisting of Stetson, Coblentz

[1] *Memoirs R. A. S., 64,* 105, 1927.
[2] *Popular Astronomy, 34,* 349, 1926.

and Arnold, measured the intensity of coronal radiation visually and photographically. Their results seemed to show that the total coronal light was 40 per cent greater in 1926 than in 1925. However, their work was done at the site of the Swarthmore party and it is possible that the haze may have affected their results. (See Chapter XXI.)

On the other side of the mountain range from Benkoelen was the expedition from the U. S. Naval Observatory including Anderson of Mount Wilson with a 21-foot concave grating for the flash spectrum. The eclipse was practically lost through clouds covering the sun for nearly the whole of totality. Freundlich of Potsdam located at Benkoelen on account of haze secured few photographs of value. In East Africa, Horn d'Arturo[1] at the head of a completely equipped Italian expedition was more successful with the weather. The photographs of the corona taken by him were compared with those obtained in Sumatra. Many changes were noted giving evidence of motions in the coronal structure, particularly in the domes which are such interesting phenomena at each eclipse.

On June 29, 1927, the eclipse track passed over England, Norway, Northern Sweden, the Arctic Ocean and Northeastern Siberia. Unfortunately the eclipse was a very brief one. At Liverpool, totality took place at 5:24 A.M. and lasted only twenty-three seconds. At Fagernes, Norway, totality occurred at 5:34 (Greenwich time) and had a duration of 34 seconds. To make matters worse, the probability of clear weather was not very great. In England, at the early hour of the morning that the eclipse took place, the most optimistic estimated that the chance of clear weather was no more than one out of three. In Norway, away from the coast and farther along the eclipse track in Lapland, the chances of clear weather appeared to be greater but at best were no more than an even chance. It seemed almost foolhardy to attempt to observe this eclipse with such a short duration of totality and with the chances of clear weather reduced to the minimum. Still nothing venture, nothing gain! It was certain beforehand that no great astronomical

[1] *Publ. d. Oss. Astron. di Bologna, 1*, 227, 1926.

discoveries would be made from coronal investigations with this short duration. Still it was very important that a record should be made of the eclipse. For photographing the flash spectrum the short duration of totality was no great drawback.

The popular enthusiasm in England was unbounded. England had not witnessed a total solar eclipse for more than two centuries (see page 58) and the populace was determined to attempt to see the wonderful phenomenon no matter how dismal were the chances of success. Nearly all of the British astronomers were in the eclipse track in England with well-planned expeditions. A large and well-equipped party from the Solar Physics Observatory of Cambridge, however, forsook England for the greater promise of clear skies in Norway and set up their equipment at Aal under the direction of H. F. Newall. The only large expedition from the United States was located in Norway at Fagernes, fifty miles across country from the Cambridge party. The McCormick-Chaloner party from the University of Virginia had a very complete equipment to photograph the flash spectrum with three concave gratings, each used without slit in the attempt to secure the flash spectrum from the extreme ultra-violet to the far red, the plates for the region of long wave lengths being stained with neocyanine. Stetson was a member of the party at Fagernes with the instrumental equipment he had used at the eclipses of 1925 and 1926 for measuring the radiation of the corona. The astronomers from Oslo were also at Fagernes. Rosseland had a 21-foot concave grating for photographing the flash spectrum, Lohse attempted to observe the times of contacts visually and by photography, Störmer had a number of small cameras for photographing the corona. A scientific party of about forty, of ten different nationalities, were assembled at Fagernes. Other parties were located farther along the eclipse track, mainly at Ringebu in Norway, and at Gällivare and Jokkmokk in Lapland.

Both in England and in Scandinavia the weather before the eclipse was much worse than had been expected. At Fagernes, instead of being clear for half the time, it had been cloudy for fourteen days before eclipse day and rain had fallen on nine of these days. Decided optimism was required in order to expect a sudden break in the weather conditions. In England in addition to the cloudy weather, there were very high winds that threatened to carry away the temporary installations. At this date of the year, June 29, and at the high northern latitudes, the adjustment of the eclipse instruments had to be accomplished without the stars. At Fagernes (latitude 61°) it did not get dark at night, while Jokkmokk was in the land of the midnight sun.

Eclipse day dawned everywhere with the weather a continuation of the cloudy conditions of the previous fortnight. Apparently only a miracle could save the situation and bring clear skies. The miracle actually did take place in England at Giggleswick, where was located the party from the Greenwich Observatory under the direction of the Astronomer Royal,[1] Sir F. W. Dyson. Here the sky was cloudy throughout the whole of eclipse day. However a hole appeared in the cloudy sky and lasted for a space of two minutes only. The eclipsed sun appeared in this blue patch of sky and successful photographs were secured. The photographs show exquisite detail in the inner corona but the outer corona was lost through thin haze. At Jokkmokk where was located the expedition from Hamburg under the direction of Schorr, a similar miracle took place, the sun at totality being surrounded on all sides by clouds. Schorr was more successful than Dyson, the sky being without haze. The photographs taken at Giggleswick and Jokkmokk exhibit many brilliant prominences with the hooded forms overtopping the prominences showing as conspicuous features. A comparison of the two photographs shows well-marked changes taking place during the time elapsed. Schorr's photographs show the corona to be of the circular type that was expected at the time near maximum of sun-spots. The brilliant prominences and extended corona must have made a gorgeous spectacle for those few fortunate souls who were lucky enough to behold the eclipsed sun under good conditions. Elsewhere along the track thin haze or heavy clouds interfered with the work or

[1] *Monthly Notices R. A. S.*, 87, 657, 1927.

made the efforts of the astronomers of no avail. At Gällivare[1] in Sweden, there were several expeditions, the party from Upsala securing photographs of the corona and of the flash spectrum. These observations were secured through thin clouds. The flash spectrum was photographed with success by Pannekoek and Minnaert,[2] by Vegard, and by Baade.

At the 1925 eclipse the attempts which were made to photograph the phenomenon in color met with partial success. At the 1927 eclipse a color film was taken at Giggleswick making a brief exposure every second and lengthening these to about 0.5 second as totality approached. The results are described by the Astronomer Royal.[3] At Giggleswick the duration of totality was 23 seconds. For 27 of the exposures by the color film it may be said that the eclipse was total, and for an additional two seconds at each end Baily beads are shown. The corona and prominences can be traced on the limb of the sun opposite to the crescent for a duration of 30 seconds both before and after totality. The corona shows on the film to be somewhat bluish in color during totality. Outside of totality this bluish color appears rather reddish, the difference in color being due to the process employed. The chromosphere appears quite red in color while the prominences vary from a deep to a whitish red.

Before totality the corona appeared in some places bluish and in others reddish. When totality commenced there was a continuous reddish arc of the chromosphere on the east limb, extending for about 130° with red prominences. The chromosphere was visible on the film for 9 seconds after the beginning of totality and reappeared again 6 seconds before the end of totality. The colors on the film do not agree with those seen by the eye.

On account of the short duration of totality the bright inner corona was left exposed beyond the dark limb of the moon for a longer space of time than is ordinarily found at a total eclipse of the sun. The result of this has been that the corona was seen and photographed for a much longer stretch of time outside of totality than is usually the case.

[1] *Kienle, Die Himmelswelt, 37,* 225, 344, 1927.
[2] *Verhand. der Kon. Akad. van Wetenschappen te Amst'dm., 13,* No. 5, 1928.
[3] Dyson, *Monthly Notices R. A. S., 88,* 142, 1927.

References to other features of the 1927 eclipse will be found at the appropriate places in the following pages.

The next eclipse, May 9, 1929, was visible in northern Sumatra, the Malay States, Siam and the Philippines. Many expeditions were attracted to the eclipse track in spite of the great distances from home that must be traveled and the uncertainty of the tropical weather. Professor J. A. Miller, a veteran of many eclipse expeditions, tells [1] of the generosity of the Dutch in caring for the work of the astronomers. All of the heavy equipment was carried free of charge by the Dutch steamship lines from the United States halfway round the world to Sumatra and return, while the official personnel of the expedition was given passage at half the regular rates. Unfortunately the Dutch could not control the weather in Takengon in northern Sumatra where was located the Swarthmore party and also the Potsdam expedition under Freundlich. Miller describes the weather as follows: "We had not seen a clear sun for days preceding the eclipse. It was at that season of the year known as the change of the Northeast to the Southwest Monsoon. The Southwest Monsoon was to bring clear weather, and it did, five days after the eclipse.

"The eclipse occurred at 12:47 o'clock. In the morning the sky was entirely overcast. About nine o'clock a round patch about 20° in diameter appeared in the southeast sky. It persisted, and seemed to drift slowly toward the completely overcast sun. At eleven o'clock one could tell where the sun was in the sky. At the time of first contact the sun was invisible, but the blue circular patch still drifted slowly on. Sometimes during the partial phase one could tell by using field glasses that the sun was eclipsed. Ten minutes before second contact we took our places at the instruments, having decided to make all exposures regardless of the clouds. At that time one could not tell with the unaided eye that the sun was in partial eclipse. Five minutes before totality it was considerably brighter, and at totality the sun was in the center of the circular blue patch, which to the naked eye seemed perfectly clear. It remained clear during

[1] *Popular Astronomy*, 37, 495, 1929.

the entire period of totality, then the clear patch moved on, and half an hour after totality the sun was again obscured by clouds."

The Swarthmore party secured successful photographs with the 63-foot tower camera, with a pair of 15-foot Einstein cameras and with smaller instruments. The German party secured excellent photographs for the Einstein effect and also spectra of the corona in fine definition. To obtain check plates the party stayed behind at the eclipse site. Observers elsewhere had varied success.

Two British expeditions were located at Alor Star in Kedah and at Pattani in Siam. The former photographed through haze, the latter was blotted out by clouds. Stetson, Arnold and Johnson at Alor Star with an illuminometer found at mid-totality a radiation of 0.15 foot candles. The U. S. Naval Observatory expedition was at Iloilo in the Philippines. Good photographs of the corona were secured through light clouds. In Cebu the Hamburg party had thin clouds; Dutch, French and Japanese expeditions were partially successful.

The annular-total-annular eclipse of April 28, 1930, presented special problems. In consequence of the usual deviation of the moon from its predicted place, would it be possible for an expedition to succeed in locating itself inside the path of totality scarcely more than a half mile in width? Evidently, at an eclipse of this character the most fruitful field of investigation would be the flash spectrum.

Heavy rain and clouds along the track in California and Nevada greatly interfered with the observations. With their customary good fortune, the Lick observers had a clear sky at the time of totality. Two prism spectrographs were employed each with moving plate according to the method developed by Campbell. The photographs by both instruments [1] recorded the lines of intermediate and high levels but the timing was too late to obtain the reversal of the lines of lowest level. The expedition from Mount Wilson was hampered by clouds. Photographs of the flash spectrum were obtained by two concave grating spectrographs, one of

[1] *Publ. A. S. P.*, *42*, 131, 1930.

10-foot radius used with moving plate and the other of 21-foot radius with minified images of the crescents. The latter arrangement designed by Anderson was used at the 1926 and 1927 eclipses. The detailed results of the spectra from this brief eclipse will be watched with interest.

The path of the recent total eclipse started at sunrise on October 22, 1930, in the North Pacific and ended at sunset off the coast of Patagonia on October 21. The observation of the eclipse was dramatic and spectacular in all of its details. The only available site was a small, isolated, volcanic island not far from the International Date Line and at 15° south latitude. The U. S. Naval Observatory secured a small appropriation from Congress for an expedition. As a result of their commendable policy, frequently put into effect in the past, astronomers with eclipse experience were invited to become guests of the expedition. The equipment consisted of 115 cases of scientific instruments and supplies, 60 tons of stores and 11,000 feet of lumber. A naval vessel, U.S.S. *Tanager,* was detailed to assist in the work. The members of the party were on the island two months before the eclipse.

The island of Niuafoou in the Tonga group is familiarly known as "Tin-Can Island." The island being both small and volcanic there are no bays nor landing places where a boat could run alongside. Once each month there is great excitement on the island; the mail steamer arrives! Unfortunately, the only sure method whereby the mail can be brought ashore is to seal it up in a tin can and lower it over the ship's side; then one white trader and two natives swim out for the tin can — at the same time carrying the outgoing mail. But the island is in the path of the trade winds, and if the water is too rough to permit the swimmers to go out, the mail steamer does not stop, and the island gets along without news from the outside world until another month rolls by.

The island is shaped like a gigantic signet ring. It has a diameter of five miles, with an interior lake three miles in diameter whose surface is about seventy-five feet above sea level. The inhabitants, 1100 in number, are natives of the

Polynesian race. When we were on Niuafoou the white male population consisted of three persons, two white traders and a Catholic priest.

The island boasts of the largest cocoanuts in the Pacific Ocean. Copra or the dried ripe cocoanut is the only article of commerce. When a boy reaches the age of eighteen he is given eight acres of land by the Government; when he marries he is given a plot of ground in the village for his home. The meat of the young green cocoanuts provides food for man, dog, chicken and pig, and the water is used for drinking and cooking, and also for the preparation of the ceremonial drink of kava.

As the cocoanuts grow readily without any attention, life for the South Sea Islander is very simple. The government taxes must be paid before the end of September and hence in the early days of the month there is great activity in gathering the ripe cocoanuts, drying the meat, and then selling the supply to one of the two trading companies. Then comes the tramp steamer from the outside world to load the copra to take it to the United States for making food supplies and soap. When the taxes are once paid and a receipt obtained, the native may take it easy for the balance of the year. In contrast to life under the benefits of white civilization, the natives lead a care-free existence. No wonder that Stevenson and others have written such fascinating tales of life in the South Seas! We enjoyed greatly the friendly intercourse on our brief sojourn on Niuafoou. At four different times we were the honored guests at a grand banquet of native delicacies followed by a dance when we were given the privilege of dancing, American fashion, with the native belles. The daughter of the head man of the village where I had been adopted as "friend," rejoiced in the euphonious name of Vaicima — which translated into English means "cement water-tank."

The erection of the big cameras was no easy task. There was the 63-foot tower telescope, the camera of 65-foot focus placed horizontally with coelostat, a pair of Einstein cameras of 15-foot focus on a separate equatorial mounting, together with smaller cameras. All of this was under the direct supervision of Professor Ross W. Marriott of Swarthmore. I had two concave grating spectrographs used at former eclipses. Fortunately, Commander Keppler and Dr. Kellers of the U. S. Navy had been at the eclipses of 1926 and 1929, and eclipse problems were to them no novelty. Two petty officers and eleven enlisted men from the Navy helped in the erection of the apparatus. A month before the eclipse the Americans were joined by a party of seven from New Zealand under the direction of my old friend, Dr. C. E. Adams.

Recent eclipse expeditions to the tropics, in 1926 and 1929, had been treated rather badly by the weather. What was in store for us? Weather records kept by us during the eight weeks preceding the eclipse had shown that only one day out of three was clear at the eclipse hour. On eclipse day, October 21 by the American date and one day later according to local reckoning, conditions before the great event were even worse than usual. Totality was to occur at 8:51 in the morning. Two hours before this a light rain was falling; first contact was observed through clouds. The clouds soon after began to get thinner and fifteen minutes before totality they had entirely cleared away. The total phase was observed under nearly perfect conditions. Possibly there was a slight haze but this in no way seemed to affect the photographs. One half hour after totality, clouds again began to gather! On the night following the eclipse, Marriott started on the development of the coronal photographs. On the following night the dark room was given over to my exclusive use. As soon as it had cooled somewhat I started, and with the help of ice brought for the purpose, I was able to keep developing, fixing and washing baths at a fairly satisfactory temperature. The washing of the plates was difficult on account of the scarcity of water. On the island the only available fresh water came from rains. Unfortunately there had been practically no rain during our residence on Niuafoou and all of the cisterns in the vicinity of the eclipse camp were dry. Fresh water had therefore to be brought by the *Tanager* in eight-gallon breakers and it was necessary to conserve this in every possible way. That used for washing the photographic plates was saved and was

The Corona of January 24, 1925
From the painting by Howard Russell Butler, N. A.

utilized again for personal use and for dish washing. Lava dust permeated everywhere and it was difficult to dry the plates, especially when large in size, without their being spoiled by dust particles.

When all of the photographic plates, direct corona and spectrographic, were developed, the eclipse camp was a cheery place. We had been fighting against tremendous odds — but we had won out. I have never seen more exquisite detail than is found on the coronal photographs taken under the direction of Marriott. My own spectra left little to be desired, with good focus from wave-length 3200 in the violet to 7800 in the red and with the region from 4650 to 6800 in duplicate. Measures of coronal radiation and observations of shadow bands will be discussed in later pages.

I had gone to Tin-Can Island, and then around the world on the return journey, to work for ninety-three seconds of time. On August 31, 1932, I shall observe my ninth total solar eclipse at a location in Magog, Quebec — comparatively speaking, almost at home. There also will be my friend Stratton at the head of an expedition from Cambridge, England, and Minnaert from Utrecht, Holland. Farther north along the track, at Parent in northern Quebec, will be an expedition from the Greenwich observatory. In northern New England there will be expeditions from the two Great California observatories, Mount Wilson and Lick, and also parties from Swarthmore, Van Vleck and other American institutions. The eclipse will be total for about one hundred seconds of time and the path will be approximately 100 miles in width. The chances of clear weather appear to be about fifty-fifty.

By referring to page 55 it will be seen that if I am ever to observe a tenth total eclipse of the sun it will be necessary for me to travel many thousands of miles from my home in Virginia.

CHAPTER XIV

THE STRUCTURE OF THE ATOM

DURING the past two and a half centuries, the astronomer and the mathematician have worked in close cooperation investigating the distances and motions of the bodies forming the sidereal universe. Under the magic wand of their combined labors, the complexity of the Ptolemaic system has given way to great simplicity and beautiful order, revealing motions obeying the inverse square law of Newton, or the very slight modification of this law of gravitation demanded by the theory of relativity. During the progress of these investigations, our conceptions of distances and dimensions have been gradually modified, so that the sidereal universe appears to be vastly greater than was formerly thought. Our own Galaxy probably has a radius exceeding 10^{21} meters or one hundred thousand light years, while beyond extend other universes at distances of hundreds of millions of light years.

While the astronomer has thus been reaching out to greater and yet still greater distances in the direction toward the infinite, the physicist and the chemist, on the other hand, have found solar and planetary systems of nearly infinitesimal dimensions within the realm of the chemical atom. The radius of the electron we seem to think we know is equal to 1.9×10^{-15} meters. The astronomer has contributed much information concerning the atom because the celestial laboratories of sun and distant stars provide high temperatures and minute pressures transcending any available in the best-equipped terrestrial laboratories.

The amazing development of our conception of the atom has come within the past three and a half decades, the beginning taking place with the study of streams of negative corpuscles or electrons. On its discovery, radium seemed superficially to exhibit a contradiction of the laws of conservation of energy. Heat and light were spontaneously emitted without any apparent changes in the radium, and thus a continuous supply of energy seemed evolved which set at naught that fundamental law of physics. But it was soon found that with the giving off of energy in radiation, the radium itself did utterly change and here the philosopher's stone seemed at last to have been discovered, for it was found that one chemical element actually changed into another.

And to think that after all the whole science of radioactivity was more or less the result of a happy accident! The year following the discovery of X-rays in 1895 by Röntgen, Becquerel of Paris wished to test the phosphorescent action of certain substances by wrapping a photographic plate in black paper, and placing on it the substance to be examined, which was then exposed to sunlight. By good fortune a preparation of uranium was chosen and the photographic plate was darkened. The Becquerel rays were thus discovered, and it was soon found that, like the X-rays, these rays penetrate substances impervious to light, even passing through thin plates of metal. The experiments were always made by placing the phosphorescent substance in the sunlight on top of the black paper enclosing the photographic plate. But one day the sun was clouded, and the plate and the phosphorescent substance were placed away in a desk and were left there for several weeks. Becquerel for some reason developed the plate, and was surprised to find the plate darkened as before, thereby showing that probably neither sunlight nor phosphorescence had anything to do with the action on the photographic plate. Thus was born, in 1896, the new science of radioactivity!

Besides the effect on the photographic plate, radioactive substances manifest themselves in three different manners: first, by exciting phosphorescence and fluorescence; second, by causing the air near them to become conductors of electricity; but most startling of all, by the continuous generation of light and heat.

Mme. Curie recognized that radioactivity was a property

of the atom and starting with this in view she found that the residues from the mine at Joachimstal, Austria, were three to five times more radioactive than uranium. From this residue she separated out a new substance, far more active than uranium which she called polonium, in honor of the place of her birth. Later she discovered radium. This appears to be an element with atomic weight 226, and it is found in excessively minute quantities, there being only one part in five million in the best pitchblende. In 1899, Rutherford showed that the radiation from uranium was complex, consisting of (1) the α rays, which are absorbed by a sheet of paper or a few centimeters of air (2) of a hundred-fold more penetrating β rays, capable of passing through several millimeters of aluminium, and (3) of still more penetrating γ rays, capable of passing through quite a thickness of iron and lead. The β rays are deflected by a magnetic field. Becquerel and Kaufmann showed that the β rays were negatively charged particles projected with a velocity approaching that of light. The very penetrating γ rays are not deflected in a magnetic or electric field, and are probably closely connected with X-rays.

Though the α rays are the least penetrating, they are much the most important of the three types of radiation. They are deflected much less by a powerful magnetic field than the β rays, and in the opposite direction, showing that the α rays consist of a stream of positively charged particles. Alpha rays, therefore, will affect a gold leaf electroscope, and this old instrument gives one of the most sensitive methods of measuring the amount of radiation. In fact, Rutherford has shown that it is not difficult to measure with certainty the presence of radium in a body which contains as small a quantity as 10^{-11} grams of radium!

The maximum velocity of the α rays is 12,000 miles per second. These α rays thus move with velocities hundreds of times greater than the fastest moving meteor. Everyone is aware of the enormous energy possessed by a meteor moving, say, at 30 miles per second. But energy varies as the square of the velocity, and thus the α particle of radium possesses a quarter of a million times more energy, mass for mass, than a swiftly moving meteor. In this enormous energy of the rays lies the secret of the surprises of radium. From whence comes this enormous store of energy?

In addition to its power of sending out radiations, radium possesses another important property, shared in by the radioactive substances actinium and thorium, namely, that of continuously emitting a radioactive " emanation " or gas. This property is rendered very striking if a specimen of radium bromide is dissolved in water and the liquid evaporated down to dryness again to get the solid substance. This simple process has caused the radium to lose the greater part of its radiation. Strangely enough the radium slowly regains its activity, and if left entirely to itself, at the end of a month it is as radioactive as ever. Rutherford has showed that the solution in water causes the radium to give off a gas called " radium emanation." This emanation has all the properties of a true gas, it can be liquefied at a temperature of —150° C, but it is 100,000 times more radioactive weight for weight than radium. It does not combine with any known substance, and is not acted upon by any chemical reagent. It is not a radium compound, but it is a new element with an atomic weight which appears to be 222. It takes its place along with the rare gases of the air, argon, helium, neon, etc., and it gives a characteristic bright-line spectrum which shows neither the radium nor helium lines. It seems, therefore, that the element radium has been transformed into another element, radium emanation, or radon. If, after a month, the radium is again dissolved in water and evaporated to dryness as before, the radium loses its activity, and a fresh crop of emanation is produced. This same process may be repeated as often as possible with the result always the same, and we are perforce compelled to assume that the radium is continually manufacturing emanation, continually changing itself into a new element. This is really only the first of a series of changes, for radium emanation changes into radium A, and this in turn to radium B, and so on. This change is an atomic change going on within the atom. But how does this change progress?

When the radium has given off the emanation, it still

gives out α particles, but only about one-fourth as copiously as before the radium was put in water. The α particles are produced by the same change as makes the emanation, and the radium atom is therefore divided into emanation and α particle.

Observations of the velocity and mass of the α particle made by Rutherford indicate either that the mass of the α particle is twice that of the hydrogen atom, or if the charge carried by the α particle is twice that of the hydrogen atom, then the mass of the α particle is four times that of the hydrogen atom and must therefore be an atom of helium. Hence each atom of radium apparently breaks up into one atom of helium and one of radon.

All that was necessary to complete Rutherford's proof that helium was actually given off from radium, was to show experimentally that helium was thus produced. This was accomplished in 1903 by Sir William Ramsay and Frederick Soddy. A tube was filled with radium emanation which was separated from all other gases by condensing it with liquid air and removing by a pump the gas not condensed. This spectrum tube was sealed and the spectrum of the gas could be examined at will. At first no helium lines were visible, but after a lapse of three or four days, when the radium emanation had disintegrated, the spectrum of helium gradually made its appearance, and finally the whole helium spectrum was complete. Similarly, Debierne has found by the spectroscope that helium is produced from the radioactive substance actinium, and Soddy has produced helium from uranium and thorium. Helium, therefore, has been found experimentally to be produced by the radioactive substances radium, thorium, uranium and actinium. These substances are alike in that each emits α particles. Hence, α particles are atoms of helium. Rutherford and Royds, however, have given a still more conclusive proof that the α particle is an atom of helium. These α particles are capable of penetrating a certain small but definite thickness of glass. Glass may be blown very thin but yet retain its ability to remain air tight. Radium emanation was stored in such a thin-walled vessel and this enclosed in a second vessel. Alpha particles given off from the radium emanation thus could penetrate through the very thin glass walls, but were stopped in the outer vessel and were there collected. At first the gas in the outer vessel was found to contain no helium, but after some days, helium lines appeared in the spectrum, proving beyond a question of doubt that radium gives off helium.

It is even possible to measure the rate of growth of the helium, which measures show that in a year, 168 cubic millimeters of helium are spontaneously manufactured by each gram of radium. Rutherford and Geiger in this connection achieved one of the greatest triumphs for experimental science in being able to count the number of helium atoms or α particles that are ejected per second from one gram of radium. Indeed two different methods were devised which led to the same results. Both methods depend on the fact that each atom of helium as it is ejected gives a small flash like a meteor. By an electrical method, these flashes were counted by Rutherford and Geiger and it was found by them that thirty-four thousand million (3.4×10^{10}) atoms of helium are ejected every second from each gram of radium. This number is in exact agreement with that obtained by noting with a microscope the number of scintillations on a given area in a given time by the spinthariscope, invented by Sir William Crookes. Thus at the same time there was measured the amount of helium produced from radium, and likewise was given the number of molecules present in matter, information which was needed to complete many theories in physics.

Investigations in radioactivity accordingly have given an entirely new conception of the atom. The atom is no longer one and indivisible, but certain atoms at least are transformed into other atoms, each radium atom being changed into one atom of helium and one of radium emanation. These atoms are continually changing, no less than thirty-four thousand million atoms of helium being produced each second of time from each gram of radium. As the atoms disintegrate, enormous stores of energy are let loose, and this energy manifests itself as light and heat. The heating effect of this energy has been measured and has been found to be 118 gram-calories per hour per gram of radium. A specimen of a grain of radium bromide would evolve about four calories per hour. In four years about 140,000 calories would have been evolved. An equal weight of coal would during complete combustion give out about 500 calories. Hence the radium in four years would give 280 times as much heat as if it had been coal and had been completely burned, and yet the radium in this time would diminish so very little in weight that it would be absolutely impossible to detect this diminution by the most sensitive balance known to modern science. The energy of radium comes from the disintegration of its atoms. The average life of a radium atom is 2280 years, so that in the complete life of one grain of radium about 100,000,000 calories are set free. This is 200,000 times more energy than if it were pure coal and entirely burned!

Helium, being permanent and not transitory, must accumulate as the result of radioactive changes. In these changes, Soddy has shown the remarkable sensitiveness of the spectroscope in detecting slight quantities of helium for he has proved, in numerous special experiments, that the D_3 line of the helium spectrum can be detected with certainty, if only one millionth of a cubic centimeter, or one five-thousand-millionth part of a gram of helium is present.

An achievement of far-reaching importance in all theories of physics was the discovery by Sir J. J. Thomson of a body having a mass much less than that of the lightest known atom, hydrogen. This body, called by its discoverer a corpuscle, but now known as an electron, is 1/1845th part of the mass of the hydrogen atom. A further study of the electron showed that it is always associated with a negative charge of electricity and in fact carries a unit charge of negative electricity. The physicist has now become convinced that the atom is an electrical structure made up of nearly equal amounts of positive and negative electrical charges. The atom is believed to consist of a central group of elementary positive charges, or protons, with a smaller number of negative charges, or electrons, called the nucleus, and about this nucleus there is an outer system of negative electrons, varying in number from one to ninety-two. These outer electrons can be expelled from the atom by a number of different methods, such as the application of heat, impact of ions, exposure to ultra-violet or X-rays, or they may be emitted by radioactive substances; in fact, the β rays consist of a stream of negatively charged particles.

The discovery by C. T. R. Wilson that the charged ions produced in gases by α and β rays become the centers for the condensation of water vapor paved the way to experimental work of a remarkable nature. Millikan has secured extraordinary results by utilizing tiny drops of oil in place of water, and as a result of his experiments, he has been able to prove conclusively that the electrical charges carried on ions "all have the same value or else small exact multiples of that value." This fundamental unit is the same, both for positive and negative electricity, and is numerically equal to the charge carried by the negative electron. This unit charge of electricity was measured by him to an accuracy of one part in 1000. With this information, it was possible to estimate with greater accuracy the mass of the electron in grams and the number of molecules of any gas per cubic centimeter at 0° C and 760 mm pressure.

Since we know therefore the size of a molecule and the number of molecules per cubic centimeter, it is possible to compute the number of molecules through which the α or the β rays emitted by radium must pass in going a given distance. The extraordinary fact revealed (by the photographs of Wilson, referred to above) is that the swift-moving β particles pass, on the average, through as many as 10,000 atoms before coming close enough to an electron to detach it from its system and form an ion. In fact, it has been shown by Eddington[1] that when an electron encounters an ionized atom it will be captured if, and only if, it actually hits the nucleus of the atom. The electron must therefore form but a very minute portion of the space enclosed within the atomic system. The α particle being an atom of helium with a mass more than seven thousand times that of the negative

[1] *Monthly Notices, R. A. S., 83,* 32, 1922.

electron, it cannot be deflected from its course by an electron which is of very minute mass, but only by some ponderable mass at least comparable with that of the helium atom. This heavier mass is found at the nucleus of the atom. As a result of Geiger and Marsden's experiments on the scattering of α rays, it was found that when these rays passed through very thin metallic foils, the deflections witnessed could be explained only by assuming a very close approach to a small but massive charged particle. Rutherford[1] was accordingly led to assume that the typical model of an atom consisted of an exceedingly minute and comparatively massive positively charged nucleus, about which is collected a number of electrons. Each and every one of the electrons forming the outermost parts of all atoms are exactly alike and each carries a unit amount of negative electricity. Since each atom is electrically neutral, the charge on the positive nucleus must be equivalent to the sum of those carried on the N electrons. The value of N for each of the atoms is a fundamental constant, for on it depends the size of the electric field surrounding the nucleus and the peculiar arrangement of the external electrons, which in turn determine the physical and chemical properties of the atom. Experiments in 1911, by Barkla on the scattering of X-rays indicated that the number, N, of electrons in an atom was approximately half the atomic weight of the element. This conclusion was abundantly verified by the magnificent work of Moseley. He found that the X-ray spectrum was similar for all elements, and that when he plotted the square root of the frequencies of the characteristic X-ray spectra, all the elements examined arranged themselves upon nearly perfectly straight lines. The atoms were then numbered in the order in which these spectra placed them to give these straight lines.

The ordinal number corresponding to the place occupied by each element in the periodic table (p. 306) has been termed its atomic number. The work of Moseley showed that all the chemical elements took their proper places in the periodic table, and as uranium is the heaviest atom

[1] *Phil. Mag.*, *21*, 669, 1911, and *27*, 488, 1914.

known (atomic number 92) there can be only 92 species of elements. With elements 85 and 87 now thought to be known, all gaps have been filled in the periodic table.

These discoveries of the physicist have been the greatest boon to the work of the chemist. The latter has always had great faith in the principles underlying the Mendeléeff table, but in this table, arranged in order of increasing atomic *weights,* certain discrepancies appeared; for instance, argon, of atomic weight 39.88, from its properties was compelled to find a place in the table before potassium having a smaller atomic weight of 39.1. The system of atomic numbers places *A* with number 18, in its rightful place before *K* with atomic number 19. In a similar way, the system of atomic numbers places *Co* in the table before *Ni,* instead of after it, if arranged according to atomic weights.

With a knowledge of atomic numbers, the difficulties of classification presented by radioactive substances were now cleared away. Soddy has shown that when an α particle is emitted, the position of the element in the periodic table is shifted by two numbers to the left towards smaller numbers, while if a β particle is emitted the atomic number increases by one unit, and the place is shifted one to the right. Since the α particle is an atom of helium with positive charge 2 and mass 4, while the β particle is a negative electron with no appreciable mass, it is evident that the emission of an α particle will diminish the atomic weight by 4, but the emission of a β particle will cause no change in the atomic weight. Consequently, if the emission of an α particle by a substance is followed by two successive changes in which β rays are set free, the net result will be that the element, after experiencing these three changes, will move back again into the position in the periodic table it had held originally. These changes have not altered the size of the atomic number, but have diminished the atomic weight by 4, and consequently it is possible for two or more elements to have the same atomic number but differ in their atomic weights. Such elements are known as *isotopes.* Isotopes are indistinguishable from each other by any chemical tests, or by any spectroscopic tests since the spectra are identical. Radium (at. wt. 226), *Th X* (at. wt. 224) and *Ac X* (at. wt. 222) are examples of isotopes, each possessing a nuclear charge or an atomic number of 88.

Since all masses are nothing more than electromagnetic manifestations, and since the mass of the electron is very minute and negligible compared with the mass of the nucleus, it should be possible to compare the masses of different elements by subjecting them to successive electric and magnetic fields. This method of "positive ray analysis" developed by J. J. Thomson consists in measuring the ratio of charge to mass. This was made possible by a beautiful photographic method, which in the capable hands of Aston has greatly improved our knowledge regarding the atoms. Although demanding technical skill of a very high order, the chemical examinations can be carried on by simple methods leading to definite results. A small supply only of the gas to be investigated is required; it need not be chemically pure nor need any special care be taken to wash away from the vacuum tube all traces of the last gas investigated. According to Aston, it is impossible to remove from a tube "all visible traces of a misspent career." Until quite recently it was thought that the experimental measurements proved that the elements whose atomic numbers are whole numbers, with oxygen assumed at 16.00, are *pure* elements, while all other elements with fractional atomic weights are *mixtures* of isotopes, each of the isotopes however having a whole number for its atomic weight. *B* (at. wt. 10.9) is a mixture of two isotopes of masses 11 and 10; *Ne* (at. wt. 20.2) consists of two isotopes, masses 20 and 22; *Mg* (at. wt. 24.32) is a mixture of three, of masses 24, 25 and 26; *Cl* (35.46) consists of two, of masses 35 and 37; while *Xe* and *Hg* are each made up of no less than 6 isotopes. This conclusion has been greatly modified by the discovery that each of the elements *C, H* and *O* has isotopes.

Our view is that the nuclei of all atoms are made up of multiples of hydrogen nuclei each carrying unit positive charge, the combination being bound together by the external electrons. Hence the mass of the atom is confined to the nucleus, but the size of the nucleus is very minute compared with the whole volume of the atom. In fact, the radius of an electron cannot be larger in comparison with the radius of the atom than is the radius of the earth compared with the distance from earth to sun. Each atom therefore forms a miniature solar system, the external electrons being held in place and compelled to perform their orbital motions by the comparatively massive nucleus. Since there may be as many as 92 external electrons, it is evident that modern mathematics cannot furnish a general solution of the motions of the electrons, except in the case of the very simplest of the atoms.

Since the chemical and physical properties depend on the distribution of the electrons of the outer atom, there have been many attempts to formulate a structure for the atom. From such attempts have gradually evolved atomic models of many different types. The Lewis-Langmuir atom has been very successful in explaining the chemical properties, particularly the valence. This atom is not based on any dynamic principles. Valence, which measures the power to combine, may be positive or negative depending on whether the atomic system has too many or too few electrons to make a stable combination. The inert gases — helium, neon, argon, krypton, xenon and radon — are elements which have no power to form compounds. Such atomic systems under ordinary conditions cannot capture an electron from another atom, nor can they get rid of one of their own. The inert gases have atomic numbers of 2, 10, 18, 36, 54 and 86, and hence we may imagine the atoms as if made up of concentric shells of electrons, the shells containing 2, 8, 8, 18, 18 and 32 electrons respectively.

Except for the hydrogen system, all atomic structures have the two electrons forming the system of helium as their innermost shell. Since helium is inert and cannot take up another electron; the third electron which forms lithium must be a single electron in a shell exterior to the helium system. The lithium system is therefore not very stable, and readily gives up its external electron. It has a positive valence of one. The system of fluorine, with seven electrons in the second shell, may be regarded as being in a position readily to capture an electron from an atom in its neighborhood, and so it has a positive valence of seven or negative valence of one. Likewise oxygen may be regarded as having positive or negative valences of 6 and 2 respectively. The Lewis-Langmuir atom is very successful in explaining the relative positions of the elements in the periodic table. The electrons forming the atom of this model are relatively fixed in position, for though each electron may be in motion, it is confined to a small portion of the space occupied by the atom. With the advance in knowledge this atomic model has been superseded by others.

Great difficulties, however, appear when discussing atomic models on the principles of physics, especially when there is a transference of energy from place to place, or when the motions of the atoms make them depart from the steady state thereby causing emission or absorption of light. After repeated failures to explain these matters by means of the accepted theories, Planck made one of the most startling proposals ever presented to the scientific world, from which developed his celebrated *quantum theory*. According to Planck, it was assumed that the transference of energy can only take place in definite but very small units, and that the total energy transferred is always an integral multiple of this small unit, called the energy quantum. This can be expressed simply in mathematical terms. If E_1 and E_2 represent the energy of a system before and after radiation has taken place, then the energy spent in radiation is $E_1 - E_2 = h\nu$, where h is Planck's constant and ν is the frequency of vibration of the body concerned. This mathematical equation gives expression to the simple statement that the total amount of energy emitted or absorbed by a radiating body is always proportional to the frequency of vibration, which in turn is inversely proportional to the wave-length of the light emitted or absorbed.

The very radical nature of Planck's hypothesis may be estimated when it is stated that it stands at variance with all the previously known laws of mechanics developed to explain the motions of material objects of large dimensions. As a matter of fact, there is no absolute necessity that the same laws that apply to ponderable material in a gross state must also be applicable to simple atoms. This duality of laws, a topic of much heated discussion by physicists in recent years, has given rise to the expression " classical dynamics " which explains the motions of matter in obedience to the law of gravitation. One fundamental conclusion of the quantum theory is that motion is not the continuous process that we have accustomed ourselves all our lives to believe; but the motion takes place " steadily by jerks," the jerks however being so small that the process is to all practical purposes continuous.

Of the many atomic models proposed, the most successful in explaining the physical facts, and more particularly the spectroscopic data, is the Bohr-Sommerfeld atom. This was first proposed in 1913 and it is based on Planck's quantum theory of energy. In explaining the motions of the external electrons, evidently little hope can be expected from the classical dynamics, for this would require the solution of the problem of n bodies when n is large. For the present, assumptions must be made for the purpose of securing results, and in spite of much inconsistency, the quantum theory of spectra is the most satisfactory attempt so far made toward interpreting spectral series. In the simple case of the hydrogen atom and ionized helium, each with one external electron, the Bohr-Sommerfeld method has been very successful in reproducing many of the details of the spectra both in electric and magnetic fields. On the basis of this theory, the problem of atomic structure consists in building up each atom in such a way that the passage of an electron from one stationary state to another will give the observed wave-lengths. Although the acceptance of the principles of wave mechanics gives a better explanation of the theory of spectra, nevertheless the Bohr atomic model is still useful in explaining the facts.

In the investigations of spectra, the first of the lines to show an arrangement in series were those due to hydrogen, discovered in 1880 by Huggins in the spectra of white stars. This series is an extension of the four lines visible in the solar spectrum. A new era in spectroscopy started in 1885 when the law underlying the hydrogen series was discovered by Balmer. The thirty-five lines found in the flash spectrum are represented by the formula

$$\lambda = 3646.125 \frac{m^2}{m^2 - 4}$$

where λ is the wave-length on Rowland's scale and m takes the values 3, 4, 5. . . .

Shortly afterwards, Kayser and Runge, and Rydberg independently, began the publication of their splendid researches. Rydberg's investigations are of the greatest importance since they have laid the foundation for all future work on spectra in series. He began by sorting out doublets and triplets and thus ascertaining the lines which belong together in a series. He was able to distinguish three chief kinds of series, as follows:

Principal, including the strongest lines
Diffuse, of intermediate intensity
Sharp, including the weakest lines.

Each of these three series may consist of single, double or triple lines. Each and every series always converges towards a limit at short wave-lengths, and the lines at the same time diminish in intensity. A fourth so-called " fundamental " series with lines mainly in the infra-red has been discovered by Bergmann.

Many and varied are the mathematical formulae employed to represent the series of spectra. The most satisfactory formula, which is due to Rydberg, takes the following form:

$$\nu_m = A - \frac{N}{(m + \mu)^2}$$

where A is the limit of the series, N is the " Rydberg constant " for hydrogen, and the wave numbers ν_m are obtained by assigning successive integral values to m. μ may be regarded as a decimal part to m, though it is sometimes greater than unity. For the details of the investigations of series spectra, one would do well to read Fowler's excellent *Report on Series in Line Spectra,* 1922.

Of special interest in dealing with the flash spectrum are the investigations regarding enhanced lines. Fowler has shown [1] that these lines form series entirely similar to those of the ordinary lines. The formula representing the enhanced series, however, differs from that of the ordinary series in that the Rydberg constant N is multiplied by 4. This has a simple explanation from Bohr's theory. The ordinary series lines are emitted when an electron of charge e returns to an atom from which it has been displaced. The enhanced lines, on the other hand, are produced when an electron returns to the atom which has already lost another electron through ionization; consequently, two electrons are detached from the atom, each electron carrying a charge e, or a total charge $2e$. The formula for the Rydberg constant N, involves the square of the charge e, and hence for the enhanced lines the multiple 4 appears. Enhanced lines therefore belong to the ionized atom, or one which has lost a negative electron and hence carries an excess of positive charge. According to a suggestion by Saunders, $He+$ and $Ca+$, refer to ionized helium and calcium, the addition of the + sign following the chemical symbol signifying that the atom is not electrically neutral but carries a unit + charge. After an atom has lost one electron and thus becomes ionized, it may lose a second electron and become "doubly ionized." The symbol adopted for calcium under these conditions is $Ca++$, the atom carrying two extra positive charges. According to Bohr's theory, the charge concerned in the production of spectrum lines by such an atom is $3e$, and hence the Rydberg constant N must be multiplied by 9.

This is not the place to give the mathematical theories underlying the formation of spectrum lines, but a synopsis may be given of the more important developments found in Fowler's *Report on Series in Line Spectra,* 1922, and in Russell, Dugan and Stewart's *Astronomy,* 1927. Any accepted theory must explain why the frequency of any line in a spectrum appears always as the difference between the terms

[1] *Phil. Trans.* A. *214,* 225, 1914.

of a quantity, neither of which represents a spectral line, and must furnish an explanation of the physical meaning of the two terms, and must further explain how an emitted frequency comes to be the difference of two of these.

Adopting the idea of Rutherford, outlined above, that each atom consisted of a heavy nucleus carrying a positive charge which was surrounded by negative electrons, Bohr was able to give a satisfactory theory in the case of the simplest type of atom. The hydrogen atom is such a unit since it consists of a single electron in orbital motion around the nucleus, and equally simple are the atoms of enhanced helium and doubly enhanced lithium. According to the Bohr theory, the single external electron is free to traverse certain specified orbits, which are determined in the simple case of circular orbits by the condition that the angular momentum is an integral multiple of $h/2\pi$, where h is Planck's constant, derived from the quantum theory. When the motion of the electron is confined to one of these stationary orbits, there is no radiation. Emission occurs only when the electron passes from one stationary orbit to another. Without attempting to explain the mechanism which causes the electron to pass from orbit to orbit, Bohr supposes that the transition is followed by the emission of light, the frequency of which can be determined from the quantum theory. In fact, the energy radiated is equal to the differences of the energies of the electron in the two orbits concerned, and is assumed to be one quantum of energy, $h\nu$, where h is again the Planck constant and ν is the frequency. At a given instant, any one electron falling from an external to an internal orbit causes one line only in the spectrum, and it is the summation of the actions of a large number of electrons that causes the whole series of spectrum lines.

By the elementary laws of mechanics, it is possible to derive the necessary equations for orbital motion. Taking E and M as the charge and mass of the nucleus, e and m as the charge and mass of the electron, c the velocity of light, and ν the wave-number of the line, then for the case of hydrogen

$$\nu = \frac{2\pi^2 E^2 e^2}{ch^3} m \left(\frac{1}{t_1^2} - \frac{1}{t_2^2}\right)$$

where t_1 and t_2 are integers.

This formula is of exactly the same form as that which represents the Balmer series of hydrogen, the quantity outside the brackets representing the Rydberg constant. In fact, as a splendid confirmation of Bohr's theory, the Rydberg constant, N, calculated from Millikan's data, furnishes a value which agrees with that found from spectral series within an accuracy of one part in 1000.

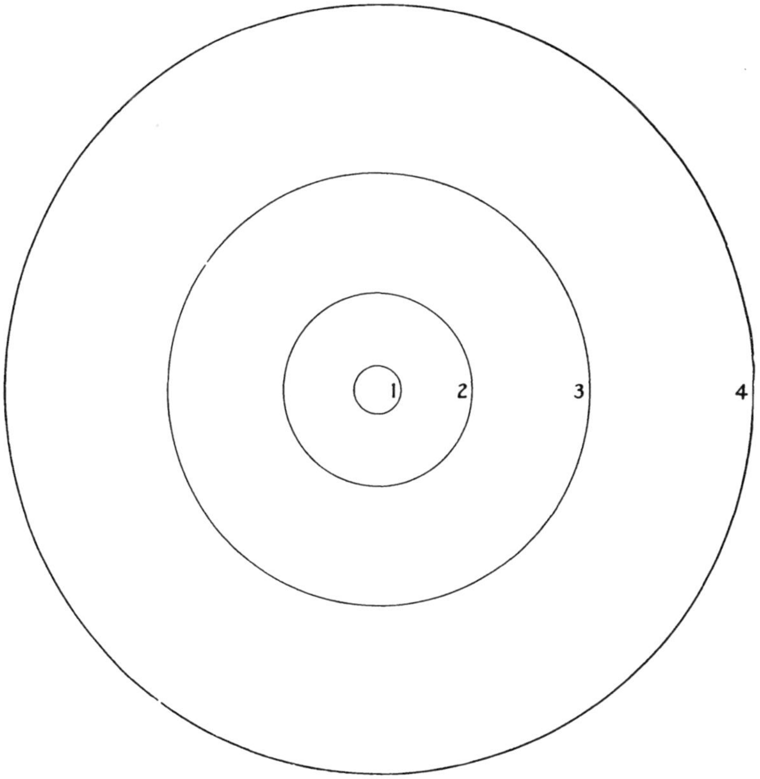

FIG. 4 Orbits of the hydrogen atom.

The successive orbits of hydrogen are in the ratios 1^2, 2^2, 3^2. . . In the normal state of the atom the single electron revolves in the innermost orbit. When the atom is disturbed by an electrical charge so that the electron is removed to a great distance from the nucleus, the electron will successively occupy different orbits on its return to its neutral position. The first line of the Balmer series, $H\alpha$, corresponds to the fall of the electron from orbit 3 to orbit 2, the second line, $H\beta$, to the fall from orbit 4 to orbit 2, and the following lines to the fall from orbits 5, 6, 7 . . . in each case to orbit 2. In the formula above, the Balmer series is represented by making $t_1 = 2$, and $t_2 = 3, 4, 5$ . . . The series discovered by Theodore Lyman of Harvard in the Schumann region in the extreme ultra-violet corresponds to falls of the electron from orbits 2, 3, 4 . . . to the innermost orbit 1, while the lines in the infra-red series are caused by falls from orbits 4, 5 . . . to orbit 3 . . . and the wave-lengths can be represented by placing $t_1 = 3$, and $t_2 = 4, 5$ . . . in the above formula.

Only two lines of the infra-red series were discovered by Paschen with wave-lengths of 18,751 A and 12,818 A, respectively. Taking advantage of a very long hydrogen tube by means of which Wood at Johns Hopkins University had photographed in the laboratory twenty lines of the Balmer series, Brackett[1] was able to measure the wave-lengths of five members of the infra-red series, two of which had been observed by Paschen, while three of them were new. These five lines represent the fall of the electron into the third orbit from orbits 4, 5, 6, 7 and 8. Two lines of an entirely new series were likewise discovered by Brackett corresponding to an electron falling into the fourth from the fifth and sixth rings of the hydrogen atom. The wave-lengths of these lines are 4.05μ and 2.63μ, respectively. These four series of hydrogen, the Lyman series in the extreme ultra-violet, the Balmer series in the visible spectrum, the Paschen series in the infra-red and the new Brackett series in the far infra-red are in splendid agreement with Bohr's theory. The four series represent falls from outer orbits of electrons to the first, second, third and fourth rings, respectively.

On Bohr's theory there was no place in the hydrogen series for the lines discovered in 1896 by E. C. Pickering in the star ζ Puppis which appeared to converge to the same limit

[1] *Astrophysical Journal 56,* 154, 1922.

as the Balmer series of hydrogen and were closely represented by substituting $(t_2 + 0.5)$ for t_2 in the Balmer formula. The line 4686 A is also of the very greatest astronomical interest. It is found in many Wolf-Rayet stars and gaseous nebulae and is likewise found in the flash spectrum. When this line was discovered in the laboratory in 1912 by Fowler, in experiments on a helium tube in which hydrogen was present as an impurity, the line was ascribed to hydrogen. This identification, however, was not satisfactory. The theory of Bohr settled the question by demonstrating that the line 4686 A, and also the Pickering series, are not due to hydrogen but to ionized helium.

The neutral atom of helium consists of a nucleus carrying a net charge of two positive units and two outer negative electrons, the mass of the nucleus being four times that of the nucleus of the hydrogen atom. When the gas is subjected to discharges of moderate intensity, one of the outer electrons is displaced, and the spectrum of neutral helium is emitted when this electron falls in to different orbits on its return towards the nucleus. The theory of the production of the spectrum is not complete since the mathematical discussion involves that of the problem of three bodies. However, when the action of the stimulus is sufficiently strong, both electrons are supposed to be removed from the normal orbit. An entirely different spectrum is emitted when one of the two electrons returns, this being the spectrum of ionized helium. The formula for the series of lines is identical in its mathematical form with that of hydrogen. The factor 4 however enters into the constant for the reason that the nuclear charge is $2e$. Bohr's theory therefore shows that the 4686 line, supposed originally to be due to hydrogen, is in reality due to ionized helium. The Pickering series is also due to ionized helium, and as members of the same series there ought to be lines nearly coincident with the Balmer series of hydrogen. Some of these lines have been observed in the laboratory by Evans and by Paschen, and as the lines have been observed when all traces of hydrogen were absent, there is abundant support for the belief that the lines are actually due to helium and not to hydrogen. These lines of ionized helium have been measured by Plaskett in certain O-type stars. At the position of $H\alpha$ the difference in wave-length from the hydrogen to the helium line is 2.63 A, while at $H\theta$ this diminishes to 1.54 A.

It has thus been found that the enhanced spectrum of helium resembles that of neutral hydrogen, and in exactly similar manner it has been concluded that in all details (even in showing doublets or triplets in their spectra) the enhanced spectrum of an alkali earth (like *Mg*) resembles the arc spectrum of the alkali metal of next lower atomic number (like *Na*). It has further been concluded that this relation exists for all alkali earths and alkali metals. In fact, this similarity in spectra between neighboring elements in the periodic table appears to be a general rule, namely, that the enhanced spectrum of any element resembles the arc spectrum of the element of next lower atomic number. If an element loses two electrons, and is therefore doubly ionized, its spectrum, for the same reason, should be similar to the arc spectrum of the element of second lower atomic number. Fowler has succeeded experimentally in discovering doubly- and also trebly-ionized silicon. Interesting comparisons have been made by him between the spectra of trebly-ionized *Si*, doubly-ionized *Al*, singly-ionized *Mg* and neutral *Na*, i.e., between the spectra of four succeeding elements in the periodic table of page 306. In the laboratory Millikan almost at will has been able to strip external electrons from atoms.

Although the mathematical analysis for a generalized theory has been too difficult to follow through to completion, nevertheless the Bohr theory has been successful in placing spectrum analysis on a very firm foundation. Further developments will be treated in the following chapters.

CHAPTER XV

PHOTOGRAPHING THE FLASH SPECTRUM

HALF a century of remarkable progress in solar physics since Young first observed the flash spectrum at the eclipse of 1870 saw a vast diversity of opinion regarding conditions of temperature and pressure in the envelope causing the dark Fraunhofer lines. On the one hand appeared Young's original view that the reversing layer is a thin shell but a few hundred miles in thickness, while on the other extreme we had Lockyer's opinion that such a reversing layer does not exist and that the corona is merely the cooler and rarer portion of the chromosphere. Within the past decade, however, the enormous increase in knowledge concerning the structure of the atom, explained in the preceding chapter, coupled with the publications of discussions of the spectrum of the chromosphere obtained at solar eclipses, has resulted in the solution of many of the outstanding problems relating to the sun.

In spite of the enormous improvement in recent years in instrumental equipment and in technique, it is still no exaggeration to state[1] that it is more difficult to secure a perfectly successful photograph of the spectrum of the chromosphere than it is to obtain an excellent photograph of any other single phenomenon attacked by astrophysical science. Witness the fact that since its first observation in 1870, the flash spectrum has continually been and still remains one of the most important of all problems taken up for solution at each succeeding solar eclipse. It is now nearly forty years since the first photograph of the flash spectrum was obtained, but in this interval of time, comparatively long when judged by the attainments of modern science, there have

[1] *Cf. Handbuch der Astrophysik, 4,* 275, 1929.

been more than one hundred attempts to photograph the flash spectrum; and yet there are not more than a half dozen photographs which may be considered to rank as first quality. Of course, it is perfectly true that many of the attempts might have succeeded had it not been for clouds, or thin haze, or poor atmospheric conditions, yet the fact remains that the great majority of those photographs which escaped the clouds have suffered from lack of perfect definition or from inaccurate timing of the exposures.

At an eclipse, the flash spectrum may be observed visually or photographically, either with or without slit, and by means of prisms or gratings. The researches of the present day demand large dispersion. The type of spectrograph depends very largely on the particular problem which the eclipse observer desires to solve. The great advantage of the prism is the greater light-gathering power, the light being concentrated in one spectrum, but on the other hand the grating possesses many points in its favor. The lines in the prismatic spectrum are crowded together at the red end and widened out at the blue part of the spectrum, thus entailing much difficulty in the determination of wave-lengths. The grating gives a normal spectrum, permits of higher dispersion, and gives higher resolving power, a larger extent of spectrum and probably better definition. Gratings either plane or concave may be used, but with a flat grating a lens becomes necessary to bring the spectrum to a focus, and such a lens introduces aberrations and absorption of light, and consequent loss of definition. In the past, plane gratings have been used at eclipses, but for the future it is safe to predict that whenever gratings are employed for the chromosphere they will be concave gratings and not plane.

The spectrograph which lends itself most readily to the easiest and sharpest focus under the temporary conditions of eclipse observations is undoubtedly one used with slit. For this purpose it is possible to take bodily into the field a spectroscope of the type that has been thoroughly tested and tried out on stars in the regular work of the observatory. This instrument is universally a prism spectograph with prisms of the very highest quality. It is readily possible to

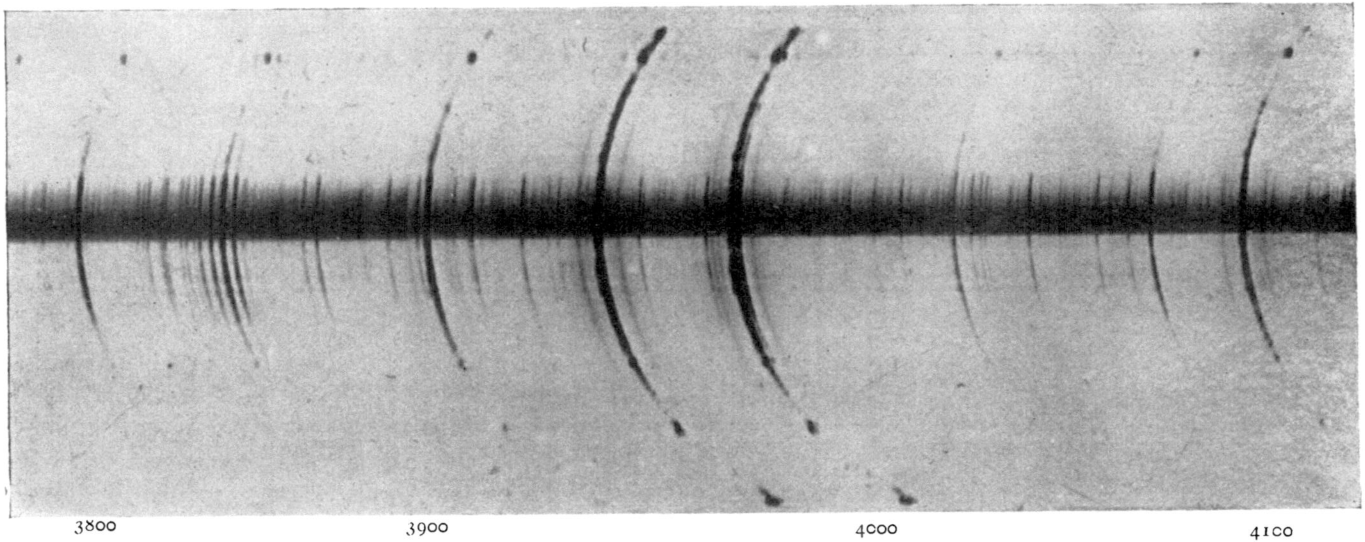

THE SPECTRUM OF THE CHROMOSPHERE IN 1905
Enlarged fivefold from photograph with concave grating without slit.

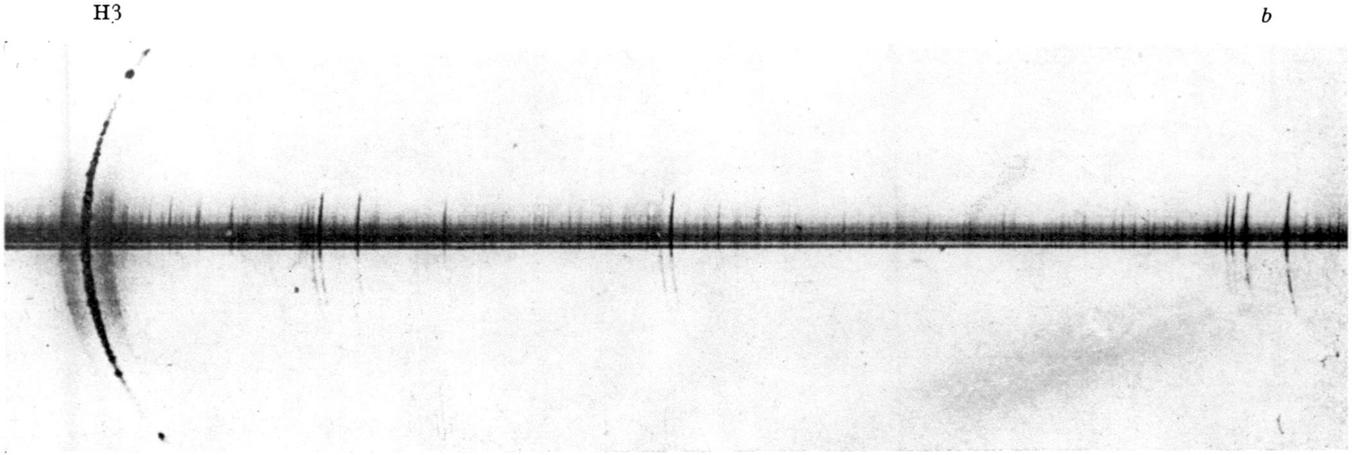

SPECTRUM OF THE CHROMOSPHERE, AUGUST 30, 1905

Region from Hβ to *b*-group in green.

test thoroughly the spectograph at home and then to mark carefully the positions of the focus for collimator and camera before dismounting. It is the work of a short while to again assemble the spectrograph at the eclipse site, and since the whole spectrograph is comparatively light in weight it can easily be picked up and set down anywhere. For instance, it is readily possible to test the focus thoroughly by the use of the electric arc, or by taking a focus plate of the daylight sky. At the time of the eclipse such an instrument is ordinarily used with a heliostat or coelostat mirror, and with the employment of a lens or concave mirror for forming an image of the sun on the slit of the spectrograph. With this type of spectrograph an eclipse observer should be absolutely certain to secure spectra of the chromosphere in excellent definition at each and every eclipse where clouds do not interfere. The only chances of failure with such an instrument are poor focus of the spectrograph and inability to keep the portion of the image of the sun, selected for observation, on the slit at the time of the eclipse. Any observer skilled in observatory manipulation through long years of handling a slit spectrograph should be ashamed to admit that the focus was poor due to the lack of careful methods of adjustment. There is however some excuse if a skilled astronomer fails to keep the slit of the spectrograph filled with light for the reason that the few moments of an eclipse are very tense and excited, the light at totality is feeble, and after all, even an efficient observer must at some time in his career see his first total eclipse — and he will have few opportunities of witnessing others.

By the use of slit spectrographs, the British observers have obtained excellent photographs of the flash spectrum. At the 1926 eclipse observed in Sumatra, Davidson and Stratton [1] used a quartz spectrograph designed to photograph to the extreme limit of the ultra-violet. The prism train consisted of four double quartz prisms of 60° angle, each prism being composed of two half-prisms of right- and left-handed quartz. The collimator and camera objectives were each a single quartz lens of 3-inches aperture and 36-inches focus.

[1] *Memoir R. A. S.*, *64*, 105, 1927.

Excellent definition was secured between wave-lengths 3066 A and 4200 A.

The greatest dispersion ever used at an eclipse, before that of 1932, was employed by Pannekoek and Minnaert[1] in 1927. Their optical system consisted of many elements each of unusually large dimensions. After reflection from the coelostat mirror, the rays passed first through an image lens 10 inches in diameter. After passing through a slit, a totally reflecting prism turned the beam through 90°, while next in the train came a triple collimating lens of 6-inch diameter. After traversing three prisms of 45° angles and large enough to receive the full beam of light, the rays were reflected from a plane mirror, and then they passed in reversed order through the prisms and 6-inch objective, which now became the camera lens for forming an image on the photographic plate. Good definition was obtained in the region of wave-lengths 4153 to 4751 A.

Although it is easier to photograph the flash spectrum by using a slit, nevertheless a slit is entirely unnecessary, the crescent arcs of the chromosphere providing narrow curved lines in the photographed spectrum. Good reasons for discarding the slit will be given later. However, it is a much more difficult task to secure perfect focus of the chromospheric spectrum without a slit, either by the use of objective prisms or grating. Slitless spectrographs of the power essential for the flash spectrum are practically never used in observatory or laboratory. In stellar work, a prism is placed in front of the objective, forming what is universally called a "prismatic camera." For many long years this type of instrument has been used in routine observatory work, but ordinarily the dispersion is small and therefore unsuited to making any contribution to our present knowledge of the flash spectrum. The dispersion of such a combination may be increased by two different methods: (1) by increasing the focal length of the objective; or (2) by increasing the number of prisms. At the 1926 eclipse, Davidson and Stratton used an ingenious device, the camera of 19-foot focus being utilized in a dual capacity. With a direct-vision slipped

[1] *Verh. Kon. Akad. te Amsterdam, 13,* No. 5, 1928.

in front of the objective the flash spectrum was photographed, while with the prism removed, direct photographs of the corona were obtained. Even with the moderate focal length of 19 feet, necessary to secure the desired dispersion, the crescent arcs were two inches in diameter on the photographs. As will be explained later, the average conditions of seeing experienced at eclipses may cause these crescents forming the lines of the spectrum to lose their clear-cut character in spite of the fact that the focus of the instrument may be perfect. None of the half dozen photographs of the flash spectrum which the writer regards as of the highest class of excellence have been obtained by this type of instrument, where the increased focal length of the camera has been relied upon to increase the dispersion.

On the other hand, excellent spectra have been obtained by objective prisms without the use of slit, through the employment of additional prisms to increase the dispersion. At the eclipse of 1905 in Spain, Campbell[1] secured exquisite definition with two 60° prisms in front of an objective of 2-inch aperture and 60-inch focus. Lines were recorded between wave-lengths 3820 and 5300 A, though the definition began to fail in the neighborhood of 4650 A. The dispersion at Hγ was 5.7 A per mm. A moving plate was used in forming the spectra, a method which will be explained later. At the Flint Island eclipse of 1908, Campbell increased the dispersion by the use of three 60° prisms and an objective of 60 inches focal length, the dispersion being about double that obtained by him in 1905. At Hγ this amounted to 2.9 A per mm, almost as great in scale as in the 1927 spectra of Pannekoek and Minnaert.

In the observation of eclipses, my own work has always been spectroscopic. I have always employed gratings and have used them without a slit. My record as an eclipse observer is probably unique. I have observed eight solar eclipses, and yet neither the Leander McCormick Observatory nor I, personally, owns a single piece of eclipse apparatus; at each eclipse everything must be borrowed. At my latest eclipse, in 1930 at Niuafoou Island, I had two concave grat-

[1] *Publications of the Lick Observatory*, *17*, 1, 1931.

ings. All of the equipment used at this expedition belonged to the U. S. Naval Observatory with the exception of one grating of 4-inch aperture kindly loaned by my good friend Saunders of Harvard University. This grating has been with me at every eclipse since 1905. Photographs of the flash spectrum were obtained in all but two of the eclipses, in 1923 in California and in 1927 in Norway. The results from the photographs obtained under clear skies in 1905 and 1925 have been published in *Astrophysical Journal, 71,* 1, 1930, and *72,* 146, 1930. The other concave grating used in 1930 was of 6-inch aperture. Both gratings are of 10-foot radius of curvature (or 60-inch focus) and each has approximately 15,000 lines per inch. The dispersion is 10.9 A per mm. At this recent eclipse, excellent definition was obtained from 3200 in the violet to 7800 A in the red, the region 4650 to 6800 A being photographed by both spectrographs. The discussion of these spectra has not been completed. The same two gratings will be employed at the 1932 eclipse.

The arrangement for using the concave grating without a slit is one of the greatest simplicity. The light from the sun falls directly on the coelostat mirror and then after reflection the beam of parallel light falls on the grating where it is diffracted and brought to a focus on the photographic plate. If the grating and the photographic plate are each perpendicular to the line joining their centers, the spectrum is normal, or to speak in more exact terms, the spectrum departs very little from a uniform scale of wave-lengths.

Used in the ordinary Rowland form of mounting in the laboratory, one of the well recognized advantages of the concave grating is the property of "astigmatism," whereby the spectrum lines are increased in length. If the astigmatism should be of approximately the same amount when the grating is used objectively without slit, then as a result of lengthening out the chromospheric lines, which are necessarily curved, the definition would be ruined. Consequently, in making plans for 1900, when concave gratings were used for the first time at an eclipse, the Naval Observatory party did not dare attempt to use such a grating without slit.

The successful photographs of stellar spectra [1] secured by concave grating used objectively showed however that these fears were groundless. Moreover, Runge's discussion of the theory of the concave grating, in Kayser's *Handbuch der Spectroscopie*, *1*, 450, 1900, proved that the amount of astigmatism for a concave grating used in the objective form would be so minute that it could have no harmful effects on the definition of the spectra.

It should hardly be necessary to add that the grating spectroscope must be very firmly mounted on solid piers of masonry or heavy timbers in order that the tremors of the apparatus caused by the wind or by the changing of the plate holders may quickly subside. It is manifestly difficult to mount such an instrument of large dispersion on an equatorial mounting or on a polar axis, with the grating in consequence directly exposed to the sun's rays. This method would indeed get rid of the coelostat mirror with its possible change in figure, but if this plan were followed, it would probably be a case of "out of the frying pan into the fire."

It might not be out of place to call attention to the very great difficulty always experienced by eclipse observers in securing sharp focus with their spectroscopes; as stated, one very prominent feature of eclipse spectra in the past has been the continued succession of photographs poorly focused. One method of securing focus frequently made use of has been to apply the final adjustments by utilizing the spectrum of the disappearing crescent of the sun a few minutes before totality. During the excitement and nerve-racking tension of these moments, a perfect adjustment can be obtained in this manner only as the result of a happy accident — and this method should never be resorted to under any circumstances. For instruments of small dispersion, the light from a star may be utilized for securing focus, if happily some bright star is conveniently located, but for spectrographs of the greatest dispersion the spectra even from the brightest of the stars are too weak. For large instruments, there is left only the option of securing focus on the sun itself or the electric arc several days before the eclipse by the em-

[1] *Astrophysical Journal*, *10*, 29, 1900.

ployment of some sort of collimating device which will give a parallel beam of light coming from a slit source. For the use of the U. S. Naval Observatory party at the eclipse of 1905, Jewell constructed a collimator consisting of a slit at the common focus of two concave mirrors, lenses not being used because of their chromatic aberrations. Several methods of placing the slit at the common focus of the mirrors will at once suggest themselves to any ingenious eclipse observer, one of the simplest being to utilize a telescope of medium size (say of five inches aperture) accurately focused on the stars.

The author has always made it a habit to be at the eclipse site a month or more in advance so that there may be plenty of time to secure exact focus without too much rush and excitement immediately preceding the eclipse. The collimator described above has always been utilized. When electric power is available, it has been found easier and more efficient to use the electric arc as the source of light for adjusting rather than the sun. This method was followed in the eclipses of 1923, 1925 and 1927. At the "Tin-Can Island" eclipse there was no electric power, and hence there was nothing to do but use the tropical sun as the source of light during the weeks of adjustment. The difficulties underlying this process are at once evident. In the optical train while adjusting, there are four mirror surfaces: the coelostat of silver-on-glass, the two metal mirrors of the collimator and the grating itself. In the adjustment, the slit becomes the source of light, and hence it is of minor importance if the coelostat mirror is not plane or if the slit is not in the exact focus of the first collimator mirror which forms the image of the sun on the slit.

Unfortunately, mirrors alter their focal lengths with changes in temperature, a fact well known to every astronomer with experience with a reflecting telescope. Excellent focus of the flash spectrum requires, among other things, that on the eclipse day the coelostat mirror be plane and not warped by the direct rays of the sun. (The mirror is always screened from sunlight until a few minutes before totality.) No matter how much time has been spent nor how much care in the determination of focus, the great fear is always present lest the grating alter its focal length owing to changes of temperature between the time of adjustment and the time of the eclipse. This requires that the wooden boxes and plate holders of the spectrograph should not expand enough to deteriorate the focus. And all of these nice adjustments had to be made and relied upon at the recent eclipse on a tropical island, with clouds always interfering with the work, with the dark-room during the daytime with temperatures that felt like inferno, and with no ice and no running water for development of the photographs. The trials and tribulations of an eclipse astronomer are many, especially when it is necessary to work with apparatus which is temporary and always borrowed!

With the collimator placed in the beam of light between coelostat mirror and grating, the focus was first determined visually with as great care as possible. For improving the focus, about a hundred and fifty focus plates were taken. The extreme range of focus in the last hundred of these films was but a fiftieth of an inch together with slight changes in the tilt of the photographic film.

The best type of spectograph to plan for the photography of the flash spectrum will unquestionably be the one that is able to make the greatest contribution to the solution of solar problems. With the gigantic equipment of great observatories working assiduously on the dark-line absorption spectrum of the sun, with the knowledge of the structure of the atom gleaned in recent years from refined researches in observatory and laboratory, how can the bright-line emission spectrum of the sun photographed with comparatively small dispersion best contribute to these investigations by supplementing daily researches? The chromospheric spectra are investigated in order to ascertain: (1) wave-lengths; (2) intensities of the lines; and (3) the heights in miles or kilometers that the vapors extend above the photosphere.

It will be shown later that the *intensities* of the lines of the flash spectrum depend on a great many different considerations, the timing of the exposures, the heights and whether the spectrograph is with slit or slitless. For refined work where the intensities are measured by a microphotometer, all of the various factors must be taken into account. If, however, the intensities are estimated on an arbitrary scale in the manner of Rowland's Tables, these estimates may be carried out with about equal facility no matter what the type of instrument. Consequently, in making a decision regarding the best type of spectrograph, the question of intensities may be laid to one side. One has a choice that is accordingly much limited. Are exact wave-lengths of the chromosphere of the highest importance?

If the decision is made in the affirmative, then it might seem that we should adopt a slit spectrograph. With equal dispersion and equally sharp focus, the wave-lengths derived from a slit spectrograph will however be superior in accuracy to those with slitless instruments in the measurement of the strongest lines of the spectrum only. If the observer is careful to place the slitless spectrograph in such a position that the tangents at the centers of the chromospheric arcs lie parallel to the lines of the grating or to the edge of the prism, or in other words, so that the spectral lines are perpendicular to the length of the spectrum, then these spectra are sharp and clear-cut and wave-lengths can be determined with high precision. With the strongest lines in all spectra there is always a spreading of light on the photographic plate by irradiation. If a slit is employed this spreading is fairly symmetrical, while without a slit the spreading is not symmetrical due to the interposition of the moon and to the elevation of the heated vapors above the surface of the photosphere. Slitless spectra of the chromosphere consequently demand greater experience and judgment in the person who measures these spectra with the result that the precision of the wave-lengths will depend to a large measure on the skill of the measurer in his ability to allow for the effects of irradiation.

With the small amount of light available at eclipses and brief exposures of a few seconds, it is possible only to utilize a dispersion small in size when compared with the much greater dispersion in the every-day solar investigations, such as are carried out with the 150-foot tower telescope and 75-foot spectrograph of the Mount Wilson Observatory. A dispersion of about 2 or 3 angstroms to the millimeter is about the maximum possible with the flash spectrum and this permits an accuracy of wave-lengths of about 0.02 angstroms. Such an accuracy is much inferior to that employed in ordinary solar work, and is not sufficiently high to permit the determination of systematic differences between eclipse wave-lengths and those taken under ordinary solar conditions. Accordingly, chromospheric wave-lengths practically can serve no other purpose than the identification of lines for accurate comparisons with Rowland's Tables in order to determine the sources whence the spectral lines originate.

As a matter of fact, no photographs of the flash spectrum taken with a slit have furnished any more accurate wave-lengths than those of the 1905 eclipse taken by concave grating without slit. It therefore appears open to question whether spectrographs with slits will give wave-lengths of greater accuracy than those of the slitless variety.

If, therefore, we assume that both wave-lengths and intensities of the spectral lines may be determined with equal accuracy with slit and slitless instruments of about the same dispersion, the choice of the best type of spectrograph for the flash spectrum must be decided by the ability of the two different types of instruments to give information regarding the *heights* to which the chromospheric vapors extend above the surface of the photosphere. The decision can now be made in the easiest possible manner. Spectrographs without slits give vastly more information about levels than can possibly be derived from any form of slit instrument.

It is, accordingly, the firm opinion of the writer, who has confined his attention at eclipses exclusively to the photography of the flash spectrum, that an observer who chooses a slit spectrograph to photograph the flash spectrum deliberately thrusts to one side the most valuable information that can come from the chromospheric spectrum, namely, the heights to which the vapors extend above the sun's surface. On the other hand, it is well not to forget that a spectrograph with slit readily permits a more accurate and reliable determination of good focus under the temporary conditions of eclipse observation than is possible with any form of slitless instrument. It would therefore perhaps be good advice to say that if an astronomer is to observe his first eclipse and wishes to do some work of value, he had better try to photograph the flash spectrum using a slit spectrograph with fair dispersion, for instance, about equal to that of the Mills spectrograph of the Lick Observatory or the Bruce of the Yerkes Observatory. For such work it will be well for him to confine his attention to the blue end of the spectrum on account of the greater dispersion of his spectrograph in this region and on account of the greater number of lines in the flash spectrum.

It is impossible to exaggerate the importance of securing the photographs of the flash spectrum at the proper instants of time so as to secure the spectra of the layers of the chromosphere as close to the photosphere as possible. At an eclipse, there are two manifestations of the flash, one at the beginning and one at the end of totality. Before the beginning of the total eclipse, the Fraunhofer lines persist as long as there is any portion of the photosphere visible, but when the moon entirely covers the sun's surface, or at the very instant of the beginning of totality, there is the sudden reversal of the Fraunhofer spectrum to that of bright lines. If one watches the phenomenon visually with some form of spectroscope, he will see many of the high level lines reversed many minutes before totality, particularly at the cusps.

The plan for securing photographs at the proper times always followed by the author is an old familiar one. A pair of old-fashioned binoculars is used and over one lens there is a direct-vision spectroscope. The one used in recent eclipses is a replica transmission grating of 15,000 lines per inch. With this it is possible to observe any particular portion of the spectrum one wishes. With a pair of binoculars and such an attachment, it is possible with the left eye to watch the disappearing crescent of the sun, shielding the eye by smoked or colored glass, while with the right eye the emission lines can be watched as they appear one after the other with the approach of second contact. Armed with this, the first flash can be observed and the exposure started with great nicety.

On account of the uncertainty in our knowledge of the exact duration of totality still persisting, one will be wise to decide that the exposure for the second flash should begin five seconds before the calculated end of totality, and should terminate with the first trace of the reappearing sun. A delay of one-tenth of a second in ending the exposure may readily bring ruin to this exposure. In photographing the flash spectrum at an eclipse it is evident that the important photographs are two, one at the beginning and one at the end of totality. Ordinarily, additional photographs are made, just before and immediately after totality, for the Fraunhofer and any emission lines. During totality, several short exposures are given, just after the first flash and again before the second flash, for the vapors of greater elevation, with a long exposure or two at mid-totality to obtain the spectrum of the corona which appears as a series of complete rings.

My little pair of binoculars just described gave me at the 1930 eclipse a most gratifying view of the flash spectrum, both at the beginning and at the end of totality. One great handicap in eclipse work is that in one's lifetime there are so few opportunities to rehearse. The 1930 photographs show that my reaction time (like that of everybody else who has ever photographed the flash spectrum) is not instantaneous; the first flash does not record the chromospheric lines of lowest level. At the end of totality, I kept the exposure going as long as I dared, and the photograph shows almost perfect timing.

It need hardly be added that the times of first and second flash, recorded preferably on the chronograph, will furnish excellent observations of the beginning and ending of the total eclipse. On account of the rough character of the moon's edge, these times will not agree with those recorded from the last appearance and first reappearance of Baily's beads — which brings up anew the time-honored question of what one means by the "beginning" and "ending" of a total eclipse.

The above descriptions of slitless instruments and methods of obtaining the flash spectrum all refer to a stationary photographic plate where the lines of the spectrum consist of a series of crescents. At the eclipse of 1898 in India, Campbell tried out an ingenious innovation of placing a narrow slot directly in front of the photographic plate in the manner shown in Figure 5; and then moving the photographic plate uniformly in a direction perpendicular to the slot. The same procedure was followed at the eclipse of 1900 in Georgia, 1905 in Spain and 1908 in Flint Island. The best definition of the four eclipses was obtained in 1905. After a delay of

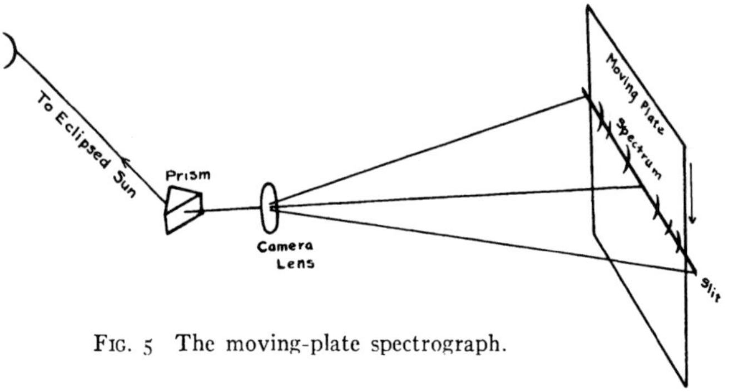

FIG. 5 The moving-plate spectrograph.

more than twenty years, a discussion by Menzel, together with an introduction by Campbell, has recently appeared in handsome form as Volume XVII of the *Publications of the Lick Observatory*. This volume forms a magnificent contribution to the study of solar problems, quite in keeping with the high standard of excellence set in the past forty years in the observation of eclipses by the Lick Observatory and by Dr. Campbell, its director for much of this period. The author wishes to express his highest praise of the beautiful Lick spectra and of the splendid character of the discussion.

If the moving plate is set in motion about twenty seconds before the beginning of totality, it will automatically record the gradual appearance as bright lines, first of the lines of highest level, then of medium level lines and then the spectacular appearance of the flash spectrum in the manner described by Young at the eclipse of 1870. At the end of totality the plate should be started more than five seconds before the calculated end of totality in order to photograph the second appearance of the flash spectrum in reversed or-

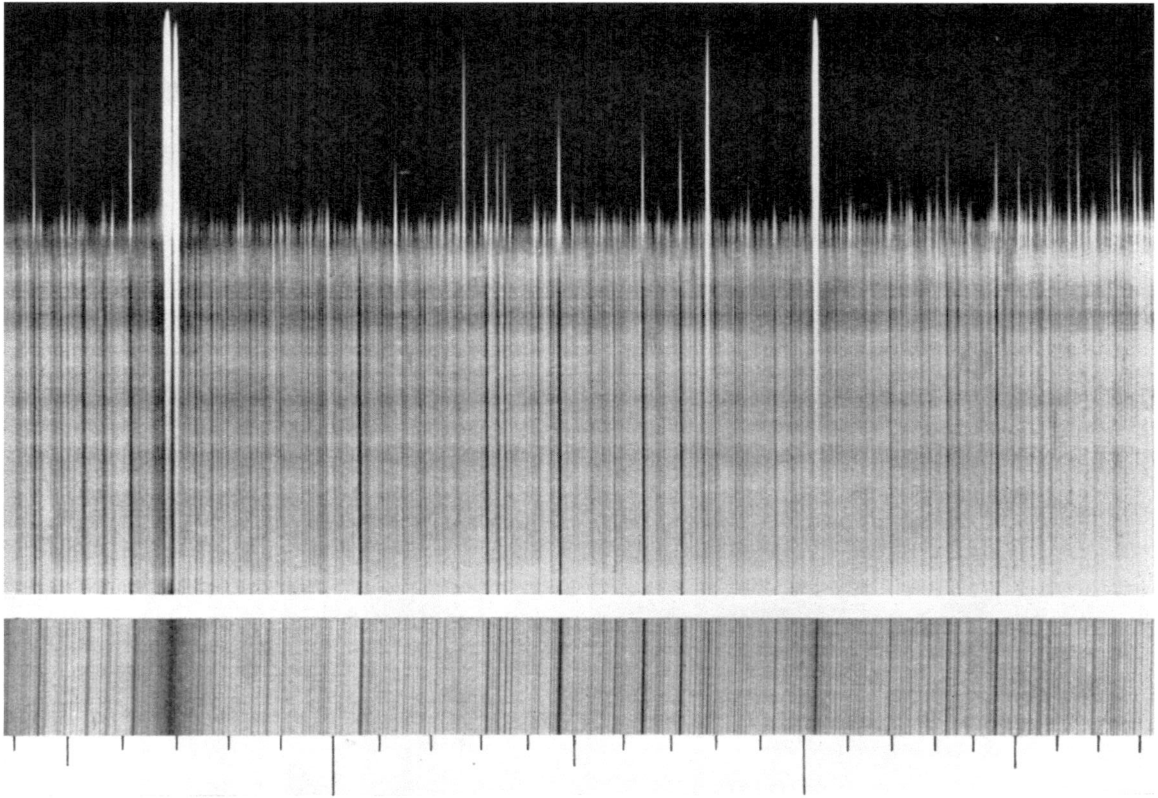

FLASH SPECTRUM, 1905 LICK MOVING PLATE, REGION 3940 TO 4180 A

The ordinary solar spectrum with the same apparatus is given below for comparison.
(An enlargement of $2\frac{1}{2}$ times the original.)

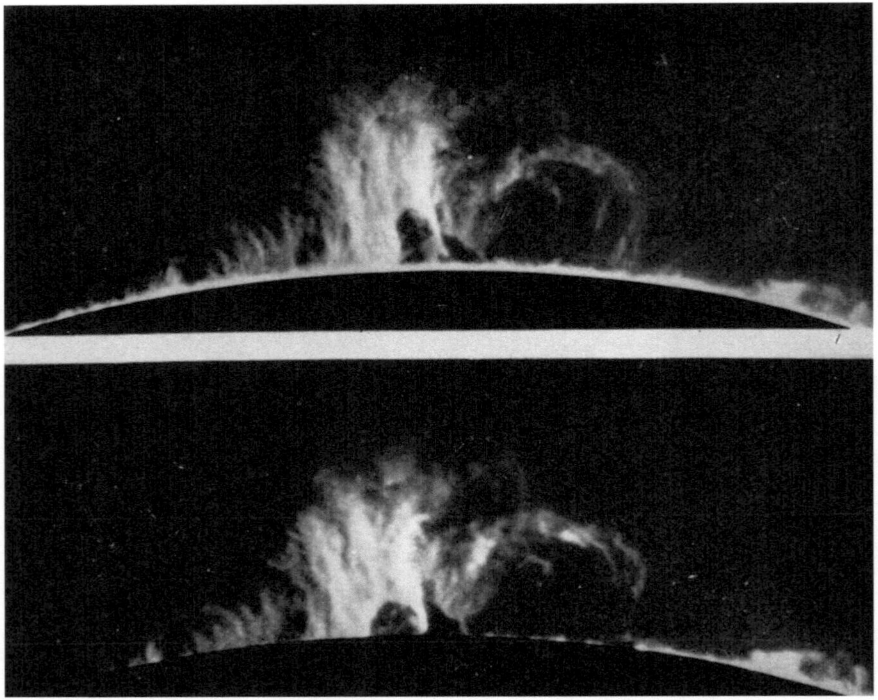

Calcium Spectroheliograms of Solar Prominences on October 10, 1910
These photographs secured by Slocum with the Yerkes refractor show great changes taking place in ten minutes of time.

der. Facing page 260, there is a reproduction of the Lick moving plate of the second flash of the 1905 eclipse. The strip of normal Fraunhofer spectrum of the sun appended below for comparison was taken by Menzel with the original 1905 apparatus reassembled after more than twenty years and with the addition of a slit and collimator lens. The streaks running lengthwise throughout the spectrum are caused by the non-uniform motion of the photographic plate.

It is interesting to compare the details of this moving plate spectrum with that taken by the author with fixed plate, as shown facing page 248. The spectra taken by the two methods refer to the same (1905) eclipse and both were taken at the end of totality. In the original spectra, in the region near the strong H and K lines shown to the left on page 248, the Lick moving plate had three times the dispersion of the fixed Mitchell plate.

One obvious and important advantage of the moving over the fixed plate is that the former makes less drastic demands on the ability of the observer to time his exposures so as to reach the lowest possible levels. Another advantage is that the moving plate records the gradual changes from the dark- to the bright-line spectrum at first flash, and in the reversed order at second flash. The fixed plate makes the record in discontinuous exposures, the moving plate has a continuous record.

Now that the Lick results have been published, it is possible to compare the relative merits of the fixed and moving plates for the photography of the flash spectrum. When the performances of any two spectrographs are compared, it is almost the universal custom among skilled spectroscopists to draw conclusions only when the two instruments have nearly identical optical powers, but more particularly when the dispersion is nearly the same for each. By disregarding this basis of comparison, many mistakes have been made in interpreting the results of the flash spectrum, the most outstanding probably being that by Lockyer, referred to on page 162, who compared the number of lines in the flash spectra photographed in 1893 and 1896 with the number found in the great Rowland Atlas obtained with vastly superior definition and dispersion. One is therefore surprised to find Campbell (*loc. cit.*) comparing the number of spectral lines of the 1905 and 1908 moving plates with those given by the 1905 fixed plate with concave grating, and then drawing conclusions in spite of the fact that the 1905 moving plate had twice the dispersion, and the 1908 plate four times the dispersion in the region of wave-lengths common with the concave grating photograph.

Investigations of the flash spectrum are attempted in order to photograph the solar spectrum under conditions differing radically from those of every-day observation without an eclipse. The heights to which chromospheric vapors extend are of the very greatest value in furthering solar research. Hence, if comparisons are made of fixed and moving plate spectrographs having the same dispersion, then it should be easy to decide which method will furnish heights with the maximum of accuracy and which will be most efficient in recording the greatest number of lines. Of course, it goes without saying that the dispersion of both fixed and moving plate spectrographs should be as great as possible. A comparison is not now being made of the relative merits of prism or grating spectrographs for the reason that either type of instrument may be combined with either fixed or moving plate.

Under ideal conditions of perfect timing and circular edges of both moon and sun (the moon without mountains and valleys, the sun without prominences), the flash spectrum appears at the instant that the whole photosphere is covered up by the moon, hence with edges of sun and moon tangent as in Figure 6. The fixed plate records the crescent arcs. A knowledge of the angular diameter of the sun and the augmented diameter of the moon furnishes a ready method of deriving from the spectra the heights to which the chromosphere vapors extend. Measurements are made of the angular length of the different arcs, or more simply by the measurement of the half-chord of this arc y_0 in Figure 6. Under the ideal conditions outlined above, the chromospheric arcs at their middle, or at the horizontal line in Figure 6, represent in each spectral line the radiations from *all* levels of the chromosphere lying above the photosphere. At a height of y_0 equal to 2.0 mm on the original spectra of 1905 (with a focal length of 60 inches) each of the crescent arcs represents the chromosphere at levels that are *above* 700 kilometers, a half-chord of 3.0 mm in each case gives the conditions at levels above 1600 km. At a value of y_0 equal to 5.0

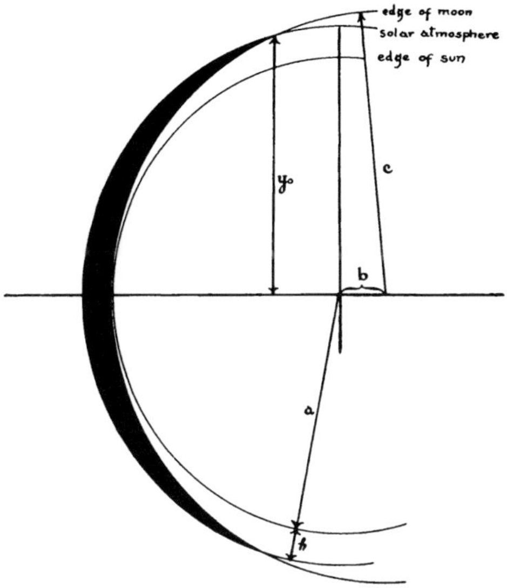

FIG. 6 Measuring the heights of solar vapors.

mm, the tip of a high level cusp is reached and this tip furnishes a detectable amount of radiation impressed on the photographic plate. At the tip of this chromospheric arc, therefore, only those radiations reach the photographic plate which come from emitting atoms *above* the 5000 km level. Hence each chromospheric arc, both above and below the middle of the arcs as recorded in the spectrum, gives a record of the radiations at the photosphere and at higher and still higher levels as the tips of the arcs are approached. If the tip of the chromospheric arc comes at a half-chord measured at 2.0 mm, then the radiations above the 700 km level are sufficiently active to give a blackening of the photographic image strong enough to be detected. The " height " is then said to be 700 kilometers. Under the actual conditions at the time of the eclipse, neither the edge of the moon nor that of the sun is circular in outline, nor is the exposure of the flash spectrum an instantaneous one, and the scale of heights is accordingly modified. The moon has mountains and valleys projected against the edge of the sun, which has prominences. The chromospheric arcs will present a picture of these effects. If these arcs are very irregular, then manifestly it would be difficult to get reliable information regarding heights, particularly if a large and active prominence is projected near the tips of the chromospheric arcs.

With the moving plate method, all features of the chromospheric arc are exactly as for the fixed plate outline above, the only difference being that the slot used with moving plate concentrates attention on a definite section of the chromospheric arc. With the arrangement used at the 1905 eclipse, the slot width uncovered a layer of 300 km at one time. As the eclipse progressed the moving plate furnished an integrated picture of what was happening within this layer 300 km in height.

In the earlier editions of this book, and also in the memoir in Volume IV of *Handbuch der Astrophysik,* the author has stated that one decided objection to the moving plate method is that at the time of an eclipse the slot might readily be superposed on a section of the sun where there was an eruptive prominence. At the eclipse of 1918, the " Heliosaurus " shown facing page 196 was exactly at the position of third contact. At the eclipse of 1919, another large prominence group is shown facing page 272.

The Lick observers recognize the force of this criticism as is shown by the following (*loc. cit.* p. 249), " One valid objection to the moving plate is that, if the image of a prominence falls on the plate, the resultant heights would be difficult to interpret." Hence at the latest eclipse observed with moving plate, that of April 28, 1930, the Lick observers examined the solar image just before the beginning of totality in an effort to set the slot on a portion of the sun devoid of large disturbances. One unfortunate feature resulting from this attempt [1] was the failure to photograph the layers of

[1] *Publ. A. S. P., 42,* 131, 1930.

lowest level owing to the plate not being put in motion until after the total phase was well started.

Regarding the relative number of lines photographed by fixed or moving plates, it seems easy to reach conclusions. In the region of wave-lengths, from 4200 to 4500 A, where all three spectra had the best definition, the 1905 and 1908 moving plates recorded 425 and 292 lines respectively, while in the same region of wave-lengths, the 1905 concave grating spectra had 437 lines; and this in spite of the fact that the moving plates had average dispersions twice and four times, respectively, that of this 1905 fixed plate. Between the region 3900 and 4100 A, the Lick 1905 moving plate recorded 413 lines and the Lick 1905 fixed plate 454 lines in spite of the better definition of the former and a dispersion more than twice greater than that of the latter. Hence with equal dispersion and definition the fixed plate will always show more lines than a moving plate.

Another disadvantage of the moving plate is the difficulty of ascertaining what zero to adopt as the initial level for the scale of heights. This zero-point depends on the reversal from dark to bright lines on the photograph (at first flash) or the reverse at second flash. As already stated, at the beginning of totality, the high-level lines become bright many seconds before those of lower levels. The reversal of the lines of lowest level forms the "flash" spectrum. With the fixed plate, measures of the lengths of the arcs give the heights directly, but with the moving plate, the photographs themselves must furnish the zero-point from the reversal of the lines. This is a difficult task which is further complicated by the fact that the slot causes an effect integrated over a region stretching over 300 km in height. It would seem therefore that this integrated effect would not permit a high precision in the determination of heights by the moving plate which must also be subject to a zero-point correction far larger than with the fixed plate.

In making comparisons of the reliability of fixed and moving plates, the best conclusion can be reached by confining attention to the low-level lines only, those that reach heights of 700 km or less, these lines being vastly greater in number than those of higher levels and their heights are known with greater precision. The Lick 1905 fixed plate gave average heights for these lines of lowest levels which differed about 50 km from those of the 1905 Mitchell plate. On the other hand, the average heights from these same low-level lines from the 1905 moving plate were 400 km greater than those from the Lick 1905 fixed plate. At one and the same eclipse, the simplest method of interpreting this systematic difference is the zero-point correction of the moving plate. Moreover, the 1905 moving plate gave average heights for the low levels more than 500 km greater than the 1908 moving plate.

Furthermore, with the moving plate, the scale of heights depends on an accurate knowledge of the speed of motion of the plate and also on the assumption that this motion is absolutely uniform.

Considering therefore the difficulties outlined above and the various causes of errors that may beset the moving plate method, the author is in hearty agreement with the opinion expressed (*loc. cit.*, p. 2) that "it is doubtful if any other astrophysical instrument is as difficult to set up and operate successfully as an objective-prism eclipse spectrograph, when used with a continuously moving photographic plate including a slit immediately in front of the plate." One might also add that it is doubtful if the photographs of any other astrophysical instrument are so difficult to interpret in order to reach conclusion free from systematic errors.

If the flash spectrum were an *exact* reversal of the Fraunhofer lines, both as regards wave-lengths and intensities, the eclipse observations could add but little to our knowledge of solar physics. In such a case, the precious moments of a total eclipse should be devoted to the investigation of other lines of research. In fact, the flash spectrum is interesting and important only in so far as it differs from spectra taken under ordinary conditions.

The *intensity* of a spectrum line depends both on the width and on the blackness of the photographic image of the line. It is unfortunate that in all spectra, whether of dark or bright lines, whether determined in the laboratory or in the observatory, it is ordinarily impossible to have a scale for the designation of relative intensities which is other than arbitrary. With such a scale the strongest line in any spectrum may be represented by 10, by 100, or even by 1000, while the weakest line receives the number 1, or 0, — 1, — 2, or even — 3, as in the revised Rowland's Tables. Such scales being arbitrary may not be uniform, and it is consequently very difficult to compare the values of the intensities of one spectrum assigned by one observer with those investigated by another.

For reasons outlined above, with chromospheric arcs photographed by the fixed plate, it is readily possible to make estimates of intensities or, better still, measures of intensities by the microphotometer at different positions along the arcs. At the middle of the arcs the intensities refer to the radiations of all atoms above the photosphere. At positions 2 mm from the middles of the arcs the intensities are thus derived from the radiation from all atoms above 700 km in height. By estimating or measuring the intensities at different places along the chromospheric arcs, from their middle to the tips of the cusps, a knowledge is obtained of the radiations of atoms at higher and still higher levels above the photosphere. In similar manner with the moving plate, estimates of intensities may be made or measures with the microphotometer be carried out, as was done by Menzel (*loc. cit.*). With the moving plate, as with the fixed plate, the intensities, estimated or measured, refer always to the radiations from atoms *above* certain levels as determined by the edge of the advancing moon.

As has been stated, with the fixed plate, the interpretation of heights is complicated by the Baily's beads which make knots in the curved lines of the spectrum. However, in estimating or measuring intensities on the fixed plate at different heights above the zero-point, the observer has it quite in his power to make measures at any particular location along the whole of the curved arc. Hence the Baily beads need be of no particular concern unless the solar disturbance is great in size and extent. With the moving plate, on the contrary, the slot has a fixed definite position and measures may be made of conditions affecting the chromosphere at this and at no other location.

The most characteristic difference between the chromospheric and the Fraunhofer spectra is found in the relative intensities of the lines. The system of intensities adopted for the chromospheric spectrum is purely an arbitrary one, in which 200 represents the strongest lines like K and $H\gamma$, and 0 that of the weakest line. Naturally the intensities depend on the character of the photographic plate used, but partial allowance may be made for the decrease in sensitiveness of the plate in the green and yellow regions. In estimating intensities, one is unconsciously influenced by the breadth of the lines, so that the values for intensity give a somewhat combined appraisal of the blackness and breadth of a certain line. These at best are but estimates, but they are perhaps comparable in accuracy with estimates of intensities by others, for instance, as in the Rowland's Tables. Estimates in Rowland and in the flash spectrum may[1] be calibrated.

The reasons for these pronounced differences in intensities between the dark line Fraunhofer spectrum and the chromospheric spectrum will be evident on a moment's reflection. Consider a Fraunhofer line coming from the center of the sun's surface, and assume that the absorption is caused by a reversing layer 500 miles in thickness. The light coming perpendicularly from the photosphere can be absorbed by the atoms in this 500-mile layer. At the time of an eclipse, the light of the chromosphere comes *tangentially* from the sun's surface, and not perpendicularly, with the result that the chromospheric light is affected by a depth of 20,000 miles of atoms. (The line of sight from a layer 500 miles in thickness passes through 20,000 miles of solar atmosphere when tangent to the sun's surface.) This has an important bearing on Saha's theory, as will be explained in Chapter XVII.

Researches on the structure of the atom, as shown in Chapter XIV have demonstrated the important fact that the intensity of a line in any spectrum depends primarily on the number of emitting or absorbing atoms. In the Fraunhofer

[1] *Astrophysical Journal, 68*, 1, 1928.

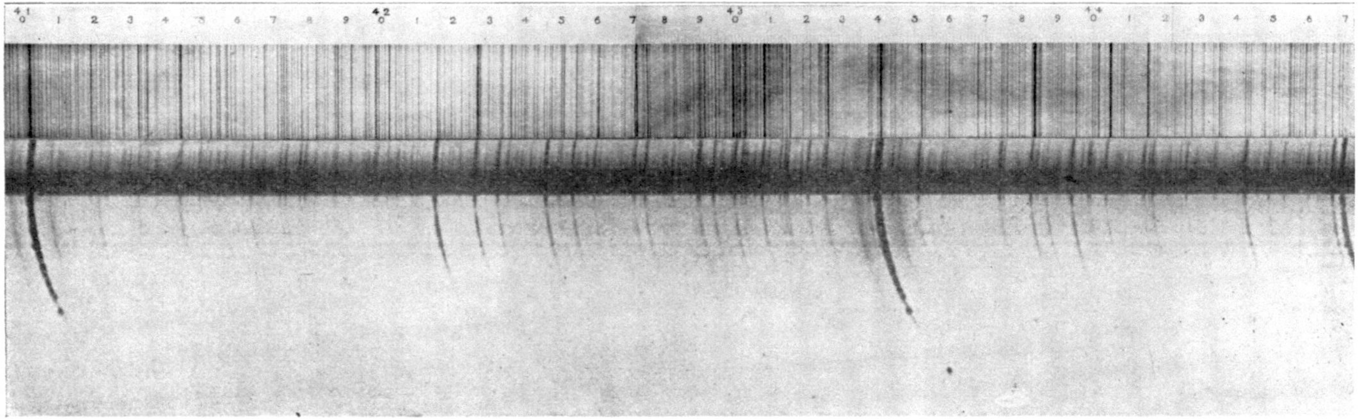

Comparison of Fraunhofer and 1905 Chromospheric Spectrum

Above: Rowland's Atlas reduced sixfold.

Below: Negative from eclipse enlarged fivefold.

THE CITADEL CROWNING THE CITY OF DAROCA

OFFICERS AND SAILORS OF THE U. S. NAVY ASSIST THE ASTRONOMERS IN OBSERVING THE TOTAL ECLIPSE IN SPAIN

spectrum the atoms are absorbing radiation but in the chromospheric spectrum they are emitting radiation. Let us now consider two different elements in the sun's envelope; one of these elements has a low density but extends high in miles above the photosphere; the other element is much heavier and its atoms are confined to layers lying much closer to the sun's surface. It is easy to imagine in the case of absorption of radiation that the total number of atoms involved in producing a line in the spectrum of the lighter vapor might be identical to the number contributing to the dark line in the heavier vapor on the condition where the light passes radially through the gases, for instance, when coming from the center of the sun's disk. If, however, at the time of a total eclipse when the moon covers the photosphere, the atoms of the chromosphere emitting radiation come tangentially from the sun's envelope. Under these circumstances, the greater height attained by the atoms by the rarer of the two vapors adds enormously to the number of atoms encountered *relative* to the number encountered in the low-lying and denser layer when the atoms are excited and radiate light. Hence, it is readily seen that although the two gases may give lines of equal intensity in their absorption spectra, they will not necessarily do so in their emission spectra; the low-lying heavy vapor will give in the chromospheric spectrum short arcs, while the other assumed vapor will give longer arcs of greater relative intensity. Though there are many contributing causes, the main reason for the great differences in intensities between the dark and bright line spectrum of the sun is the differences in heights to which the vapors extend. The H and K lines of enhanced calcium and the lines of the hydrogen series are the strongest lines in the chromosphere mainly for the reason that the atoms of enhanced calcium and hydrogen are detected at greater elevations than are attained with any of the other elements.

As a matter of fact, there are such enormous differences in the intensities of the Fraunhofer and flash spectra that placed side by side, as they are on page 268, the spectra seem to belong to stars of two different types rather than to the same object under different conditions. It is these differences that make observations of eclipse spectra of the greatest value in widening our knowledge of solar physics. The chief differences in intensity for the stronger lines are found in the elements helium and hydrogen. As is well known, no helium absorption lines are ordinarily found in the sun, whereas in the eclipse spectrum the helium lines are conspicuous by their great strength. In the Fraunhofer spectrum there are only four hydrogen lines visible, while in the flash spectrum there is the whole Balmer series, no less than thirty-four lines being measured on the 1905 plates.

One can therefore see at a glance the very great importance attaching to a knowledge of heights in the chromosphere. The following chapters will show how this information not only has been the basis of Saha's Theory of Ionization but in addition has had an important practical bearing on many solar and stellar problems.

This close comparison of the solar and chromospheric spectra side by side helped make the identifications of sources rather certain. But the Rowland spectrum had a dispersion fully ten times the dispersion of the chromospheric spectrum (21-foot radius in the second order compared with 5-foot focus in first order, the gratings having nearly the same number of lines per inch). Naturally, lines which appear single in the chromospheric spectrum may be a blend of two or more lines with the greater dispersion. But lines which appear as a close pair or a blend in the chromosphere must be the result of the blend of corresponding lines in Rowland. On account of the great differences in intensity of the chromospheric and Rowland spectra, it was difficult to be always sure of identifications until photographs were compared side by side. The original photograph of the flash spectrum was enlarged five times. Rowland's great Atlas was reduced six times. Since the flash spectrum was nearly normal, it was possible to procure both spectra on a comparatively close approximation to scale. This comparison of spectra will perhaps speak more strongly, than any words or comparison of wave-lengths, concerning the sharpness of the original spectrum of the chromosphere. On account of the small variations from the normal spectrum (noted above) it was impossible to obtain an exact match in scale in the two photographs. Those who are interested sufficiently will be able to carry the comparison along line for line.

Repeated attention must be directed to the fact that the heights of the chromospheric vapors determined from the measurement of the angular length of the cusps, can afford no great accuracy in the determination of the *absolute* heights in kilometers to which the various layers extend above the photosphere. The method depends on the visibility of the ends of the cusps. It is quite possible, and probable, that vapors extend in detectable amounts to elevations beyond the limits visible in the cusps. The heights derived by this method can therefore only represent a mean height and cannot be expected to furnish the maximum heights to which the vapors in detectable amounts extend. Attention should likewise be called to the fact that the heights measured in this manner cannot give the elevations above the photosphere, but rather above the average level of the layer photographed in this particular flash spectrum. With these limitations, the method is capable of furnishing the relative heights of the layers producing the individual spectrum lines with a fair degree of accuracy.

This is not the place to go into greater technical details which may be found in *Publications of the Leander McCormick Observatory*, Vol. 5, part 7, 1932, or in the *Astrophysical Journal*. However, a brief summary may be made here of the advantages and disadvantages of the moving over the fixed plate for the spectrum of the chromosphere. It is assumed that the two types of spectrographs have about equal dispersion and light gathering power. Wave-lengths are obtained with equal accuracy by the two methods and hence may be left out of consideration.

The moving plate is a much more complicated instrument and demands greater technical skill in the erection and adjustment at the eclipse site. In photographing the flash spectrum, the moving plate has distinct advantages in the proper timing of the exposures. The moving plate shows the gradual change from dark to bright lines at the first flash (or the reverse on second flash). The separate exposures with the fixed plate make a similar but less complete record. With equal dispersion and definition, the fixed plate records more lines than the moving plate.

If the chromosphere is greatly disturbed near the positions of second and third contacts, the heights recorded from the lengths of the arcs on the fixed plates are difficult to interpret, especially if a solar prominence occurs near one of the cusps. With the moving plate, the slot which defines the region of investigation may be superposed on an active prominence region and therefore the photographs will not represent average solar conditions. For the first flash, the slot might possibly be set immediately before totality so as to avoid disturbed regions in the chromosphere. This however could not be done for second flash. The present writer would never under any circumstances attempt such an operation. It is far too difficult a task to undertake in the few excited seconds available. It is far easier to observe the appearance of the flash spectrum with the binoculars and direct-vision spectroscope, and then to start and close the exposures accordingly.

If successful photographs are obtained with both fixed and moving plates, it is much easier to interpret the results of the former. With the arcs recorded by the fixed plate, any portion of them may be adopted for discussion. The levels that lie closest to the photosphere manifest themselves by the appearance of a strong continuous spectrum on the photograph. Hence with the fixed plate, the zero-point correction should be small, probably of the order of 50 kilometers. At the 1905 eclipse, the Lick fixed plate at second contact and the Mitchell fixed plate at third contact gave low-lying levels which differed about 50 km, those of the latter recording the greater heights.

With the moving plate, the zero-point correction depends on the estimation of the location where on the photograph the spectral lines reverse from dark to bright (or bright to dark). The zero-level should be that of the vapors closest to the photosphere, those of lowest level. If there are many enhanced lines in the spectrum, as in the region of 4300 and 4400 A, the problem becomes complicated and more diffi-

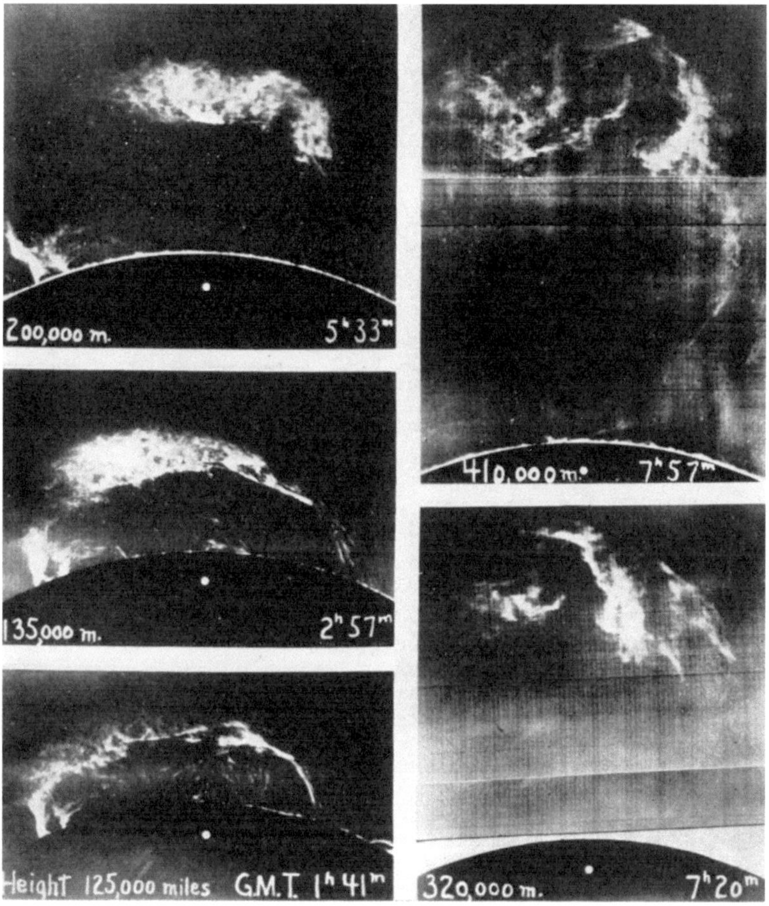

THE GREAT PROMINENCE OF MAY 29, 1919

Photographed at Yerkes Observatory by Pettit with the spectroheliograph on the day of the total eclipse. The greatest height attained on that day was 760,000 kms.

British Photograph of Total Eclipse of May 29, 1919

The same prominence is shown facing page 272 from photographs taken at Yerkes Observatory where there was no eclipse.

cult. As already stated, the Lick 1905 moving plate gave the heights of the vapors of lowest level 400 km in excess of the Lick 1905 fixed plate. This systematic difference indicates zero-point corrections which are larger for moving than for fixed plates.

The 1905 Lick moving plate had a slot width of .05 inches which permitted radiation from 300 kms of vapors to pass through and be recorded on the photograph. The motion of the plate causes progressive exposures to be made along the spectral line. The height of the vapor is derived in each case from the total length of the line measured from the assumed zero-point up to the vanishing tip. But this total length depends on the integrated effect of the radiation passing through the slot of the mechanism. Hence it would seem that the moving plate might readily have systematic differences between the low-level heights and those of greater elevation, and this is further complicated by departures of the motion of the plate from a uniform character. "An examination of the 1905 eclipse plate indicates that in the neighborhood of 1800 kms, the plate had experienced a slight acceleration" (*loc. cit.*, p. 258).

In regard to the photometric measures, a quotation may be made (in its corrected form) by Campbell (*loc. cit.*, p. iv). "My *moving-plate method* gives the *average* or *integrated* effect of a relatively *long segment* of the chromospheric crescent itself, say 300 km long, the inner edge of which is serrated by lower mountain peaks. The *fixed plate* method records the serrated crescent as a whole (Mitchell), or in part (Pannekoek and Minnaert); and it is interesting to note, a *long segment* of the serrated photographic crescent was *averaged* or *integrated* upon a moving plate of the recording microphotometer by Pannekoek and Minnaert as the basis of their photometry of the chromospheric spectrum. In their photometric applications, therefore, the two methods seem to have an element of equivalence." However, we might remind Dr. Campbell that in the measurement of the fixed plate spectra, it is not necessary to use a long slit with the microphotometer. A short slit may be used in order to integrate the measures over a relatively short range in heights.

In the actual operation of measuring the intensities of the spectra of fixed or moving plates there should be practically no difference in procedure.

In the foregoing, there has been stressed some few of the difficulties besetting the eclipse observer. No matter how carefully and precisely the focus of the spectrograph is determined, no matter how accurately the exposures are timed, either by fixed or moving plate, the photographed spectra may not be of finest quality for causes entirely outside the control of the observer, namely, the "seeing." As an example, the experience of the writer may be cited. At the 1925 eclipse the flash spectrum was photographed with the same equipment as at the 1905 eclipse. At the 1925 eclipse, the sun was only 17° above the horizon, and in consequence the seeing was poor. The 1905 eclipse was observed under excellent conditions of seeing. The 1905 spectra show the faintest lines of the spectrum in much greater numbers than the 1925 photographs, the lines in the former eclipse being stronger and more clearly defined. The explanations for the differences in the two spectra cannot be found in differences of focus, or of exposures, or of levels photographed. Under the poor conditions of seeing of the 1925 eclipse, the light of the bright-lined chromospheric arcs was rendered ill-defined by being spread over a larger area. With the strong lines of the spectrum, this effect made the edges hazy but diminished the total intensities of the lines but little. With the weak lines, however, the spreading of the light over a larger area on the plate caused a marked diminution in intensity and sharpness of each line with the necessary result that the weakest lines in the 1925 spectrum were practically obliterated by the poor seeing and only those lines of a certain minimum strength survived. Astronomers who are engaged in the observational work of stellar photography, particularly with telescopes of great focal length, are entirely familiar with an exactly analogous problem. Under the best conditions of seeing, the star images are hard and sharp and with well-defined edges. Under poorer and poorer conditions of seeing, the star images increase in size and the edges become more and more fuzzy. The larger stellar images with poor seeing require an increase of exposure to photograph a star of any given magnitude. The length of exposure becomes greater and greater as the seeing deteriorates more and more. Under poor conditions, the star images on the plates are large and ill-defined and the accuracy of measurement, when compared with plates taken under good conditions, is much diminished.

If photographs of the flash spectrum are actually secured in spite of all difficulties to be overcome, then comes the much more important problem of the discussion. The size of this task may be surmised from the fact that Dr. Menzel spent about five years in preparing for publication the handsome *Lick Publication,* Volume 17. On page 2, it took one-half a quarto page for him to render thanks to those individuals or bodies who had given scientific or financial assistance. In similar manner it has also taken the present writer, although working almost alone and with his time interrupted by many duties, five years to prepare the original contribution in *Publications of the Leander McCormick Observatory,* Volume II, parts 2 and 3 and the revision in Volume V, parts 2, 3, 6 and 7.

It will not be necessary here to explain in further detail the methods followed in deriving wave-lengths and heights. This information may be obtained from the sources just cited.

When the spectra of the 1905 eclipse were being discussed by the writer, comparatively little was known of the differences between the conditions existing in the Fraunhofer spectrum and in the chromosphere. The main difference known to exist was that pointed out by Fowler from the 1898 eclipse spectra, namely, that the enhanced lines played an important rôle in the flash spectrum by showing increased intensities over those in the solar spectrum. But how was one to find the wave-lengths of the lines which had greater strengths in the spark than in the arc spectrum? Lockyer had given lists of some of these lines, and these were good as far as they went. But the known lines were few in number and entirely inadequate for a discussion of the problem. What was to be done? Kayser's important lists had not yet appeared. There was nothing to do but to follow the example set by Kayser and others, namely, that of going to the original sources one by one. There seemed no other method, and this meant many long and tedious hours of patient comparisons.

Within the past decade, however, the combined attack on the structure of the atom by the astronomer, physicist and chemist has resulted in a stupendous increase in knowledge of conditions underlying the production of emission and absorption lines in different spectra. The outstanding achievements have been the Bohr-Sommerfeld atom, Saha's theory of ionization, investigations of series in spectra coming mainly from the hands of Fowler, Russell and Meggers, and the publication from Mount Wilson Observatory of the Revision of Rowland's Table of Solar Wave-Lengths. As a result of this greatly increased information now available, particularly in the proper identifications of the sources causing the spectral lines, the discussion of the flash spectrum has become one of greatly increased simplicity.

CHAPTER XVI

DISCUSSING THE FLASH SPECTRUM

THE depths of the various layers of gases surrounding the sun have an importance which is very vital in all theories of solar physics. Eclipse spectra furnish the only means yet known of directly measuring these depths. Comparisons of the flash spectrum with the solar spectrum taken under ordinary conditions reveal many important conclusions, the first of which is that wave-lengths are identical in the flash and in the Fraunhofer spectrum. This statement can be true only within the limits of accuracy attainable in the measurements of wave-lengths of the flash spectrum, or expressed in other terms, it may be said that no differences between the two spectra exceed 0.02 angstroms. The flash spectrum must therefore be regarded as a true reversal of the ordinary spectrum of the sun since every strong line in the solar spectrum, without any exception, is changed to a bright line at the beginning and at the ending of totality.

Although the wave-lengths in the two spectra are identical, this is far from being true with the relative intensities of the lines. In the flash, many lines appear which are not found in the ordinary solar spectrum; furthermore, some strong lines in the Fraunhofer spectrum appear as weaker lines in the flash, and *vice versa,* weak lines in the sun are strengthened in the spectrum of the chromosphere. Differences in intensity signify differences in level and in electrical, thermal and pressure conditions of the vapors producing the lines, and consequently an intimate study of such dissimilarities will give valuable information regarding the distribution of atoms in the solar atmosphere.

The problem, however, is a complicated one for the reason that the intensities depend so intimately on the level of the chromosphere photographed, all spectral lines being most intense at levels closest to the photosphere. Hence, at one and the same eclipse and with exactly the same timing, different intensities are found on the photographs depending on whether the spectrograph is used with or without slit. Without a slit, all levels contribute their emission to the spectral line, while with a slit only those levels covered by the slit give up their light to the line in question. Manifestly, the position of the image of the slit with respect to the photosphere is of prime importance, and also whether or not there is a prominence at the sun's edge where the slit is projected. Hence with a slit, there are many reasons why the intensities of the flash lines estimated or measured from the photographs of any one eclipse may vary from photograph to photograph or from instrument to instrument. Even on the same photograph there are differences of intensities for the same line. In their publication dealing with the 1926 eclipse, Davidson and Stratton give the estimates at two different positions in their photographs, called respectively "flash" and "prominence." From their photographs taken at the 1927 eclipse with slit, Pannekoek and Minnaert note similar differences in intensities.

For the same reasons, eclipse spectra taken without slit show wide ranges in intensities depending on the levels photographed. For instance, the author secured spectra with the same equipment in 1905 and 1925. The intensities differ materially in the two years, and in turn they both differ in most details from the intensities observed at the 1926 and 1927 eclipses, and also from the Lick results at the eclipses of 1898, 1900, 1905 and 1908. One should, therefore, be very guarded when comparing one eclipse with another before coming to the conclusion that differences in intensities between two eclipses mean differences in solar activity. Manifestly one should not push too far any deductions based on the intensities at any one eclipse.

Since the date of the second edition of this book, several important discussions have appeared dealing with the observations of the flash spectrum. The instrumental equipment has been described in the preceding chapter. A list of the more extended of these publications follows, arranged in order of the dates of the eclipses observed. As already stated, Campbell with the moving plate obtained photographs in 1898, 1900, 1905 and 1908. In 1905, with the fine seeing of the Spanish eclipse, the spectrum also was photographed on a fixed plate. The discussion by Menzel makes a veritable storehouse of valuable information. Accurate wave-lengths derived from the weighted mean of the different spectra make possible a close comparison with Rowland. The identification of the individual lines are given in splendid detail together with their multiplet designation. The intensities were estimated on the Rowland scale. For certain of the stronger lines, the intensities were measured by the microphotometer at different heights above the photosphere. On the observational side, the most important information is contained in the heights from both fixed and moving plates. As shown in the preceding chapter, the heights from the moving plates are subject to larger systematic errors than those from fixed plates. After tabulating the material in order of wave-lengths, a further tabulation rearranges the material according to the source of the line, a grouping being made according to the multiplet structure. A summary of the various recent theories underlying the problems of solar physics makes this Lick publication an epoch-making contribution. As an illustration of the tremendous activity since 1923 (the date of the first edition of this book) caused by researches on the structure of the atom, there may be cited the number of references to scientific publications made by Menzel in footnotes. Before 1923, references were made to 18 articles; in the years 1923–25 there were 22 references; while in the interval 1926–30, no less than 100 publications appeared, of which number there were 69 citations in 1928 and 1929.

Following the eclipse of 1908 observed by Campbell, the next eclipse to provide a photograph of the flash spectrum was 1914, when Abetti secured spectra of small dispersion. At the eclipse of 1918, on account of widespread clouds, the only photograph of the flash spectrum in good detail was obtained by H. C. Wilson with concave grating, but also with small dispersion. No attention was paid to the spectrum of the chromosphere in 1919 and 1922, the Einstein problem being of paramount importance. Little progress was made at the 1923 eclipse.

At the 1925 eclipse, Curtis and Burns[1] with a concave grating of short focal length obtained photographs at the red end of the spectrum on plates stained with dicyanine. About 100 lines were measured between the D lines and 8807 A, the dispersion being small, of 80 angstroms per mm. At this same eclipse, Anderson with a 21-foot concave grating without slit secured spectra on which the definition was vitiated by the poor qualities of seeing. At the same eclipse, with 10-foot concave grating without slit, Mitchell obtained photographs of the flash spectrum. For publication, the results were combined with those of the 1905 eclipse photographed with the same equipment. At the extreme ultra-violet, the measures of the 1926 spectra of Davidson and Stratton were included in the discussion by Mitchell. This publication, like that of Menzel, involved the newer ideas of the structure of the atom. In fact, both Menzel and Mitchell, through the kindness of the Mount Wilson astronomers, had free access to the valuable compilations on multiplet structure. The region of wave-lengths included in Mitchell's discussion was from 3066 to 7065 A, while that of Menzel included the region 3229 to 5328 A.

In 1926, Davidson and Stratton[2] secured beautiful photographs with a quartz slit-spectrograph giving excellent definition from 3066 to 3968 A. With a camera of 19-feet focus and direct-vision prism, the region photographed was extended to Hβ, and with camera of 40-feet focus and prism of 40°, the region was further extended to beyond Hα. The photographs with the two prismatic cameras were taken without slit and furnish heights for the stronger lines of the spectrum. The results were grouped according to multiplets shown by the different chemical elements.

Davidson and Stratton compared the heights derived from their prismatic cameras with those of the 1905 eclipse with concave grating. Differences were noted which depend

[1] *Publications of Allegheny Observatory, 6,* 95, 1925.
[2] *Memoir R. A. S., 64,* 105, 1927.

mainly on definition but there was no systematic error in zero-point between the heights from the two eclipses.

At the 1927 eclipse, Pannekoek and Minnaert secured excellent photographs of the flash spectrum at Gällivare in Lapland, one of the few places where clear skies greeted the observers of this eclipse. The spectrograph was one of three prisms and slit, the photograph of the first flash was in splendid definition between the wave-lengths 4154 and 4751 A. The dispersion of 1 mm equal to 3 angstroms is the largest ever successfully employed on eclipse spectra.

The most interesting part of their discussion is the photometry of the flash spectrum, the Moll registering microphotometer being used for the measurement of intensities. In spite of the large dispersion employed, no attempts were made to measure exact wave-lengths. The positions of the spectral lines were read from the photometer sheets and wave-lengths were derived from a Hartmann formula. Comparisons were made for each line with Rowland and with the intensities estimated by Mitchell from the 1905 eclipse, an excellent agreement being found between the 1905 and 1927 results.

These spectra were very carefully calibrated by means of three other plates cut from the same larger photographic plate as the one on which the flash spectrum was photographed, and developed together with it. On one of these plates was photographed the iron arc; on the second, the continuous spectrum of a standard incandescent lamp whose intensity distribution over various wave-lengths was known; and on the third, the Fraunhofer spectrum of the sun. In front of the slit of the spectroscope they had placed a "step weakener." By means of this they obtained a calibration curve for each region of the spectrum separately; and by means of the spectrum of the standard lamp and the Fraunhofer spectrum they obtained the relation between these individual calibration curves.

In a very thorough discussion they were able to find the number of ergs per second emitted in a unit solid angle by an arc of the chromosphere 1′ in length corresponding to each bright line of their flash spectrum. Combining this material with the theoretical intensities of lines within multiplets, they made a careful study of the effects of self-reversal in the flash spectrum. They also found the total intensities of entire multiplets relative to the total intensities of other multiplets of the same element. When the theory of atomic spectra has been sufficiently developed to predict the relative intensities from one multiplet to another, such absolute intensities in the chromospheric spectrum will offer the best data for comparison, much better, in fact, than laboratory spectra.

To derive from the measured transmissions of the microphotometer, a knowledge of the absolute units of energy of the source in the chromosphere is a problem beset with many difficulties. By the employment of their careful methods of calibration, this was accomplished in three steps: (1) From the measured transmission curve for a short interval of wave-lengths to ascertain the apparent intensity. (2) By proper allowance for instrumental causes, such as change of dispersion in the spectrograph, color sensitivity of the photographic plate, selective absorption in the apparatus, and other factors, to change the apparent intensities thus determined into real intensities. (3) By proper methods of calibration and standardization it is then necessary to find the absolute intensities. According to Pannekoek and Minnaert, " It is easy to foresee that the determination of the *apparent* intensities is the most accurate, that the determination of the *real* intensities is more difficult, and that the determination of *absolute* intensities may be liable to many sources of errors."

After making allowance for all possible factors and deriving intensities measured in absolute units, comparisons were then made with Rowland's intensities and with those of Kayser. An interpretation of these comparisons which will be reasonably free from large systematic errors is encumbered with many grave difficulties for the reason that the absolute scale of intensities derived from the microphotometer has no definite relationship with the estimates of Rowland and Kayser which are on arbitrary scales. On account of the many lines of the iron spectrum, this element gave the most complete information. Between the microphotometer readings and Kayser's estimates there was found a strong dependence on wave-length; a line at 4200 A having a three-fold greater microphotometer intensity than a line of similar Kayser intensity at 4700 A. On the other hand, a close agreement was found to exist between Rowland's solar intensities and Kayser's estimates for *Fe;* but a similar agreement is not found for other important elements in the sun, such as *Cr, Mn, Ti, Ni* and *Co,* the Rowland's intensities being much lower than Kayser's intensities of the elements.

In the calibration of Rowland's scale, referred to in a later chapter, Russell, Adams and Miss Moore find that the Rowland scale at 4200 A is seven per cent greater than at 4700 A, a result not in harmony with the findings from the 1927 eclipse. On account of this disagreement it is evident, either that the 1927 spectra were not properly calibrated or else that the homogeneous scale derived at Mount Wilson in the process of the revision of Rowland's Table must be changed by a large factor.

As a further result of their spectra, Pannekoek and Minnaert find that, " The energy emitted by the whole chromosphere in each of these wave-lengths is not proportional to the theoretical emission. In general, the observed intensities increase together with the theoretical intensities, but more slowly. This proves that the chromosphere may not be considered as an optically thin layer of small effective depth, but that there is an appreciable amount of self-absorption in the layer of gas viewed tangentially." Further discussions carried out by Pannekoek [1] lead to the conclusion that the density of the chromospheric gases being small, the effect of collisions may be neglected with the result that absorption and emission of radiation is the only method of energy transfer. The atoms forming the chromosphere being exposed to the radiation from the hemisphere below, there is no diminution of radiation at the highest levels. This implies that the chromospheric gases produce only a small part of the total absorption in the solar chromosphere with the consequent result that the Fraunhofer absorption takes place almost com-

[1] *B. A. N., 4,* 263, 1928.

pletely within the confines of the lower levels of the reversing layer. These results lead to the conclusion that the chromospheric gases are supported mainly by radiation pressure. For obvious reasons, these conclusions must be accepted with reservations.

By combining the material from six successive exposures they were able to find the density gradient of hydrogen in the chromosphere up to heights of 3000 km. They found that its density gradient was much less than would be expected in an isothermal atmosphere in which the only counterbalance to gravitation is gas pressure. By the same method they found that the density of the emitting atoms of helium is actually greater at 700 km above the photosphere than at levels closer to the photosphere. Their work is the most successful yet accomplished in exact spectrophotometry of the chromosphere.

The region investigated at the 1927 eclipse, of 600 angstroms at the red side of 4153 A, is but a small portion of the chromospheric spectrum. Menzel (*op. cit.*, p. 12) with the Moll registering microphotometer measured the 1905 Lick moving-plate spectrum between 3913 and 4466 A, the region of best definition. Runs were made in the ordinary way at six different levels, assumed to be at 170, 385, 600, 1030, 1675, and 2030 kilometers, respectively, above the photosphere.

Although the method adopted by Menzel, developed independently, for reducing the microphotograms, was very similar to that of Pannekoek and Minnaert, nevertheless he worked under severe handicaps in that he did not even know the character of the photographic plate used at the 1905 eclipse, and naturally, no standard squares were impressed thereon for purposes of calibration. No corrections were made by him for atmospheric or instrumental absorption.

As stated in the preceding chapter, the slot of the moving plate permitted an integrated radiation of 300 kms in each spectral line, and moreover, each line under consideration was defined on one side by the serrated edge of the moon, and on the other side by emitting atoms assumed to lie above the six different levels under consideration.

With so many unknown factors, and with heights subject

The Approaching Shadow of the Moon, Baker, Oregon, June 8, 1918
From the painting by Howard Russell Butler, N. A.

to systematic errors, Menzel is forced to evaluate his material from the interagreement of results from separate lines. "When the continuous background was subtracted, it was found that all of the curves for a given chromospheric height required multiplication by a single factor, inversely proportional to the ordinate at maximum, to bring them into surprisingly good agreement. This was probably due, in part, to the short spectral region studied. No correction for atmospheric or instrumental absorption has been introduced. The intensity of a spectral line is equal to the area under the 'true intensity' curve. Since all of the curves, for a given level, were of similar 'shape,' the ordinate at the mid-point determines the relative intensity of a line.

"The observed peculiarity of line shape is just the reverse of what might have been expected on a priori grounds. It is easily shown that the widening that arose from the deviation of the small segment of the solar crescent from a straight line was negligible. Almost all other causes that act to broaden a spectral line — pressure, Stark effect, self-reversal, temperature (Doppler motion) — apparently should be more effective at the lower levels. There is, of course, the chance that 'seeing' conditions during totality were not uniform, but there is evidence against this explanation, in the form of a marked exception to my previous statement that all lines, at a given level, conform to the same contour curve. The line 4026 A of helium is essentially a 'high-level' line, as is immediately evident from its great extension in relation to its intensity. Consequently, most of the intensity, recorded on the tracings at lower levels, is the integrated light from much greater heights. The lines of greatest intensity present a special problem. Small errors in drawing the characteristic curve of the photographic plate produce large errors in the resultant intensities. Rather than introduce such indeterminate intensities, I have computed the intensities of these lines on the assumption that their contours were the same as for the fainter lines. It then suffices to determine the ordinate at any point of the line contour. The intensity is measured, not at the center of the line, but in the wings where the photographic density is less, and where the characteristic curve is more certain. Intensities calculated in this manner are 'upper limits.' They are subject to correction for self-reversal."

Continuing the study by the investigation of the difficult problem of self-reversal, Menzel continues on page 261 as follows. "The density gradient of the chromosphere is closely related to the observed intensities of the chromospheric lines at the various levels studied by the microphotometer. The energy at any one level on the plate represents the integrated light from all the higher levels not yet cut off by the moon's limb. The presence of self-reversal complicates the problem. Theoretical estimates of the amount of self-reversal seem to be unreliable. We can, however, determine the amount of the self-reversal in what seems to be a perfectly direct way, by comparing the *observed* and *theoretical* intensities through the intermediary of the *estimated flash* intensities."

After going through an intricate theoretical discussion, Menzel apparently puts little weight upon his final results. This is shown by the fact that in the determination of the distribution of vapors in the chromosphere and the derivation of density gradients, he prefers the density gradients deduced from a discussion of the observed heights to those derived from his microphotometer readings.

One of the most pressing needs concerning the atom at the present time (1932) is a more accurate knowledge of the contours of lines in the ordinary solar spectrum in order to ascertain the numbers of atoms effective in producing absorption in the centers and wings of individual lines. Valuable work has been done by Unsöld and others, but this forms merely the beginning of a more comprehensive research necessary. In the Fraunhofer spectrum the atoms are absorbing energy, in the chromospheric spectrum they are emitting. To those who are not fully acquainted with the difficulties involved, it would seem an easy matter to photograph the chromospheric spectrum at different levels, and then to obtain a knowledge of the number of atoms involved at different levels. After several years of investigation on the Fraunhofer spectrum, with huge equipment and unlimited resources, we do not yet feel on very firm ground. If an eclipse came every day, and if photographs could be obtained of the flash spectrum with powerful equipment, that is, with large dispersion and large solar image, much might be accomplished. Unfortunately, eclipses are rare phenomena, the " flash " lasts but a few seconds, and it is impossible to use large dispersion with a large solar image on account of the paucity of light and erratic conditions of seeing. The present writer is always an optimist (or he would not be an eclipse observer) but he has a practical turn of mind. In the immediate future, he cannot foresee any brilliant discoveries from the measurement of eclipse spectra by the microphotometer.

A knowledge of conditions in the sun's atmosphere are of such great importance to solar physics that attempts are made to observe or photograph the chromospheric spectrum at times when there is no total eclipse. This is possible either during a partial eclipse by observations at the cusps, or when there is no eclipse at all. Observations were secured visually by Fowler in London during the progress of the *partial* eclipse of the sun, April 17, 1912. At its maximum phase, this eclipse was 0.91 total (the sun's diameter = 1.0). The 6-inch equatorial was employed in conjunction with the Evershed spectroscope, the observing conditions being very favorable. The number of bright chromospheric lines that were seen vastly exceeded expectations. As early as thirty-five minutes before maximum eclipse, high-level chromospheric lines were noted at the cusps and the number of lines increased with the progress of the eclipse. During the maximum phase, hundreds of Fraunhofer lines were reversed, in fact the number of lines seen was so great that it was impossible to record all of the lines. The appearance greatly resembled the flash spectrum that had already been observed by Fowler at more than one total eclipse. About seventy bright lines between *b* and D were actually identified. That so many lines of the chromosphere were visible near the cusps at the time of the eclipse may be partially explained as the result of reduced sky illumination due to the fact that but nine per cent of the sun remained uncovered. A greater advantage may have resulted from the smaller effect of unsteadiness or "boiling" at the cusp as compared with the limb under ordinary conditions.

The success of these observations was so great that it seemed possible to Fowler that even better results could be obtained at a similar eclipse by the use of more powerful apparatus, and that the multitude of bright lines seen visually for half an hour, while the eclipse ranged from eight- to nine-tenths total, might profitably be photographed by suitably arranged instruments. In fact, at this same eclipse, Newall at Cambridge actually secured successful photographs. On the best photograph only about forty lines were recorded as bright, many of them exceedingly narrow with the continuous spectrum quite faint. Apparently, therefore, it is more difficult to photograph the bright lines than it is to observe them visually. In view of the success attained in photographing, Newall came to the conclusion [1] that "exceedingly valuable work could be carried out with an instrument of high power by an observer who, in a total eclipse, stationed himself to the north or south of the band of totality at such a distance that the maximum phase was about 0.99. At such a station, detailed observations could be conveniently made with a much more leisurely program and with far greater completeness than on the central line."

In view of their success in 1912, the Cambridge [2] observers prepared to photograph the eclipse of 1921 with the McClean spectroscope of the Solar Physics Observatory. This instrument uses an image of the sun 168 mm in diameter, the dispersion being caused by a 6-inch plane grating. Two photographs were secured with excellent definition — but the total number of bright lines visible was only two, due to hydrogen, not a single bright metallic line appearing on either of the plates. A similar disappointment resulted from the photographs taken with the Huggins refractor, also at Cambridge.

Observations similar in kind were also made [3] visually at

[1] *Monthly Notices, R. A. S., 72,* 538, 1912.
[2] *Monthly Notices, R. A. S., 81,* 482, 1921.
[3] *Monthly Notices, R. A. S., 81,* 485, 1921.

the same eclipse by Fathers Cortie and Rowland at Stonyhurst. A Browning spectroscope with a dispersion of eight prisms of 60° was employed, the region observed being in the green from 5167 A to 5400 A. "Every line in the field was reversed, the bright lines tapering to a point indicating decrease of pressure."

Newall (*loc. cit.*) could not offer any explanation for his failure to photograph the bright lines at the 1921 partial eclipse. The experience of the writer, both at total eclipses and at Mt. Wilson (*see below*) causes him to suspect that the conditions of seeing at Cambridge were inferior to what they were at the eclipse of 1912 in spite of the fact that at the time the observers regarded the conditions as about equal.

A quarter of a century ago,[1] Hale and Adams secured photographs of the flash spectrum without an eclipse by using the 60-foot tower telescope and powerful spectograph attached. The method employed was to allow light from the sun's limb to fall upon a diagonal prism so placed as to reflect the light horizontally to a second prism immediately above the slit. The first prism was mounted upon a slide with a screw adjustment allowing motion toward or away from the second prism.

After the sun's image has been brought to the slit, the observer selects a bright line of the chromospheric spectrum and brings it into the field of view of an eye-piece, mounted in an opening near the end of the plate-holder. During the exposure, this line is maintained at maximum brightness by guiding with the screw controlling the position of the first diagonal prism, thus moving the sun's image slightly on the slit. The second order spectrum was employed giving a dispersion of 1 mm = 0.9 angstroms; and exposures of four minutes were required in the yellow part of the spectrum, and double this amount in the red.

Two reports on the wave-lengths, etc., obtained from the spectra have been published from Mt. Wilson. The first one by Hale and Adams appeared in *Communications* No. 41, and the second by Adams and Burwell in *Communications* No. 95, the former coming before and the latter after the

[1] *Astrophysical Journal, 30*, 222, 1909; *Mt. Wilson Contr.*, No. 41.

publication by Mitchell of his flash spectrum results obtained at the eclipse of 1905. For obvious reasons, the second communication from Mt. Wilson is more complete than the first. Unquestionably the photographs at Mt. Wilson were secured at a very low level, but whether at a lower level than the 1905 eclipse spectra it is rather difficult to decide. In comparing the results obtained within and without an eclipse, there are very great advantages in favor of the latter method, the most important being the possibility of securing a much higher dispersion by the method without an eclipse than can be secured at the time of an eclipse. The dispersion of the 1905 eclipse spectra was 1 mm = 10.9 angstroms, or one-twelfth of that in the discussion in the *Mt. Wilson Communications*. For reasons stated in the foregoing, it is impossible to use large dispersion at an eclipse. Because the exposures without an eclipse can be increased at will, the dispersion can also be increased, and it may be possible for the Mt. Wilson observers to employ the 150-foot tower telescope with the 75-foot spectograph in the second order and thus secure a dispersion thirty times that of the 1905 eclipse spectra. At an eclipse only two exposures of the flash spectrum are possible, and unfortunately the best focus is secured more or less only as the result of a happy accident. Outside of an eclipse there are no such limitations, observations may be repeated until moments of good seeing are secured and until photographs are obtained of the best definition exhibiting the results of layers sufficiently close to the sun.

The comparison of the results obtained with and without an eclipse may be briefly summarized here. First, let us take up the *number* of the lines photographed by the two methods in a given region of the spectrum. Compared with spectra taken outside of an eclipse, those obtained at an eclipse have the two-fold disadvantage of the much smaller dispersion and of the superposition of the eclipse lines on a very strong continuous spectrum. In the process of measuring the eclipse spectra, it was difficult to differentiate the weakest lines from the continuous spectrum. Consequently many eclipse lines were actually measured in the flash spectrum but were not published at the time (*see Astrophysical Journal, 38,* 481, 1913), for it seemed unwise to add any lines to the publication unless those lines could be definitely identified with lines appearing in Rowland's Tables. For these reasons, therefore, it must be concluded that the number of the *weakest* lines secured at Mt. Wilson without an eclipse is probably not greater than the number found on Mitchell's eclipse spectra. Since the intensities of the lines depend on the levels at which the lines originate, it seems highly probable, therefore, that the levels photographed within and without an eclipse are not very different. For the stronger lines originating at higher levels, the intensities inside and outside of the eclipse greatly differ, such variations however being readily explained on account of these very differences in level.

Regarding the accuracy of the wave-lengths, the following may be said. On account of the spreading of the photographic images of the strong lines of the eclipse spectra it is impossible to secure accurate wave-lengths from such lines. Omitting these lines, the eclipse wave-lengths differ from Rowland on the average by 0.020 angstroms. With the twelve-fold greater dispersion at Mt. Wilson the wave-lengths have an accuracy of 0.012 angstroms,[1] being a precision about twice that obtained from eclipse spectra.

With a slit tangent to the sun's limb it is evident that the *length* of the lines in the spectrum taken without an eclipse should furnish information regarding the levels at which such lines originate. Such measures have not as yet been carried out at Mt. Wilson. The need of more reliable determinations of the levels at which the spectrum lines of different intensities originate and the great importance of such knowledge in present-day problems of solar physics cannot be overemphasized.

In summing up the problem regarding the photography of the flash spectrum as it appeared to him in 1924 while in residence for a brief time at Mt. Wilson Observatory, the writer voiced his opinions in the following paragraph which is found in the second edition.

The recommendation is therefore made to the investiga-

[1] *Mt. Wilson Contributions,* No. 95.

tors of solar rotation, and is hereby urged upon them for their consideration, that they curtail their work on solar rotation, and instead devote their energies for a short while to the flash spectrum without an eclipse. With the equipment already in hand, and under *good* conditions of seeing and with adequate facilities for proper guiding, good photographs of the flash spectrum should be possible at very low levels. The accurate measurement of wave-lengths, and particularly the determinations of the levels at which different lines of the spectrum originate, will add greatly to our knowledge concerning these lines and will supply information so sadly needed in deciphering the laws underlying the production of spectral lines. The Bohr atom and Saha's theory of ionization have made possible the identification of new spectral lines in the sun and in the laboratory, and undoubtedly we are on the eve of a very great advance in knowledge regarding the laws underlying the production of lines in the spectra of various sources, provided wave-lengths of great accuracy and free from systematic errors can be furnished.

The eight hectic years since the second edition of this book have revolutionized our conceptions of solar research. Regarding the problem of the flash spectrum, Director Adams gave the writer, while he was in residence at Mt. Wilson, every opportunity for observing with the 150-foot tower telescope. The writer then realized the important rôle played by the qualities of seeing and steadiness of image of the sun on the slit of the spectrograph. If the slit is set tangent to the limb of the sun and if the bright-lined spectrum is observed visually by the methods described above, then it is immediately noticeable that if the seeing is ragged or fair, there will be few bright lines that can be seen, such for instance as the stronger lines of the solar spectrum like the *b*-group in the green. If the seeing improves and the image of the sun becomes more and more steady, then more and more lines are seen reversed into bright lines. At Mt. Wilson, the best seeing ordinarily comes shortly after sunrise. After the sun's rays heat up the slopes of the mountain and the rising of the heated air disturbs the atmosphere, the steadiness of the image of the sun on the slit of the spectrograph becomes less and less. In the short time at his disposal for making observations, the writer came to the conclusion that for photographing the flash spectrum without an eclipse the best possible conditions are absolutely necessary, or in other words, with seeing and steadiness of quality 10 where 10 represents the maximum of perfection. If therefore one wishes to obtain such photographs without an eclipse, he must be continually on the alert to catch the few precious moments of perfect seeing when they come; and then if not too impatient, after taking hundreds or possibly thousands of photographs for this purpose, a photograph may finally be secured which will be of quality equal to, or even better than, that of Adams obtained in the early years at Mt. Wilson. These observations only serve to emphasize the importance of excellent seeing, no matter what method is employed for photographing the flash spectrum.

Anticipating somewhat the theoretical considerations to be developed in the three following chapters, it is possible to make a summary of results achieved from a study of the flash spectrum, and also to make recommendations for work at future eclipses. The following general conclusions may be drawn:

1. On account of the different conditions under which the chromospheric and Fraunhofer spectra originate, it is probable that wave-lengths from the two spectra differ systematically. However, these differences do not amount to as much as 0.02 A.

2. Every strong line in the Fraunhofer spectrum is found in the flash spectrum, and every strong line in the latter (with the exception of *H* and *He* lines) is matched by a line in the former.

3. The flash spectrum may therefore be regarded as a reversal of the Fraunhofer spectrum.

4. The " flash " is not an instantaneous appearance. At the beginning of totality the chromospheric lines of greatest elevation appear first, and at the end of totality remain the last.

5. The " reversing layer " which contains the majority of

the low-level lines of the chromosphere is about 600 km in height.

6. The " reversing layer " has no existence separate from the chromosphere.

7. It is the densest part of the chromosphere lying closest to the photosphere, and it is the cause of the greatest portion of the absorption producing the Fraunhofer lines.

8. The " Evershed-effect " measured in sun-spots, and photographs of flocculi which exhibit vastly different aspects when photographed at various elevations above the photosphere prove that the shadings of such *strong* lines as H and K are caused by absorption at different levels and pressures above the photosphere.

9. The chromospheric spectrum differs greatly from the ordinary solar spectrum in the intensity of the lines.

10. The Fraunhofer spectrum is essentially an arc spectrum. The chromospheric spectrum more closely resembles the spark spectrum, and its spectrum corresponds to an " earlier " type than that of the sun.

11. Especially prominent in the chromosphere are the enhanced lines.

12. The enhanced lines ascend to greater elevations above the photosphere than do the ordinary lines.

13. The increased elevations cause greatly diminished pressures.

14. As Saha has shown, the reduced pressures permit the ionization of the atom. As a result, the lines of the ionized atoms are specially prominent in the flash spectrum. The enhanced lines are produced by the ionized atoms.

15. The depth of the chromosphere is not constant.

Recommendations for future work on the flash spectrum may be briefly summarized. The most important contribution from the chromosphere will undoubtedly come from the investigations of heights and intensities of the spectral lines in order to gain further information regarding atomic structure. To be of the greatest value in furthering scientific research, photographs of the flash spectrum should be secured with large dispersion, they should extend as far to the violet and as far to the red as possible, the definition should be of the very best and the exposures should be timed so as to photograph the very lowest possible levels. The occupation of a station near the edge of the moon's shadow path would permit relatively long exposures on the low-lying layers closest to the sun's pole. Such spectra would afford interesting comparisons with those taken near the central line of totality which give spectra near the sun's equator. Another important investigation for the future will be to make comparisons between flash spectra taken at different phases of the sun-spot period. If we are to judge by the changes which take place in the spectra of the corona, we should expect that the flash would be richest in lines at times of maximum sun-spots. The eruptions taking place near spot zones seem to elevate the low-lying metallic vapors. If this be true, it would be natural to expect that if it were possible to photograph at a constant level above the photosphere in securing the flash spectrum at an eclipse, the lines should appear to be of greater strength at sun-spot maximum than at minimum.

The type of spectrograph is all important. Eclipses, however, are rare phenomena. After the eclipse of 1932, which will be assiduously observed, the total eclipses of the next two decades require long voyages and expensive expeditions. Few institutions, except those like the Lick, U. S. Naval and Mt. Wilson Observatories in the United States, and the British Joint Permanent Eclipse Committee, can afford the luxury of assembling spectrographs for infrequent use on chromospheric work. Hence most astronomers will be forced to use the equipment they already have, others, less fortunate, must borrow their apparatus.

As it is impossible to employ a dispersion sufficiently high to permit the detection of systematic differences in wave-lengths between the solar and chromospheric spectra, we come to the conclusion that eclipse wave-lengths are of secondary importance, serving as they do merely as a means of accurately identifying the origins of the spectral lines. Although it is far easier to get good definition by the use of a slit spectrograph, each observer must come to the decision whether to try the easier method and thereby sacrifice information regarding levels, or to try the more difficult plan of photographing without a slit in order to attempt to gain knowledge regarding heights which are now about the most important contribution that can come from eclipse spectra.

In work on eclipses, as in all other branches of scientific research, continued advances must be made, if the time and energy and money are to be profitably spent. Just because eclipses are rare phenomena is no reason in itself why an astronomer without eclipse experience can expect to make a scientific contribution of value unless he knows the subject thoroughly and has a problem of importance to attack.

Prisms have advantages over gratings in their greater light-gathering power. For work at the blue end of the spectrum, prisms may be the best means of securing the high dispersion now necessary. On account of erratic conditions of seeing at the time of an eclipse, no one should attempt to increase the dispersion by increasing the focal length of the camera to 20 feet or more. No photographs of the flash spectrum of first quality have yet been taken by this type of prismatic camera. In the future, its use at eclipses should be discontinued.

It is possible to use higher dispersion on the chromosphere than has been employed up to the present. We badly need photographs at the extreme violet end of the spectrum with large dispersion in the region which cannot be reached by glass prisms, and hence quartz must be used.

Except for the light-gathering power, gratings have distinct advantages over prisms. At the red end the prismatic spectrum is crowded together. To get high dispersion, many prisms must be used. The grating gives a "normal" spectrum and a greater region of wave-lengths in good definition. If gratings are used, they should be concave and not flat gratings. Similarly, larger dispersion than that employed to date may be used profitably with concave gratings. To increase the dispersion one cannot expect to have more lines per inch than the number of 15,000 per inch which have been used. The only other means is to work in the second order (if it is possible to rule gratings specially for this purpose), or

should be of the very best and the exposures should be timed so as to photograph the very lowest possible levels. The occupation of a station near the edge of the moon's shadow path would permit relatively long exposures on the low-lying layers closest to the sun's pole. Such spectra would afford interesting comparisons with those taken near the central line of totality which give spectra near the sun's equator. Another important investigation for the future will be to make comparisons between flash spectra taken at different phases of the sun-spot period. If we are to judge by the changes which take place in the spectra of the corona, we should expect that the flash would be richest in lines at times of maximum sun-spots. The eruptions taking place near spot zones seem to elevate the low-lying metallic vapors. If this be true, it would be natural to expect that if it were possible to photograph at a constant level above the photosphere in securing the flash spectrum at an eclipse, the lines should appear to be of greater strength at sun-spot maximum than at minimum.

The type of spectrograph is all important. Eclipses, however, are rare phenomena. After the eclipse of 1932, which will be assiduously observed, the total eclipses of the next two decades require long voyages and expensive expeditions. Few institutions, except those like the Lick, U. S. Naval and Mt. Wilson Observatories in the United States, and the British Joint Permanent Eclipse Committee, can afford the luxury of assembling spectrographs for infrequent use on chromospheric work. Hence most astronomers will be forced to use the equipment they already have, others, less fortunate, must borrow their apparatus.

As it is impossible to employ a dispersion sufficiently high to permit the detection of systematic differences in wavelengths between the solar and chromospheric spectra, we come to the conclusion that eclipse wave-lengths are of secondary importance, serving as they do merely as a means of accurately identifying the origins of the spectral lines. Although it is far easier to get good definition by the use of a slit spectrograph, each observer must come to the decision whether to try the easier method and thereby sacrifice in-

that of 10 feet,
oses. As Ander-
the best at the
heric arcs on the
ncrease of focal
the intensity of
spheric arcs will
, in consequence,
ence. To over-
ice for the 1926
of 15,000 lines
ic arcs were di-
This method has
and April, 1930,
e tried again in

e will be brave
erately to place
r to photograph
eclipse of 1900,
track as the re-
e of observation
ary to await the
al eclipse passes
erful equipment
lue will be pos-
se is more than
eration of active
e to pass within
sical laboratory
hern edge of the

the spectrum of
spersion than is
that without an
very high quali-
Hence, taken all
to rely on total

eclipses for adding to the information already available on the chromosphere.

The splendid work of Adams at the Mt. Wilson Observatory in photographing the spectra of the brightest stars of various types with very high dispersion will supplement the work on the chromosphere by giving information on series relationships and multiplet groups in spectra. The work so beautifully inaugurated with the powerful spectrograph attached to the 100-inch mirror may be carried out with greater facility when the gigantic 200-inch telescope is put into operation.

At the present time, astronomers are devoting much attention to the measures of intensities by means of the microphotometer. In the foregoing, the author has tried to point out some of the difficulties to be overcome when applying photometric measures to eclipse spectra. Unless the measures can be reduced to absolute units, fairly free from systematic errors, the measured intensities will have a reliability little higher than estimates by a skilled observer. In all of the discussions so far published, the intensities of the chromospheric lines have been compared with Rowland's values and with those of the laboratory, both of which are on estimated scales. Hence when reducing eclipse spectra, each observer must decide for himself whether it is worth while both to estimate and measure chromospheric intensities, when perforce all comparisons must be made with arbitrary estimates. However, it is very desirable that each eclipse spectroscopist should place standard squares on his photographic plates so that later these may be available for photometric purposes if desired.

Fixed or moving plates may be used for the flash spectrum with either prisms or grating. Those who disagree with the author in his opinion that the fixed plate gives the more reliable information have the opportunity of using the moving plate method. At the 1932 eclipse, both methods will have a thorough trial for there will be many American and European parties in the field with well-equipped expeditions.

CHAPTER XVII

THE IMPORTANCE OF IONIZATION

THE discussion of the results of the flash spectrum obtained in 1905 has shown the great importance of enhanced lines, or those due to the ionized atom, and the explanation offered by the writer was that the cause of the enhancement was due to the great heights attained by the vapors producing the enchanced lines and, consequently, to the reduced pressures at which these lines were found. It was also pointed out that at the time of an eclipse the light from the sun is capable of reaching us only in a direction tangential to the sun's surface. As a result, a beam of light from the bottom of a layer only 500 km in thickness would encounter no less than 20,000 km of emitting atoms in the tangential line of sight before getting out of the shallow layer. Any theory that can explain the method whereby atoms are ionized will be of the very greatest importance in all problems of modern astronomy. Such a theory has been developed by Dr. Megh Nad Saha. The sun is the nearest of the fixed stars, and on account of its proximity its structure may be examined in detail. In the sun and stars are found high temperatures, very minute pressures and electromagnetic conditions that together cause ionization to take place with great facility. The celestial laboratories thus opened to the astronomer have given him the opportunity of supplementing the work of the physicist and chemist in terrestrial laboratories in the combined attack on atomic structure.

Accepting the correctness of the Bohr-Sommerfeld theory of atomic radiation, and assuming that the general laws of thermodynamics apply equally well to electrons and to molecules of gases, Saha has been able to calculate the degree of ionization that takes place in gases under different conditions of temperature and pressure, and has derived formulas which can readily be applied to conditions existing in the atmosphere of the sun and of the stars. This theory explains both qualitatively and quantitatively many of the features observed in the spectrum of the sun and of the stars, and it likewise finds a ready application in laboratory spectra under conditions when enhanced lines appear.

In addition to the original papers by Saha,[1] valuable contributions have been made by Milne,[2] by Russell,[3] and also by many others.[4] Assuming that the decomposition of a molecule or an atom into one or more electrons and a positively charged ion is essentially of the same nature as an ordinary chemical reaction, Saha derives a simple equation to express the self-ionization of a gas at high temperatures. The equation derived is:

$$\frac{P\,x^2}{1 - x^2} = K$$

$x/(1 - x)$ is the ratio of the percentage of atoms ionized to those left neutral, and this ratio multiplied by the partial pressure of the free electrons ($P\,x/(1 + x)$) is equal to K, which is a function only of the absolute temperature of the gas and the ionization potential. This latter is a measure of the work done to ionize a single molecule, or to drive an electron from its neutral ring to infinity, and it is expressed as the number of volts through which the electron must fall to acquire this energy. Since the ionization potential is a constant, the quantity K, in the formula above, depends only on the absolute temperature. Hence for a given pressure, the smaller the ionization potential P, the more nearly x approaches unity, or in other words the more nearly complete is the ionization. For all gases where the ionization potential is known, Saha is enabled to calculate the percentage of ionization found under different conditions of temperature and pressure. The higher the value of the ionization

[1] *Philosophical Magazine*, *40*, 472 and 809, 1920; *Ibid.*, *41*, 267, 1921; and *Proceedings of the Royal Society*, A, *99*, 135, 1921.

[2] *Monthly Notices, R. A. S.* Many papers beginning 83, 1923.

[3] *Astrophysical Journal*, *55*, 119 and 354, 1921.

[4] Eddington, *Internal Constitution of the Stars*, 1926.

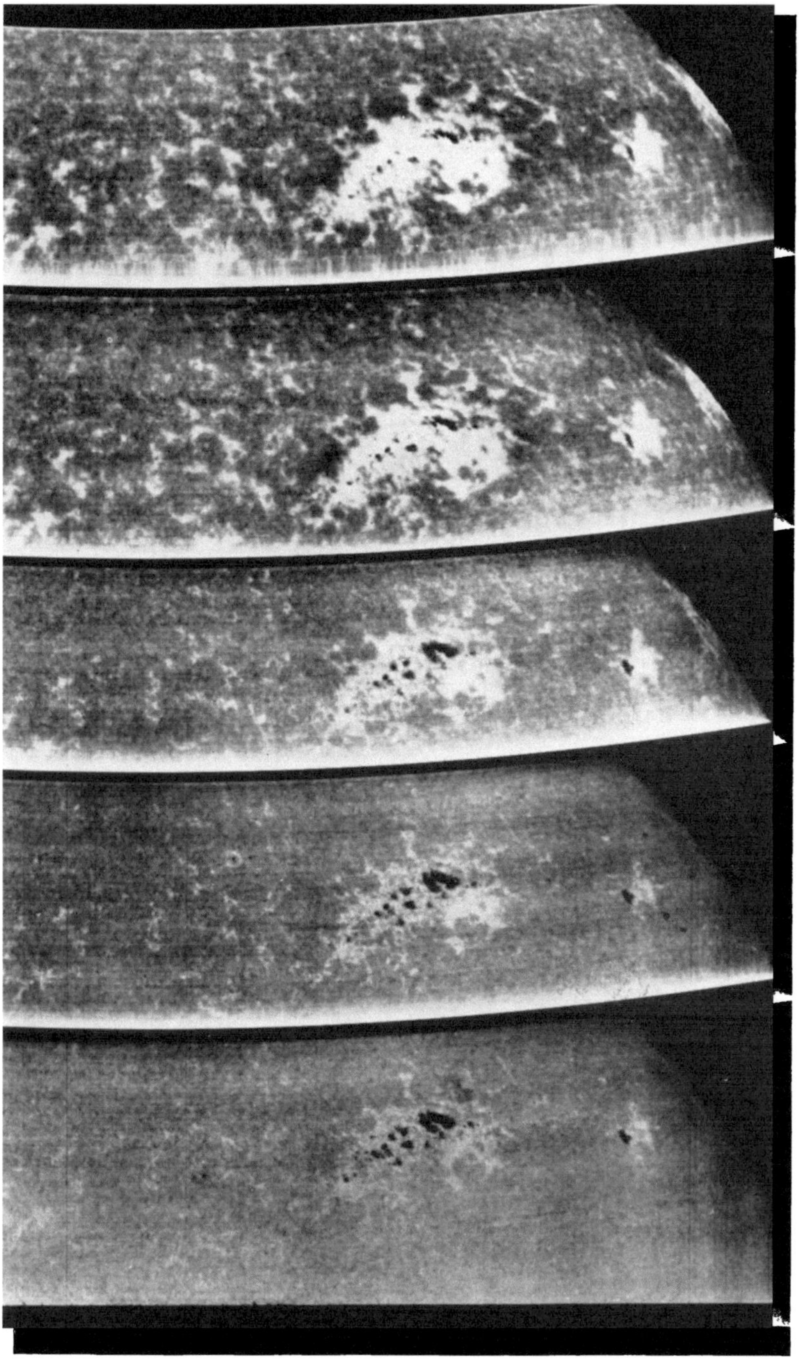

Calcium Flocculi Photographed by Fox with the 40-inch Yerkes Refractor, August 25, 1904

The slit was set at different wave-lengths corresponding to different levels above the sun's surface. Order: from lowest upwards.

U. S. Naval Observatory Eclipse Station, August 30, 1905
The 40-foot camera is placed horizontally.

Lick Observatory Eclipse Station, June 8, 1918
The 40-foot camera is pointed directly at the Sun.

potential, the higher must be the temperature to sustain a given degree of ionization. This is readily seen in the case of helium which possesses the highest known ionization potential, of 24 volts; for the series due to enhanced helium is found only in stars of the highest temperature.

Saha calculates tables giving the percentage of ionization in atmospheres at various temperatures and pressures (measured in atmospheres). The following tables are copied from his publication.

PERCENTAGE IONIZATION OF CALCIUM

Pressure (atmos.) Temp.	10	1	10^{-1}	10^{-2}	10^{-3}	10^{-4}
°						
4,000	0	0	0	3	9	26
5,000	0	2	6	20	55	90
6,000	2	8	24	64	93	99
7,000	7	23	68	91	99	100
8,000	16	46	84	98.5	100	100
10,000	46	85	98.5	100	100	100
12,000	76	96.5	100	100	100	100
14,000	90	100	100	100	100	100

PERCENTAGE IONIZATION OF STRONTIUM

Pressure (atmos.) Temp.	10	1	10^{-1}	10^{-2}	10^{-3}	10^{-4}
°						
4,000	0	0	2	5	15	45
5,000	1	3	11	32	73	96
6,000	4	13	37	78	97	100
7,000	10	32	73	96	100	100
8,000	22	58	91	99	100	100
10,000	56	90	98.5	100	100	100
12,000	82	97.5	100	100	100	100
14,000	93	100	100	100	100	100

Saha applies his theory to explain the differences between the spectrum of the sun and the chromosphere, and Russell broadens the scope of the theory by applying it to show the meaning of the differences in intensities of lines in the sun-spot and in the solar spectrum. The results furnish a complete triumph for the Saha theory.

Fortunately for the theory, the flash spectrum gives the height in kilometers (or miles), that the various vapors producing different spectral lines extend above the level of the sun's photosphere. Within the past few years there has been a great revision of ideas regarding the pressures found in the reversing layer. Where formerly it was regarded that these pressures amounted to several times that found at sea level on earth, it is now known with certainty that the chromospheric pressures are even less than a thousandth of an atmosphere. According to many researches in recent years by a number of competent investigators,[1] it is known that the solar temperature is not far from 6000° C. absolute. Schwarzschild[2] has shown that if the variation in temperature in the upper atmosphere is caused only by radiation, then the temperature should not fall below $6000°/2^{\frac{1}{4}}$, or 5000°, approximately. Hence in the solar atmosphere, where the temperatures vary between 5000° and 6000°, it is easy to see at a glance from Saha's tables the percentage of ionization. For instance, in the chromosphere at elevations where the pressure amounts to one ten-thousandth of an atmosphere, ionization for calcium is ninety per cent complete.

Saha's theory is in complete harmony with the conclusions derived from the discussion of the flash spectrum and furnishes an adequate explanation. The case of calcium is most interesting. The lines H and K are enhanced lines and are caused by the ionized atom, while the *g* line at 4227 A takes its origin from the neutral atom. In the neighborhood of the reversing layer, both normal and ionized atoms will be plentiful and the presence of *g* and the H and K lines are fully explained. At great heights above the reversing layer, however, the pressure will be very small and, as a result, ionization will be nearly complete. Under these conditions the neutral atom cannot exist, the ionized atom exhibiting the enhanced lines alone being found. In the flash spectrum, measures indicate that the H and K lines extend upwards to heights of 14,000 kms, but the *g* line only to 5000 kms. The presence of H and K above the 5000 km level shows

[1] Russell, Dugan and Stewart, *Astronomy*, 535, 1927.
[2] *Gott. Nachrichten*, *41*, 1906.

that calcium actually exists above this level, and we must therefore interpret the failure of the atom to emit the *g* line above the 5000 km level to be due to the fact that practically all of the atoms are ionized and there are few normal atoms left to produce the 4227 line. For strontium and barium, which also exhibit enhanced lines, their ionization potentials (5.7 and 5.1 volts) are lower than that of calcium (6.1 volts), and consequently complete ionization in the chromosphere is found at higher pressures or, in other words, at lower elevations above the photosphere. The strongest line of neutral *Sr* is 4607 A, which reaches an elevation of only 400 km, while the ionized atom *Sr*+ shows the two lines at 4215 A and 4077 A which are found at elevations much greater than that of the neutral atom, in fact, at 6000 kms, but this level is much less than the 14,000 km height attained by the H and K lines of *Ca*+.

In addition to *Ca* and *Sr, Ba* also belongs to the alkali earths of Group II of the periodic table of the chemical elements. In the following table, there are given the details concerning the *strongest* single line found in the neutral series of each of the three elements, and also the *strongest* doublet belonging to the ionized atom. In the second column is given the wave-length and in the last the heights in kilometers measured in the flash spectrum. In the four other columns are given the intensities in sun, chromosphere, arc and spark, respectively. The intensities for the laboratory spectra are taken from Exner and Haschek. In spite of its high atomic weight (137.4), *Ba* is a conspicuous element, both in the sun and in the chromosphere, on account of the enhanced lines. No neutral lines of *Ba* are found either in the sun or in the chromosphere.

This small table contains some of the conclusions derived from a study of the flash spectrum which, on account of their importance, will bear being repeated. The strongest line in the Fraunhofer spectrum is K (Rowland intensity 1000) and this line is the strongest in the flash spectrum (intensity 200 on a different scale). The enhanced lines which are produced by the ionized atom (designated by +) have an intensity in the chromosphere greater than in the sun, and these enhanced lines extend to greater elevations above the photosphere than do the ordinary or unenhanced lines.

In the flash spectrum the elements arranged according to the intensity of the *strongest* lines give the following sequence: *Ca, H, He, Ti, Mg, Sr, Sc, Cr, Ba, Fe, Al* and *Y*. This order differs materially from a similar table for the

COMPARISONS OF THE STRONGEST LINES OF THE NEUTRAL AND IONIZED ATOMS

Element	Wave-Length	Sun	Chromosphere	Arc	Spark	Height in km.
Ca	4227 (g)	20	40	1000	100	5,000
Ca+	3933 (K)	1000	200	500	1000	14,000
Ca+	3969 (H)	700	175	300	500	14,000
Sr	4607	1	2	1000	50	400
Sr+	4077	8	80	1000	1000	6,000
Sr+	4215	5	60	500	500	6,000
Ba	5535	—	—	100	30	—
Ba+	4554	8	50	1000	1000	2,000
Ba+	4934	7	25	100	300	1,200

ordinary solar spectrum. In lists arranged according to the number of lines, *Fe* stands first in each case. *Ni* and *Co* are second and sixth in the solar spectrum, but become sixth and eighth, respectively, in the flash spectrum. *V* is the eighth element in the Fraunhofer spectrum but fourth in the chromosphere. *Cb, Mo* and *Pd* appear in the first list but not at all in the second, while in the chromosphere is found *He* which is absent in the ordinary spectrum. The rare earths are represented relatively by more lines in the flash than in the ordinary spectrum.

Interesting results are obtained when the total number of lines identified in the flash spectrum are arranged according to the periodic tables of the elements. This is given in the table on page 306.

For each element in the table there is given: in the first line the atomic *number* and the chemical symbol, in the second line is found the atomic *weight*, and in the third line in *italics* the total number of lines in the chromosphere.

The mark (?) is found with the elements *Cb, Mo, Ag* and *Cd* to signify that there are no strong lines in the chromosphere resulting from these elements, but they may possibly be represented by weak lines in combination with stronger lines of other elements. Eight of the rare earths are represented in the flash spectrum by a total of 178 lines, *Ce, La* and *Nd* being responsible for 57, 47 and 31 lines, respectively. All of the rare-earth lines without exception are enhanced lines. The element *Ca* occupies an interesting place in the table. To the right of *Ca* in the table are found *Sc, Ti, V, Cr, Mn, Fe, Co* and *Ni,* all represented by numerous strong lines both in chromosphere and in Fraunhofer spectrum. Vertically above *Ca* is *Mg,* and below it are *Sr* and *Ba,* all important elements in solar investigations. In column I, of the alkali metals, the only element represented in the chromosphere is *Na,* where it is found with less prominence than in the Fraunhofer spectrum. *He* is the only element in column O, the inert gases, that manifests itself in the flash spectrum.

As mentioned above, the chromospheric and solar spectra agree exactly as to wave-lengths, but differ very greatly in the relative intensities of the lines. These differences of intensity are accentuated in the case of the "enhanced" lines, or those which are more intense in the spectrum of the spark than in the arc.

A comparison of the intensities of the lines in the Fraunhofer, chromospheric, arc and spark spectra forces one to the conclusion that while the Fraunhofer spectrum corresponds to the arc spectrum, the spectrum of the chromosphere more closely resembles that of the spark spectrum. The sun thus exhibits three distinct spectra under different conditions: the chromospheric spectrum, the Fraunhofer spectrum and the sun-spot spectrum. These three closely resemble the spectra of the stars γ Cygni, Capella and Arcturus, respectively; γ Cygni representing an "earlier" and Arcturus a "later" type of spectrum than that of the sun. Hence, the Harvard classification of the three spectra are: chromosphere = Fo, sun = Go and sun-spot spectrum = Ko.

A comparison of the spectra of the sun and the chromosphere reveals the following interesting facts: first, the

PERIODIC TABLE OF THE ELEMENTS

In italic figures are given the total number of lines for each element found in the chromosphere.

	O	I	II	III	IV	V	VI	VII	VIII
1.*H* 1.008 *35*									
	2.*He* 4.0 *19*	3.*Li* 6.9	4.*Be* 9.0 *1*	5.*B* 10.9	6.*C* 12.0 *140*	7.*N* 14.01	8.*O* 16.00	9.*F* 19.0	
	10.*Ne* 20.2	11.*Na* 23.00 *6*	12.*Mg* 24.32 *16*	13.*Al* 27.0 *2*	14.*Si* 28.1 *11*	15.*P* 31.0	16.*S* 32.06	17.*Cl* 35.46	
	18.*A* 39.9	19.*K* 39.10	20.*Ca* 40.1 *59*	21.*Sc* 45.1 *62*	22.*Ti* 48.1 *421*	23.*V* 51.0 *130*	24.*Cr* 52.0 *240*	25.*Mn* 54.9 *92*	26.*Fe* 27.*Co* 28.*Ni* 55.8 58.94 58.68 *1313* *84* *184*
		29.*Cu* 63.57 *3*	30.*Zn* 65.38 *3*	31.*Ga* 69.7	32.*Ge* 72.6	33.*As* 74.96	34.*Se* 79.2	35.*Br* 79.9	
	36.*Kr* 82.9	37.*Rb* 85.4	38.*Sr* 87.63 *6*	39.*Y* 88.9 *55*	40.*Zr* 90.6 *106*	41.*Cb* 93.2 ?	42.*Mo* 96.0 ?	43.*Ma*	44.*Ru* 45.*Rh* 46.*Pa* 101.7 102.9 106.7
		47.*Ag* 107.88 ?	48.*Cd* 112.4 ?	49.*In* 114.8	50.*Sn* 118.7	51.*Sb* 121.7	52.*Te* 127.5	53.*I* 126.9	
	54.*Xe* 130.2	55.*Cs* 132.8	56.*Ba* 137.37 *8*	57–71 Rare Earths *178*	72.*Hf* 178.6	73.*Ta* 181.5	74.*W* 184.0	75.*Re*	76.*Os* 77.*Ir* 78.*Pt* 190.8 193.1 195.2
		79.*Au* 197.2	80.*Hg* 200.6	81.*Tl* 204.4	82.*Pb* 207.2	83.*Bi* 209.0	84.*Po* 210	85....	
	86.*Rn* 222	87....	88.*Ra* 226.0	89.*Ac* 227	90.*Th* 232.2	91.*Pa* 234	92.*U* 238.2		

values of Rowland, giving the intensities of the lines of the ordinary solar spectrum, are comparable with arc intensities, while those of the chromosphere approximate more closely to the intensities of the spark; and second, the enhanced lines of any element are in every case stronger in the chromosphere than the lines of the neutral atom, and the heights attained are much greater. The results are a complete verification of Saha's theory that enhanced lines are caused by ionization which becomes more and more complete in the sun's upper atmosphere where the great altitudes make the pressures very small.

Not only is Saha's theory able to explain the facts regarding the enhanced lines of the ionized atom, but it makes clearer the details concerning the lines of the neutral, or un-ionized atom. Take, for example, the D-lines of sodium, so well known in the Fraunhofer spectrum. At pressures below one-thousandth of an atmosphere, *Na* with an ionization potential of 5.1 volts, is completely ionized. The D-lines belong to the principal series of the normal atom, and accordingly, they have no connection with, and are not produced by, the ionized atom. The normal atoms forming the D-lines therefore cannot exist when the pressure in the chromosphere is reduced to the thousandth of an atmosphere. It is quite in keeping with theory to find the flash spectrum photographs furnishing the information that the D-lines reach the comparatively small heights of only 1500 kms above the photosphere (H and K are found at 14,000 kms). The contrast in behavior in passing from the Fraunhofer to the flash spectrum for the D-lines of sodium on the one hand, and D_3 of helium on the other, is very marked. The sodium lines are weakened in the flash while the helium line is enormously strengthened, being entirely lacking in the ordinary solar spectrum. Furthermore, in view of the great prominence of the D-lines in the solar spectrum, it has always been a matter of the greatest surprise that the element potassium, so similar in its properties to sodium, is not found represented by strong lines in the sun. The explanation is a very simple one. The lines of the neutral atom of potassium, corresponding in its series to the D-lines of sodium, are found in the deep red part of the spectrum at wave-lengths 7664 A and 7699 A, and consequently they are not in the visible spectrum. Like the D-lines, both lines of this pair are strengthened in sun-spots. The only lines due to potassium found in the visible solar spectrum are very weak lines at 4044 A and 4047 A, of Rowland intensities 0 and 00, respectively. Russell finds both these lines strengthened in sun-spots. No enhanced lines are known for *Na* or for *K*, and consequently neither element is conspicuous in the flash spectrum.

The temperature of the photosphere is approximately 6000°, while that of sun-spots is lower and probably somewhere near 4700°.[1] The pressures found in sun-spots can differ but little from those in the lowest depths of the reversing layer. On account of the lower temperatures in the spots, however, ionization is less complete according to Saha's theory. As a result, the lines of the neutral atom, the so-called "low temperature" lines, are strengthened in sun-spots, while on the other hand, and also as a direct consequence of Saha's theory, the enhanced, or "high-temperature" lines are weakened in the spectrum of sun-spots. Since the variations in pressure in the neighborhood of the sun are much greater than the variations in temperature, it would have been more fortunate if the enhanced lines had been referred to as "low-pressure" rather than as "high-temperature" lines.

In the light of Saha's theory, Russell has investigated the sun-spot lines. His conclusions for the alkali metals (*loc. cit.*, p. 129) are here briefly given. Sodium is represented in the sun by the principal, diffuse and sharp lines of the neutral series, and all of its lines are much strengthened in the spot spectrum. Potassium is represented by the principal series only and its lines are also strengthened in sun-spots. Lithium is found in the spot spectrum only at wave-length 6707 A. Rubidium, hitherto unknown in the sun, was discovered to be present by both members of the strongest pair of the principal series, the wave-lengths being 7800 A and 7947 A.

[1] Miss Moore, *Astrophysical Journal*, 75, 222, 1932.

We are now in a position to explain some of the peculiarities regarding the appearance of lines in the spectrum of the sun and chromosphere and the heights found in the flash spectrum. The peculiarities noted (Chapter XVI) are as follows: The H and K lines of calcium of atomic weight 40 are stronger in sun and chromosphere and reach greater heights than hydrogen, the lightest gas in the sun. In the chromosphere the whole Balmer series for hydrogen is found, while only the first four members are seen in the Fraunhofer spectrum. No helium lines are found in the ordinary solar spectrum, but they are of great strength in the chromosphere. The elements, other than *H* and *He*, arranged according to the periodic table (see page 306) have remarkable progressions in the number and intensities of the lines involved. Group II, the alkali earths, represent the strong lines in the chromosphere, the strongest lines of all belonging to *Ca*. Group I, the alkali metals, have few strong lines in sun or chromosphere other than the D-lines of *Na*. None of the lines of Group O originating from the inert gases *Ne, A, Kr* and *Xe* are found in sun or chromosphere. In Group III, strong lines are found for *Al, Sc, Y,* and the rare earths, but the strength of lines is not as great as reached by the corresponding elements in Group II. In Group IV intensities are still less. The only element in Group V, found with certainty in the chromosphere, is vanadium, and in Groups VI and VII, *O* and *Cr,* and *Mn,* respectively, and in Group VIII, the three metals *Fe, Co,* and *Ni*.

It is easy to see why the metals of Group I, the alkalis, are represented by such feeble lines in the chromospheric spectrum. For reasons already stated, the enhanced spectra of the alkali metals resemble the spectra of the neutral atoms in the preceding group in the periodic table, the inert gases; and consequently, such spectra are very difficult to produce on account of the outer electrons forming part of a very stable ring or shell. As a matter of fact, no enhanced lines are found for any of the alkali metals in the visible portion of the spectrum. It is apparent, therefore, why the alkali metals cannot be prominently represented in the chromosphere since the flash spectrum is essentially an enhanced spectrum.

Quite different is the situation regarding the elements of Group II, the alkali earths, which are specially important in the chromospheric spectrum, for the reason that the strongest lines of their spectra are enhanced lines, and the principal members ($1s - 2p$) of the series lie in the familiar portion of the spectrum. This is true for the elements with the exception of *Mg,* the strongest lines of which are found in the extreme ultra-violet at 2795 A and 2802 A, in a region in fact where no light can reach the earth's surface from the sun on account of the absorption of light in the earth's atmosphere.

The great strength of the H and K lines of calcium both in the sun and in the chromosphere and the great heights to which these lines extend in the flash spectrum are now completely explained as the result of Saha's theory. In spite of the great difference in the atomic weights of the two gases, calcium and hydrogen, the atomic weight of the former being forty times the latter, the spectrum lines of calcium are seen to reach greater heights than are attained by hydrogen. The reasons for this curious circumstance are very simple. H and K are lines due to the ionized atom, and in virtue of the great elevations the ionization is greatly increased. The lines H and K are the chief lines belonging to the principal series, and in fact are the only lines of this series in the chromosphere. Only five others lines of $Ca+$ are found in the chromosphere, they are all of subordinate series and they have much smaller intensities than H and K. The hydrogen lines in the visible spectrum, on the contrary, belong to a subordinate series and not to the principal one.

In the above, attention has been called to the great differences in the spectra observed in sun, chromosphere and sunspots. To find an explanation of the observed facts, it is necessary to look more closely into the newer developments regarding the structure of the atom. In Chapter XIV it was shown that many atomic characteristics are explained on the assumption that the electrons forming the more complex atoms are arranged in successive shells. The inert gases helium, etc. in column O of the periodic table, have atomic numbers of 2, 10, 18, 36, 54 and 86, and hence it has been assumed that the concentric shells contain 2, 8, 8, 18, 18 and 32 electrons, respectively.

Lithium of atomic number 3, and sodium of atomic number 11, are elements which have one external electron outside of a complete shell. The process of removing an outer electron from an atom is called *ionization*. This may be accomplished by various electrical processes, some of which permit the measurement of the energy required in volts. As each electron carries a unit charge of negative electricity, an ionized atom, called an "ion," has a net positive charge of electricity. As we have seen, the spectrum of the ionized atom is radically different from that of the neutral atom. For reasons stated in the foregoing, the spectral lines belonging to the ionized atoms are called "spark" or "enhanced" lines.

Not only can an atom be removed from an outer shell through ionization, but a given atom or ion can be excited. That is, the position of one or more of the electrons can be changed within the atom itself. To change from the normal state to an excited state requires energy. If this energy is absorbed by the atom, the spectrum refers to the dark line or absorption spectrum. When the change takes place in the reverse direction, from excited state to normal state, the same amount of energy is emitted by the atom and the spectrum is a bright-line or emission spectrum. As already stated in a previous chapter, the transference of energy from one state to another is always in definite amounts, or quanta, of energy.

It has been found that the simple element hydrogen can exist in more than twenty excited states. More than one hundred different energy levels have already been discovered by spectroscopic means for the more complicated atom iron. The transference of energy from the excited to the normal state has been found to take place in the brief interval of time of approximately one-millionth of a second.

Planck's quantum theory is justly regarded as one of the

most important triumphs of the twentieth century. By means of it, we know that the frequency of the radiation emitted by the atom, and hence the wave-length, is exactly proportioned to the amount of energy radiated.

In the case of hydrogen, we have seen that the wave-lengths of the 35 lines which appear in the chromosphere can be represented by the difference between two spectroscopic terms. With ordinary hydrogen and enhanced helium there is only one external electron. When there are two or more electrons, the mathematical computations become very complicated and as a result the exact calculations and predictions of spectral lines are impossible. However, there are sound reasons for believing that all electrons in the outer shells of atoms, no matter how complicated these atoms may be, obey rules, called quantum conditions, that are quite similar to those underlying the simple structures of hydrogen and enhanced helium. The changes in the outer electrons give rise to radiations which produce spectral lines. The energy of the electron depends on the relation of the size, shape and inclination of the orbits to each other, and hence with the more complicated atoms, the number of different energy states, or spectroscopic terms, is much greater than in the simple case of hydrogen.

This is not the place to go into the technical details involved in the transference of energy from one spectroscopic state to another. For the more complicated atoms, it is impossible to represent the orbits as was done so simply for hydrogen in Figure 4. However, by means of a diagram suggested by Bohr and Grotrian, the different energy states of an atom can be represented quite readily. Each state is represented by a dot. The distance below a heavy horizontal line is proportional to the amount of energy (measured in volts) which must be communicated to the external electron to pull it entirely away from the atom and thence to ionize the atom. The lowest dots in the diagram represent those of least energy while the heavy horizontal line represents the ionized condition.

The rules of the game [1] laid down for the electrons to fol-

[1] See Russell, Dugan and Stewart, *Astronomy*, 548, 1927.

low are not complicated. In the diagram, states corresponding to orbits of the same angular momentum plotted by dots in the same column thus *s, p, d, f, g, h,* etc., represent conditions of increasing angular momentum. When there is a transference of energy from one state to another, two of the dots are joined by a line. An atom in the 1*s* state (the lowest energy state in the *s* column) can change to any of the *p* states but not to the others; or an atom in the 1*p* state can change to any *s* or *d* state, and so forth.

In the sodium spectrum, for instance, when a small amount of heat is applied, as in the Bunsen burner, the only lines emitted are those confined to the two lowest levels *s* and *p*. The lines emitted correspond to transitions from the *p* states to the lowest *s* state. If white light is passed through sodium vapor these lines are absorbed. This series of lines is designated by 1*s* — m*p*. The level 1*s* is the normal state of the sodium atom in consequence of the lowest amount of energy. If the atom is excited and is in the 1*p* state, it can return to the normal state only by the radiation of a quantum of energy corresponding to the transition 1*s* — 1*p*, which causes the appearance of the familiar D lines in the spectrum.

The transitions 1*s* — m*p* are called the *principal* series. If the temperature is raised above that of the Bunsen burner, by the application of electricity, the atom can exist in higher energy states, such as 2*s* and 2*d*. With increase of temperature, the fraction of atoms in the 1*p* state, though quite small, increases and hence the intensity of the lines become greater. At still higher temperatures, the fractions of the atoms in the 2*s* and 2*d* states continually increase, with the consequent emission of lines represented by the transitions 1*p* — 2*s* and 1*p* — 2*d*, etc.

If sodium is placed in the electric arc, then it is found that all the lines of the series 1*s* — m*p* are *sharp* while the lines 1*p* — m*d* are more fuzzy or *diffuse*. Hence the significance of the letters *s* and *d*.

The amount of energy required to raise an atom from the normal state to any given excited state is called the *excitation potential* which is expressed in *volts*. A simple formula connects the energy V, in volts, with the wave-length A, in angstroms as follows.

$$V = \frac{12345}{A}$$

The *p* levels of *Na,* and also of all the alkali metals are double, and hence the transitions cause double lines in all the elements in column I of the periodic table just as with the familiar D lines of sodium.

When there is more than one valence electron, the spectra are much more complex than that of sodium. There are then many different energy states and these may be single, double, triple and so forth up to eight-fold. The transitions from one energy level to another level give rise to numbers of lines known as *multiplet groups.* These may contain as many as fifteen members of varying intensities, some lines strong and some weak.

All of these important spectroscopic developments have taken place in the past ten years. The chief workers in bringing order out of chaos, in the grouping of spectral lines into multiplets and deriving the excitation potentials have been A. Fowler, Russell, Saunders, Meggers and Miss Moore. Their investigations have been greatly helped through accurate wave-lengths furnished by King of Mt. Wilson and Meggers of the Bureau of Standards.

The method of finding series relationships is somewhat similar to that used for the simple element hydrogen. Instead of working with the observed wave-lengths it is found to be easier to work with frequencies, or the reciprocal of wave-lengths. Certain rules have been evolved to which the frequencies are subjected. Then in a manner somewhat similar to that of working out a Chinese puzzle, relationships are gradually discovered between the lines of a certain element, and finally these lines can be grouped into multiplets.

Not only are the wave-lengths, or frequencies of the spectral lines found to obey the quantum conditions, but the intensities are also subject to laws. Since 1922 the progress in this work has been very rapid until now all of the important lines in the spectra of most of the elements have been interpreted.

In unravelling the puzzle, King's temperature classes have been of the greatest use. Lines of Class I appear at the lowest temperatures, that of the Bunsen flame. As the temperature is raised, first through the electric furnace and then at the higher and higher temperatures of the arc and spark, lines of Classes II, III, IV and V successively appear and the lines of the earlier classes become stronger and stronger. Class V lines appear in the arc only. Conversely, at lower and lower temperatures, all of the lines become progressively weaker and lines disappear in groups. The last lines to disappear, those that are visible even at the lowest temperatures, are called *ultimate lines.* These correspond to transitions to or from the lowest level. Lines from higher levels form subordinate series.

With increase of temperature the energy to which the atoms are subjected increases. The rate at which atoms are ionized increases with the temperature. If the electrons liberated from the atom by ionization could be kept from combining with some other atom, then the process would go on until there were no more neutral atoms left. However, the liberated electron soon meets another ion and combines, again forming a neutral atom. Evidently the relative numbers of neutral and ionized atoms depend on conditions. High temperature increases the rate at which atoms are ionized while on the other hand high density, where atoms are crowded together, increases the rate of recombination into neutral atoms. Hence it is evident that at a constant temperature, ionization increases in amount with decrease in density or pressure. At the same time it is evident that at the same temperature and pressure those elements can give up an external electron most freely when the energy required for this process, the ionization potential, is the least. Or in other words, the greater the ionization potential, the less the degree of ionization.

As already stated, from a knowledge of heights to which vapors ascend in the chromosphere, Saha was able to show that close to the photosphere where the temperatures changed but little, increase of height meant decrease of density and pressure and a greater facility whereby atoms became ionized. Hence with the lines of the spectra grouped into multiplets for which excitation potentials are known, we are enabled to explain quite simply most of the problems connected with the chromosphere.

Let us take as an example, the interesting element radium in order to see whether or not it is in the sun. Twenty years ago there was much discussion over the question, Dyson having found flash spectrum lines due to radium, while Evershed and Mitchell took the opposite stand. The strongest lines of the principal series of enhanced radium are at 3814 A and 4682 A, while three lines belonging to the diffuse series are at wave-lengths 3649 A, 4340 A and 4436 A. These are the *strongest* lines of $Ra+$, and if radium is in the chromosphere, we should unquestionably expect it to display its presence by these wave-lengths. As shown in *Popular Astronomy, 21*, 321, 1913, each of these five lines is already satisfactorily identified by coincidences with lines in Rowland's Tables, without invoking radium as a source. If therefore, we are to prove that radium is in the chromosphere, it will be possible only as the result of flash spectrum photographs with much greater dispersion than those taken up to the present.

The case of the element *Mg* is specially interesting for the reason that it is one of the alkali metals and is found in Column I of the periodic table immediately above the element *Ca*. The strongest lines of the solar and chromospheric spectra are the H and K lines of enhanced calcium. Unquestionably *Mg* is very abundant in the sun, but why then are there no very intense lines of ionized *Mg* found in sun and chromosphere? The reason is not that the ionization potential is too high, for the values of *Ca* and *Mg* differ little, being 6.1 and 7.6 volts, respectively. In the chromosphere appear the three lines of the well-known *b*-group in the green, there is a triplet in the violet at 3829, 3832 and 3838 A, all ascending to heights of 6000 or 7000 km. Farther to the violet is another triplet at 3329, 3332 and 3336 A. Each of the three triplets has an excitation potential of 2.70 volts. In addition there are many singlet *Mg* lines in the chromosphere of the higher excitation potential of 4.33 volts. These lines of *Mg* are all of them much more prominent than the corresponding lines in the spectrum of *Ca*. Hence it appears certain that *Mg* is a more prominent, and thus a more abundant element in the sun than *Ca*. But where are the ultimate lines of enhanced *Mg* corresponding to the H and K lines of enhanced *Ca?* The answer has already been given. They are far in the ultra-violet, at wave-lengths 2795 and 2802, and on account of the absorption in our terrestrial atmosphere they cannot be photographed either in the sun or in the chromosphere. The only line of $Mg+$ of any importance in either solar or stellar spectra is the line at wave-length 4481 A. In the flash spectrum this line exists but it is comparatively weak and is detected only to a height of 400 km. The reason is now clear. This line has a high excitation potential of 8.82 volts, it does not appear at all in the arc spectrum. Hence its presence only in the spectra of the hotter stars is readily explained.

Saha's theory thus interprets in a beautifully clear manner the systematic differences between the flash spectrum, the solar spectrum and the sun-spot spectrum. It goes much further, however, and furnishes the causes of the progression in type of the stars from the red stars of class M to the early types of B and O. Lockyer was the first to call attention to the change in the appearance of the lines H and K, very faint or even missing in late M stars, with a maximum intensity in the solar, or Go stars, and becoming faint again in early B stars and disappearing in certain O stars. Lockyer's interpretation, one of temperature only, was unsatisfactory. The hydrogen lines have their maximum at type Ao and are less intense in both the earlier and later types. The lines of neutral helium appear only in the stars of very early type, while the 4686 line and the Pickering series due to enhanced helium are found only in still earlier types. The conditions of appearance and disappearance of spectral lines due to ionization are calculable, and it has thus been possible to assign temperatures to stars of different types which are in substantial agreement with those derived from other lines of research. All of the difficulties have not been entirely cleared away, but there has been a great step forward.

The theory of ionization has shown the essential unity of astronomy, proving not only that the sun is a typical star but also that a study of the stars can shed much information on solar questions. The great problem of astronomy, the evolution of the stars and the structure of the universe, can find their complete solution only through an intimate study of the ultimate constitution of matter. The size of an atom can be deduced from observations of the gigantic stars. The structure of the atom and the theory of ionization are unquestionably among the most important problems of present-day physical science.

CHAPTER XVIII

HEIGHTS AND RELATED PROBLEMS

THE conclusions drawn from eclipse spectra that strong lines originate at higher levels above the sun's surface than do lines less intense have very important consequences in the problem of the period of rotation of the sun. The determination of this rotation by spectroscopic methods is one of the very greatest perplexity and still awaits an adequate solution. The information drawn from direct observation of sun-spots was positive enough as far as it went. At the equator on the sun, spots take about twenty-five days to make a complete circumference, while at latitudes north and south, the sun rotates more slowly, a day longer being required for a spot at 30° than at the equator. On the whole, the information from spots furnishes contradictory conclusions since the individual spots have peculiar motions of their own which are not representative of the general surface of the sun. Few spots appear at distances greater than 40° from the sun's equator, and manifestly it is impossible to determine the law of rotation in high latitudes where observations on spots are impossible.

With the application of the spectroscope and the photographic plate to the problem, it was confidently expected that all difficulties would disappear, since the measurements were henceforth to be made on sharp and well-defined Fraunhofer lines. All that seemed necessary for a complete solution of the problem was to utilize a spectroscope of sufficiently high dispersion so that the shift in wave-length due to rotation could be measured with an adequate degree of precision. The great advantage of the spectroscopic method was that the observations might be carried out at any time without waiting for the appearance of a spot. A still greater advantage, however, was that observations were not confined to the zones of sun-spots, but could be pushed even to the sun's poles. At the solar equator, the eastern limb advances at the rate of 2.0 km (1.2 miles) per second, while the western limb recedes at the same rate.

From observations of spots, several different empirical formulas have been devised to represent the rotation of the sun at different latitudes. Chief among them may be mentioned those of Carrington and Faye. The latter has usually been regarded as the most satisfactory formula, and to it spectroscopic observations closely conform. Faye's equation may be readily adapted for spectroscopic work and takes the form [1]

$$v + v_1 = (a - b \sin^2 l) \operatorname{Cos} l$$

where v is the velocity in the line of sight deduced from the actual observations; v_1 is a correction allowing for the orbital revolution of the earth so as to convert synodic periods to sidereal; l is the latitude of the region on the sun under investigation; and a and b are velocities measured in kilometers (or miles) per second. It has been assumed that a and b are constants, or, in other words, that the sun's rotation is not varying. If b is equal to zero, then the sun would rotate like a solid sphere with equal angular rotations in all solar latitudes.

The first spectroscopic observations for determining solar rotation are more than forty years old, and were made by Dunér of Upsala. Two lines were chosen in the red and their wave-lengths were compared with terrestrial lines which consequently show no rotational shift. The measures were carried to within fifteen degrees of the poles of the sun. Observations have been in progress continuously for more than thirty years. The values for the first decade of intensive work have furnished [2] the basis for a detailed study of the problem.

After the expenditure of so much time and energy on this research it must be confessed that the results attained

[1] Newall, *Monthly Notices, R. A. S., 82,* 101, 1921.
[2] St. John, *Publ. Astron. Soc. Pacific, 30,* 319, 1918.

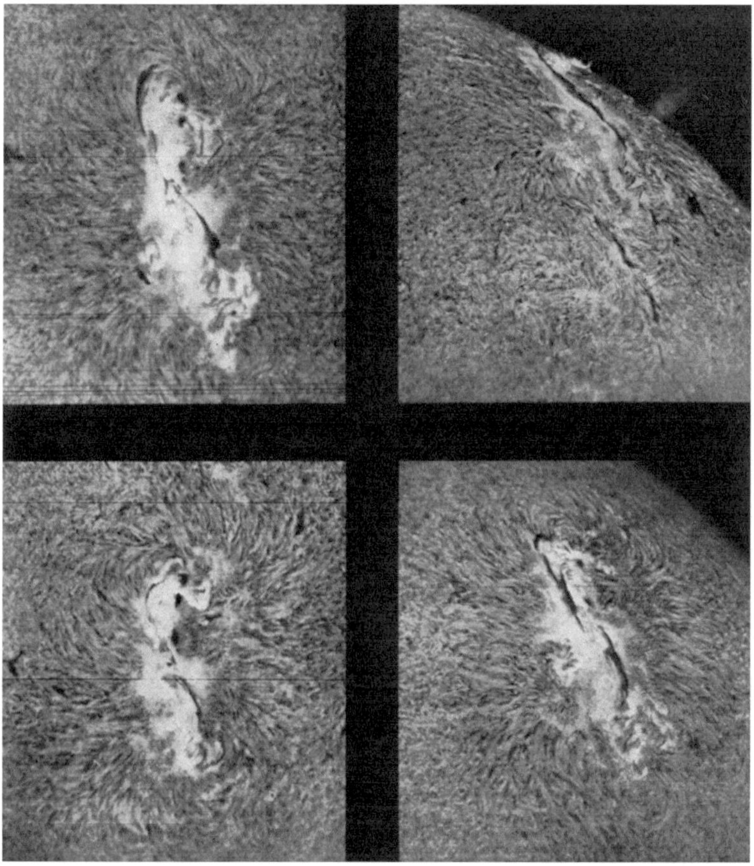

Sun-Spot Group and Prominence Photographed at Mt. Wilson
These photographs taken in the red light of hydrogen show exquisite detail.

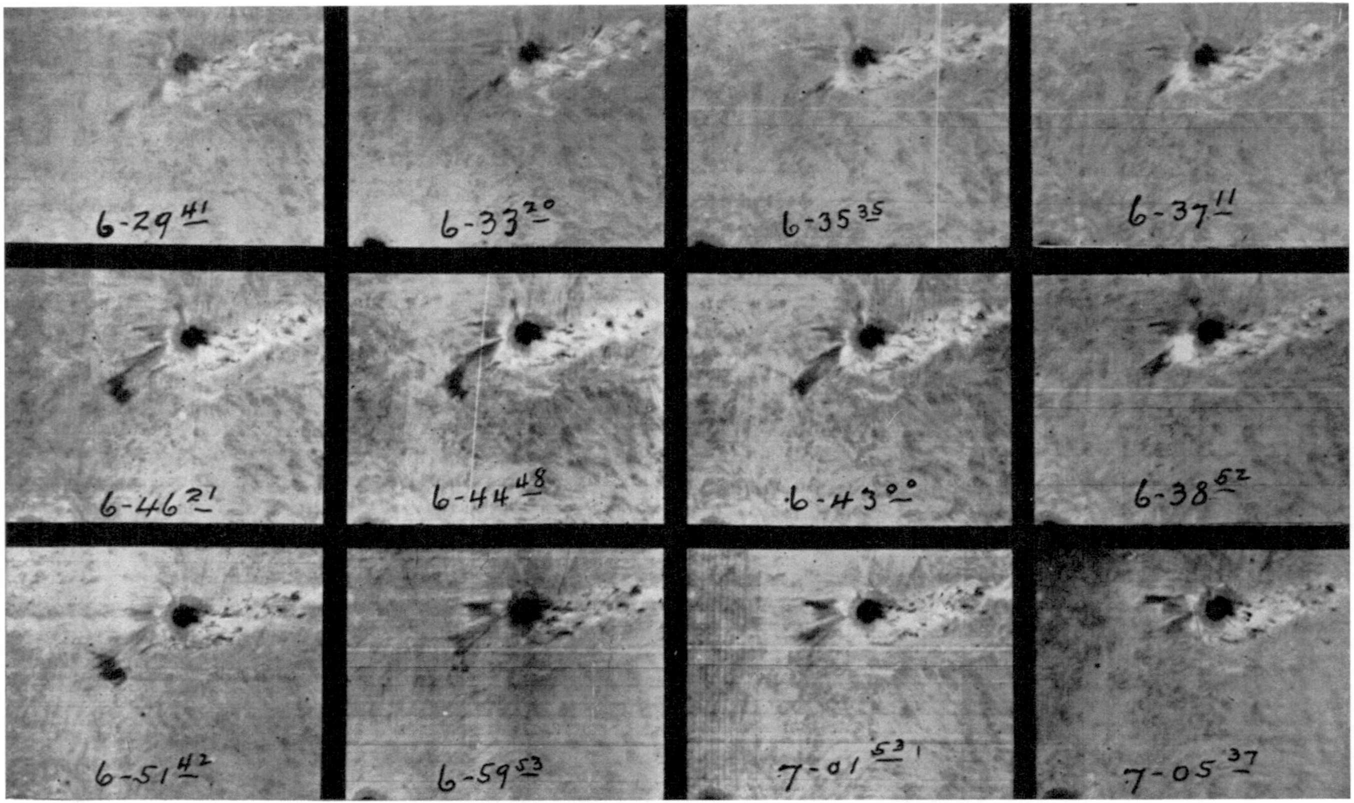

Mt. Wilson Photographs of Prominence Projected on Sun's Disk

Dark region to left of spot is the prominence showing rapid changes taking place at the Greenwich times given.

are a grave disappointment. A glance at the velocities at the equator given in the table will reveal values ranging in size from 1.86 to 2.08 km per sec. The measures were carried out very carefully, using every precaution to free them from er-

LINEAR VELOCITY OF SOLAR ROTATION AT THE EQUATOR

Observe	Location	Velocity km/sec.	No. of Lines	Region	Date
Dunér	Upsala	2.08	2	6301–6302	1900.5
Halm	Edinburgh	2.04	2	6301–6302	1904
Adams	Mt. Wilson	2.06	20	4196–4294	1907
Adams	Mt. Wilson	2.05	22	4196–4291	1908.5
Storey and Wilson	Edinburgh	2.08	10	6280–6318	1909
Plaskett, J. S.	Ottawa	2.01	19	5506–5688	1911
Plaskett, J. S.	Ottawa	2.02	15	4196–4291	1911
De Lury	Ottawa	1.97	19	5506–5688	1911
Hubrecht	Cambridge	1.86	40	4299–4400	1911
Plaskett, J. S.	Ottawa	2.01	27	4250 and 5600	1911–12–13
Schlesinger	Allegheny	2.00	20	4058–4276	1912
Evershed and Royds	Kodaikanal	1.95	—	3906 and 5624	1913
Plaskett, H. H.	Ottawa	1.98	12	5576–5628	1913
St. John and Ware	Mt. Wilson	1.94	35	4123–4338	1914
Plaskett, H. H.	Ottawa	1.95	5	5900–	1915
St. John and Ware	Mt. Wilson	1.94	26	5018–5316	1914–18
St. John and Ware	Mt. Wilson	1.95	7	6265–6337	1916–17

rors. Each of the results in the above table is the mean of a very great number of observations on many different spectrum lines (in one case numbering forty), and it would seem as a consequence, that the measures should have a high degree of precision with the final values entirely free from systematic errors. Taken as a whole the measures seem to prove conclusively that during the two decades of observation the sun's speed of rotation has gradually diminished, showing a total change of five per cent. Before accepting this result as the truth, we must not forget the fact that the wave-lengths of lines in the spectrum have not the constant values originally supposed to exist; and to the many variations already known, eclipse observations add another, viz., that the stronger spectrum lines occur at higher levels in the sun and must therefore display greater velocities of rotation. It is manifest that if the observations of one observer are to be compared with those of another, it will be necessary to subject the lines to a number of refinements as follows: (1), Only lines with well-determined wave-lengths should be employed, they should not exhibit any "pole-effect," and to be representative lines they should not be "enhanced;" (2), lines of like Rowland intensities only should be used; (3), lines of the same region of the spectrum only should be utilized since we are enabled to see into the sun to greater depths in the violet than in the red, this being due to the scattering of light. These conditions will limit very materially the choice of available lines.

Spectroscopic observations are ordinarily made at both ends of a solar diameter, the differential measures giving twice the value of the displacement. In order to eliminate any local effect at the limb due to the near presence of spots, faculae or filaments, it would be well to compare wave-lengths at the sun's edge with those at the center. The investigations of St. John [1] show that the wave-lengths of lines at the center of the sun are "constant to a surprising degree of accuracy." Newall recommends the method developed at Cambridge, of making the measures of displacements between east and west points on fixed chords, parallel to and equadistant from the projected axis of solar rotation, the east and west points being chosen in the same heliographic latitude, in either northern or southern hemisphere. This method is essentially one of comparing the law of solar rotation with the simple law of uniform rotation of a solid body. The method has some advantages over that ordinarily in use, and some disadvantages. The peculiarity of this method is that the spectra are taken on the bright parts of the sun's surface and not near the fainter limb. The work of De Lury has shown that wave-lengths determined at the limb are subject to slight uncertainties due to the superposition of the sky spectrum. If the photographs are taken through thin haze, the wave-lengths may be displaced by amounts depending on the strength of the solar lines investigated.

The discussions by Newall [2] and by Halm [3] of these spectroscopic observations have led to interesting conclusions. The method usually followed by spectroscopists is to secure

[1] *Mt. Wilson Contributions*, No. 223, 1922.
[2] *Monthly Notices, R. A. S.*, *82*, 101, 1921.
[3] *Ibid.*, *82*. 479, 1922.

observations on as many days, at as many different solar latitudes as possible. When sufficient material is obtained, the velocities in the line of sight being found, they are then subjected to a discussion by least squares in order to determine the best possible values of the constants *a* and *b* in Faye's formula above. The different values of *a*, the velocity at the equator, obtained from various series of observations are listed in the table.

It might be said, in passing, that if ever in the history of spectroscopic work positive and conclusive results were promised by any piece of astronomical research, the chances were very great that such results could be positively assured beforehand by the investigation on the rotation of the sun by spectroscopic methods.

In the series investigated by the different observatories, various spectral lines and regions of wave-lengths were employed. In spite of the possible sources of error to which the measures were subject, it would seem, if more than a dozen or fifteen lines were measured in a certain series, that any local peculiarities of a single line should be certainly averaged out. From the nature of the problem, it seemed altogether probable that lines of approximately the same average Rowland intensity would necessarily be chosen by the different observers. Hence it might well have been expected that simultaneous observations of the sun made from different observatories would thoroughly agree in furnishing the same value of the solar rotation. How ideas have been changed in the past two decades regarding the constancy of wave-lengths! In comparing the longer series of observations, we have the results tabulated below, where are given all of the observations secured since the year 1906 as listed by Halm in *Monthly Notices.* In addition to the individual series secured at Edinburgh, Mt. Wilson and Ottawa, there is given the mean of all the values secured between the years 1901 and 1913, and also the results for the hydrogen line, $H\alpha$, obtained at Mt. Wilson. The velocities are given in kilometers per second for every five degrees of heliographic latitude.

The following facts should be noted regarding the quantities in the table: (1), the values derived at Edinburgh agree perfectly with those taken almost simultaneously at Mt. Wilson; for among the seventeen separate linear velocities, only two differ as much as 0.02 km per second, showing a remarkable accord; (2), the three series at Ottawa are very consistent throughout all latitudes; (3), the three Mt. Wilson series found in the third, fourth and fifth columns agree

LINEAR VELOCITIES OF ROTATION OF THE SUN

Heliographic Latitude	1906.3 Ed.	1906.4 Mt. W	1907.0 Mt. W	1908.5 Mt. W	1911.5 Ott.	1912.5 Ott.	1913.5 Ott.	1901-13 Mean	1908.5 $H\alpha$
0°	2.03	2.05	2.07	2.06	2.01	2.00	1.98	2.04	2.11
5	2.02	2.03	2.04	2.04	1.98	1.98	1.96	2.02	2.09
10	1.98	1.99	2.00	2.00	1.95	1.95	1.93	1.99	2.05
15	1.92	1.94	1.94	1.94	1.89	1.89	1.88	1.93	2.00
20	1.86	1.87	1.86	1.86	1.82	1.83	1.82	1.86	1.93
25	1.78	1.78	1.77	1.76	1.74	1.75	1.74	1.77	1.83
30	1.69	1.68	1.66	1.65	1.64	1.64	1.62	1.66	1.73
35	1.58	1.57	1.54	1.54	1.53	1.52	1.50	1.54	1.61
40	1.44	1.43	1.41	1.40	1.40	1.39	1.37	1.42	1.50
45	1.30	1.29	1.27	1.27	1.26	1.24	1.23	1.28	1.37
50	1.16	1.15	1.13	1.12	1.11	1.10	1.09	1.14	1.22
55	1.02	1.01	0.99	0.97	0.96	0.95	0.95	0.99	1.08
60	0.88	0.87	0.85	0.81	0.81	0.79	0.80	0.84	0.93
65	0.73	0.73	0.71	0.67	0.67	0.63	0.65	0.70	0.78
70	0.60	0.59	0.56	0.53	0.52	0.48	0.50	0.55	0.63
75	0.45	0.45	0.43	0.39	0.38	0.35	0.37	0.41	0.48
80	0.30	0.30	0.28	0.26	0.24	0.22	0.24	0.27	0.34

perfectly among themselves between latitudes 0° and 45°, but show a progressive diminution of velocities at all latitudes poleward of 45°; (4), the Mt. Wilson $H\alpha$ values are greater in amount than the Mt. Wilson results on the reversing layer for the year 1906.4, by an amount which is approximately constant and equal to 0.06 km per second; (5), the $H\alpha$ values exceed the velocities of the three Ottawa series by a constant which amounts to 0.11 km per second; (6), the time of observations given in the Edinburgh and in the first Mt. Wilson column (1906.4) agrees fairly closely with sun-spot maximum, while the mean of the three Ottawa series differs little from sun-spot minimum.

On account of the splendid accord between the observations, we seem almost forced to conclude that the rotational velocity of the sun can be determined spectroscopically with great exactitude. The results tabulated above seem to prove most emphatically that at spot maximum the reversing layer revolves with a speed that is greater than the average, while at minimum of spots the rotational velocity is less than the average. This is the conclusion independently reached by Newall and by Halm.

A decade ago, at the 1922 meeting of the International Astronomical Union, the experts were forced to the conclusion that, although there is such a beautiful accord between the measures listed in the above table, it may not be impossible that some of the observations may have been affected by errors, either accidental or systematic in nature, and that in consequence, the difference in rotational values at maximum and minimum of spots may be illusory and not real. For the reasons already stated, systematic errors due to different spectral lines investigated may readily find place in any series of measurements, and the only conclusion to draw from the large differences between Mt. Wilson and Ottawa is that one or both of these series of measurements are unquestionably affected by systematic errors. In fact, investigations made at Mt. Wilson, in which simultaneous observations are made upon the two limbs and the center of the sun, indicate that temporary and local conditions frequently exist in the reversing layer which produce differences of as much as ten per cent in the rotation values obtained by comparing east and west limbs directly. Such results make it obvious that misleading values may be actually derived from a short series of observations; in fact it may be well to ask, along with the Chairman of the Committee on Solar Rotation of the International Astronomical Union, "whether the solar rotation can be determined as definitely as has been thought."

One obvious method of testing whether systematic errors affect the observational values and whether the rotation of the sun is different at sun-spot maximum from that at minimum is for one observatory with fine equipment to keep at the problem for a long stretch of years, taking great care to keep both methods and observers as constant as possible. Hence as early as 1914, a program of observations was inaugurated at Mt. Wilson Observatory by St. John and Miss Ware. Subsequent to the values given in the table above, are the following linear equatorial velocities [1] at the sun's equator in units of kms per second:

1916	1917	1918	1919	1920	1921	1922	1923	1924	1928
1.91	1.94	1.95	1.90	1.90	1.91	1.91	1.89	1.91	1.89

More recently, St. John [2] obtained values in 1929 and 1930. That in the latter year, amounting to 1.95, agreed in size with those of 1917 and 1918. Still more recently, Evershed [3] obtained the value 2.02 in 1931. It should be added, however, that Evershed made no distinction between high- and low-level lines and it is possible there may be a systematic error between his results and St. John's.

Following the method of Adams, measures of rotation have been made on the high-level $H\alpha$ of the chromosphere. At Mt. Wilson in 1907, the value was 2.11 km per sec., in 1919 this had changed to 2.00. However, in 1918, Evershed's results gave 2.07 km per sec. while at Arcetri in 1929 the value 2.07 was also obtained. Again we are faced with the conclusion that variations in instrumental equipment might readily permit different levels to be measured at two observatories with consequent systematic differences in velocities. The simplest method of explaining the large differences in the Mt. Wilson values is to assume that the measures of the bright line of $H\alpha$ in the two years were unconsciously referred to layers at different heights.

With the splendid equipment at Kodaikanal, measures have been carried out on the dark $H\alpha$ line of the Fraunhofer spectrum. The observations in the years 1918–22 were consistent and numerous and a value for the linear velocity at the equator was 2.03 km per sec. In the years 1923–29 the observations were not so numerous and they yielded a slightly increased value of 2.05. The weighted Kodaikanal value for the whole period 1918–29 is 2.03 km per sec. During the

[1] Mt. Wilson Obsy., *Annual Report of Director,* 127, 1928.
[2] International Research Council, *Third Report,* 121, 1931.
[3] *Monthly Notices, R. A. S., 92,* 105, 1931.

same interval at Mt. Wilson the value from different lines was quite constant at 1.90 km per sec. On account of the higher level at which the $H\alpha$ line has its origin, the results of the two great observatories are quite consistent in showing during a fifteen-year period, 1914–28, a solar rotation that appears to be nearly constant.

In the latest annual report (1931) of the director of the Mt. Wilson Observatory, the intricate problem of solar rotation is carefully summed up in the following words: "It has become increasingly evident that the rotation of the sun as determined from observations of the reversing layer is not constant. Observations by Halm at Edinburgh and Adams at Mt. Wilson gave a velocity of 2.07 km per sec. and a minimum period of 24.75 days for 1905.5, while a smaller velocity was indicated by observations at Ottawa, Allegheny and Kodaikanal for the epoch of 1912. For the years 1919–29, the linear velocity averaged 1.90 km per sec., and the period two days longer than in 1905.5. Since 1928 the linear velocity has slowly increased, while the period has decreased by 0.6 day. On the other hand, the velocity derived from long-lived spots has remained practically constant at 2.02 km per sec. for the last thirty years. The lower reversing layer is 200 to 500 km above the photosphere (level of spots). In 1905.5 its linear velocity was 0.06 km per sec. greater than that of the photosphere, and in 1918, 0.10 km per sec. less, thus indicating a streaming of the lower reversing layer relative to the photosphere from east to west during the spot cycle 1901–12, and from west to east during the following cycle. If connected with the direction of whirl in the spot-forming vortex, this systematic difference would indicate a counter-clockwise rotation of the vortex in the northern hemisphere during the cycle 1901–13 and a clockwise rotation during the following cycle. For the cycle beginning in 1923, the linear velocity of the reversing layer might have been expected to increase, and apparently it has increased; but whether it will continue to do so remains to be determined. A correlation between linear velocity of the reversing layer and direction of whirl in the spot-forming vortex would be of great interest in its bearing on the sign of the charge in the spot vortex."

A summary[1] is given for other solar phenomena observed for the rotation of the sun. The number of degrees per day at or near the sun's equator shown from observations are: sun-spots, 14.39; faculae, 14.49; calcium dark filaments, 14.45; reversing layer, 14.27; *Hα* (dark marking) 14.40; *Hα* (chromosphere, Adams), 15.00; K (prominences) 17.1. To find how the rotation periods change at different heliographic latitudes, one should also consult Abetti, *Handbuch der Astrophysik, 4,* 158, 1929.

During the early years of the work at Mt. Wilson, the *Hα* line had shown such different rotational values from the other lines that it seemed advisable to extend the investigation to other solar phenomena. What values would be given by flocculi? To photograph in the red required special plates sensitive to this region. When the flocculi plates were taken, the results were a great surprise to Hale and Adams, for it was evident at a glance that the hydrogen flocculi differed from those taken by means of the H_2 line of calcium in several remarkable features.[2] The photographs of hydrogen flocculi already secured showed that they did not share the same retardation in rotation assumed by spots, faculae and calcium flocculi. But the striking differences between the details of the photographs of flocculi in the *Hα* light of hydrogen and in the H_2 line of calcium, when examined side by side, showed that whereas most of the calcium flocculi are bright, those due to hydrogen are dark. Further differences were exhibited by the accentuated definiteness of structure of the hydrogen photographs which show details much smaller in size and with greater distinctness. The details brought forth by the *Hα* line were more marked than those obtained by the other hydrogen lines *Hβ*, *Hγ* and *Hδ*, the reason for the difference being caused by the greater strength of the *Hα* line in chromosphere and prominces and the greater heights attained. A most unusual series of photographs was secured by Hale on June 3, 1908, showing a dark

[1] *Kodaikanal Bulletin,* No. 113, 1931.
[2] *Mt. Wilson Contributions,* No. 26, 1908.

hydrogen flocculus which was actually seen drawn into a sun-spot vortex, the remarkable change taking place within the brief space of time of ten minutes. The day previous, the location of the flocculus was evident from local whirls. On the day following the disappearance of the mass of *cool* hydrogen into the spot, photographs were secured showing eruptions in the neighborhood of the spot due to *bright* and *hot* hydrogen gas. Apparently, therefore, a sun-spot is a vortex somewhat resembling a terrestrial cyclone. Relatively cool matter in the gaseous form, floating high above the solar surface, is sucked into the vortex with a whirling motion. After sinking into the interior of the sun, the cool gas is heated, and later in the heated state makes its reappearance outside of the limits of the spot, but in its immediate neighborhood. Meanwhile, the researches of J. J. Thomson and others had shown that electrified particles, both positively and negatively charged, must occur in vast numbers in a hot gaseous body like the sun. If there is a preponderance of positive or negative charges, their rapid rotation must give rise to a magnetic field. In other words, a sun-spot by its vortex motion appears to create its own magnetic field.

The most promising method of attack appeared to be the Zeeman effect. When a luminous vapor is subjected to a magnetic field by being placed between the poles of a powerful magnet, an effect is produced on the lines of the spectrum. If the radiation is observed *along* the lines of force, the spectrum lines appear in most cases as doublets, having components circularly polarized in opposite directions. It was found that different lines of the same element are affected to a different degree, and that the distance between the components of a given doublet is directly proportional to the strength of the field. In a field of moderate strength the distance between the two components of a doublet may not result in complete separation, the lines being merely widened, while in other lines, which are exceptionally sensitive, the separation may be complete. As early as 1892, Young with the Princeton refractor, and later W. M. Mitchell, observed lines due to iron in the sun existing as single lines in the spectra of the photosphere but double in spot spectra.

If, therefore, the observers at Mt. Wilson were to endeavor to find a magnetic field in sun-spots, the line of attack was clearly outlined. First of all, there was necessary a large image of the sun, possible only by the use of a telescope of great focal length so that the surface of the sun could be examined in regions surrounding the spot. The solar light must then be examined by a spectrograph of such great power that the separation of the doublets showing the Zeeman effect might be as great as possible. To interpret the measures of the solar photographs, investigations in the laboratory unquestionably would be necessary. Fortunately for the development of astronomical research, the resources of the Carnegie Institution were ample to provide, on the top of Mt. Wilson, the 60-foot and the 150-foot tower telescopes with powerful spectrographs attached, and to equip in Pasadena a physical laboratory with the necessary forms of refined apparatus and Hale's spectrohelioscope. As the result of twenty-five years of investigation, the scientific information garnered regarding spots and the general magnetic field of the sun has been very startling and most important.

Observations are secured in the second order spectrum of the 75-foot spectograph attached to the 150-foot tower telescope. In front of the slit of the spectograph[1] is placed a Nicol prism in combination with a "quarter-wave plate" built up of mica strips 2 mm wide, mounted so that the principal sections of successive strips make an angle of 45° with the slit and 90° with each other. If the long axis of the Nicol is placed parallel to the slit, the mica strips will alternately extinguish the red and the violet members of a doublet, and the photograph will have a dentated appearance shown in the illustration facing page 332. It is hardly to be expected, however, that the extinction of either component will be complete. A partial extinction of the red or the violet component has the effect of shifting the maximum of brightness of the doublet towards the red or the violet of the average position, depending on whether the red or the violet

[1] Hale, *Nature, 113*, 105, 1924.

component is the stronger. When the two components are not entirely separated, the lack of uniformity in the lines causes a displacement which depends on the relative strength of the two components. This unsymmetrical character of the lines renders the accurate measurement of the displacements one of the very greatest difficulty. In fact, even with the great scale of the Mt. Wilson photographs, reliable measures are possible only on a comparatively small number of lines.

The best line for the investigation of sun-spots is one in the red at 6173.553 A. This line is split into three components, the two outer ones being elliptically polarized in opposite directions, the inner component plane polarized. If the red component of the line is transmitted by the compound quarter-wave plate and the violet component extinguished, then the polarity of the spot is N, north-seeking or positive; while if the violet component is seen the polarity is S or negative. Since the distance between the components is proportional to the strength of the magnetic field, the direct measurement of a photograph by means of a micrometer readily furnishes the field-strength. In addition, the angle between the magnetic lines of force and the solar surface can be determined in all parts of a spot.

Ever since the time of Galileo it has been noticed that spots have a tendency to appear in pairs. In fact, the observations at Mt. Wilson show only that in ten per cent of all cases are the spots single or unipolar in appearance. Sixty per cent of spot groups are distinctly bipolar, while the balance of thirty per cent show a tendency towards the bipolar type, exhibiting calcium or hydrogen flocculi following or preceding a single spot. In many bipolar groups, one of the members, usually the preceding or western spot, is much larger than the other. In nearly all cases it has been found that the eastern member of a spot pair differs in polarity from the western member. When the group is formed of a number of spots small in size but without any dominant members, then the groups of spots at opposite ends of the stream are usually opposite in polarity.

According to the opinion of Hale, a spot is a vortex in

which the solar gases are expanded, and consequently cooled, by centrifugal action. If this cooling is sufficient, the gases appear darker over the vortex because of their decreased radiation and increased absorption at reduced temperatures. In the cooler regions, chemical compounds are formed, such as titanium oxide and certain hydrides, and their presence is recognized by the existence of characteristic bands in sun-spot spectra. It is quite possible, however, that the cooling may not have advanced sufficiently far for the spot to become visible as a darkened area on the surface of the sun, but nonetheless it has been found that the magnetic effects persist even after the spot ceases to be visible and also appear before the darkening of the spot becomes sufficiently great to bring it into view. Thus have been discovered "invisible spots," with the result that the life history of a spot is increased by lengthening out the times of observation both before and after the spot becomes visible to the eye.

Ever since the beginning of the work at Mt. Wilson in the year 1908, many thousands of spots have been observed with great care. It has been found that all spots, independent of size, contain magnetic fields, but that the field-strengths up to a certain maximum increase with the diameter of the spot. Since the spots show themselves to be gigantic vortices, a natural question to ask is whether the directions of the whirls follow a law similar to cyclones and tornadoes on our earth which move in directions contra-clockwise in the northern hemisphere but right-handed in the southern hemisphere. Little is known regarding the sign of the charge dominant in sun-spots, but if it is assumed that these charges are the same in all solar vortices then the observed polarities will give the directions of the whirls. It has been found that preceding spots in the northern and southern hemispheres of the sun exhibit opposite polarities and consequently opposite directions of whirl.

What is the law underlying the magnetic changes in sun-spots? From the time of the inception of the work at Mt. Wilson in June, 1908, to the minimum of spots in December, 1912, there were a total of twenty-six groups observed. In seven of these groups that appeared in the northern hemi-

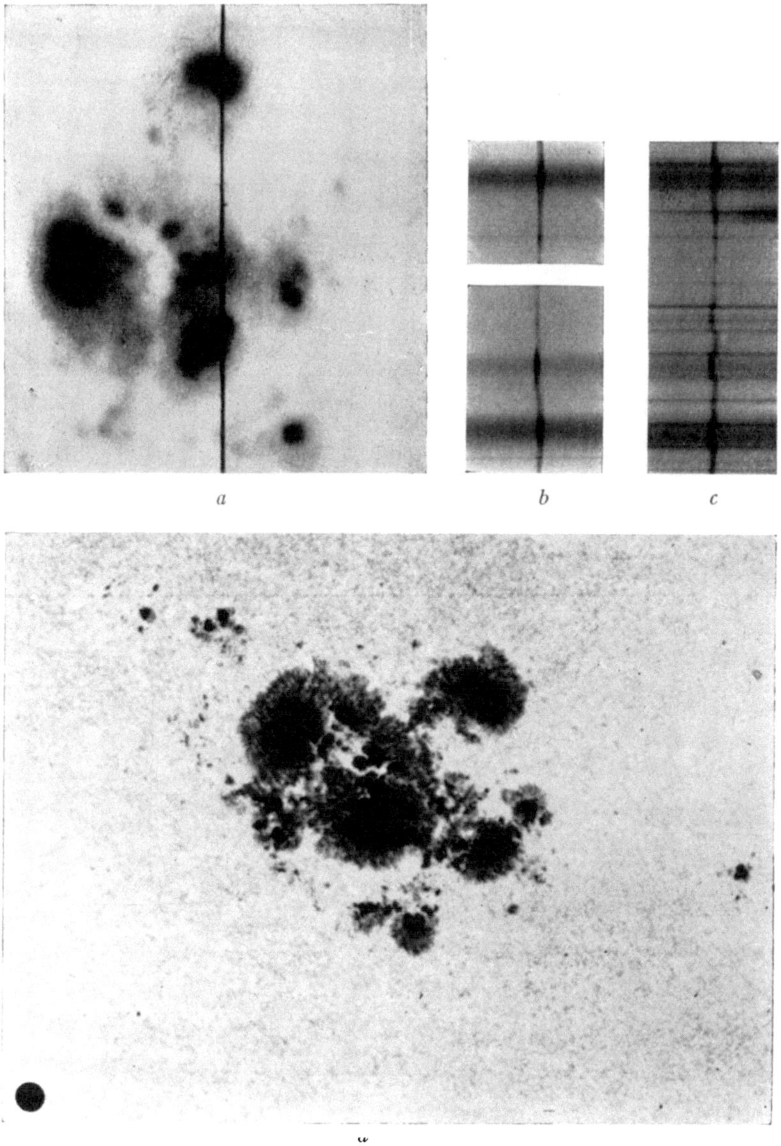

ZEEMAN EFFECT IN SUN-SPOTS

a, b, c Photographic observations of a multipolar sun-spot group on August 8, 1917 at Mt. Wilson Observatory; *b* and *c* photographed with Nicol and quarter-wave plate show the Zeeman effect with the iron triplet 6173 A. *d* An enlarged direct photograph of the same spot group on August 8, 1917 with the 60-foot tower telescope. (The black disk in corner shows the size of the Earth.)

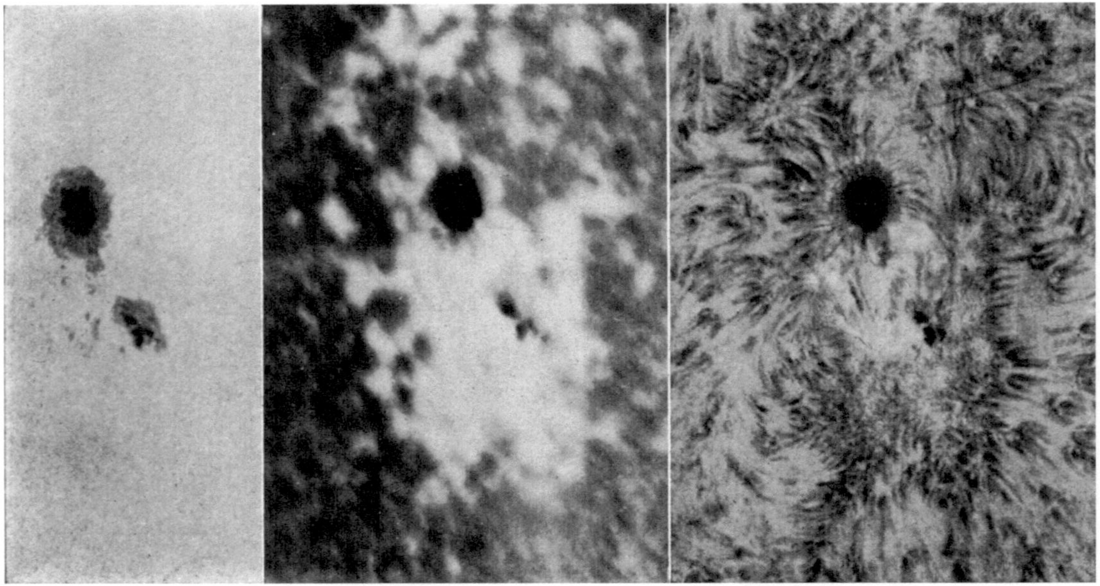

BIPOLAR SUN-SPOT, JULY 12, 1914 (MT. WILSON)

From left to right: Direct photograph; K_2 spectroheliogram; $H\alpha$ spectroheliogram.

sphere, the polarity of the preceding member of the pair, characterized by the violet member of the 6173-line, was south-seeking or negative, and that of the following spot was north-seeking or positive. In the southern hemisphere, for seventeen groups the polarities were the reverse of those in northern latitudes, N or positive for the preceding member, and S or negative for the following member of the pair. Two southern spots out of the total of twenty-six observed showed opposite polarities to those expected, thereby furnishing exceptions to the rule.

The spots which were observed were the dying members of the spot cycle and had an average latitude of 9°. Would

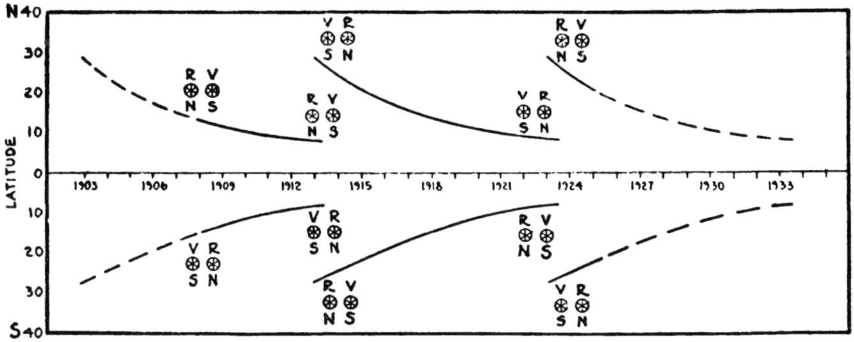

Fig. 7 Sun-Spot Polarities observed at Mt. Wilson between 1908 and 1925. The curves represent the change in latitudes of the spots; the preceding spot is shown on the right.

the spots appearing in the new cycle, in high northern and southern latitudes continue the same persistence of polarities that the spots had already exhibited? There appeared no reason for a change, but when the observations were obtained they furnished a great surprise in showing a complete reversal of polarities from those observed in the previous cycle.

During the eleven-year period which closed in 1923 more than two thousand spot groups were observed at Mt. Wilson. Unipolar and bipolar groups, with only four per cent of exceptions, continued to have the polarities reversed from that of the preceding cycle and showed south-seeking or negative polarities in the northern hemisphere, and north-seeking or positive polarities for the preceding spots of the southern hemisphere.

Observations after the spot minimum in 1923 were anxiously awaited to know whether a reversal would again take place with a resumption of polarities that had been the rule in the second preceding cycle. The answer was a very conclusive one, the reversal which was *now* expected had actually taken place.

The law of sun-spot polarities is expressed in the two appended figures. At sun-spot minimum the new cycle originates in high northern and southern heliographic latitudes.

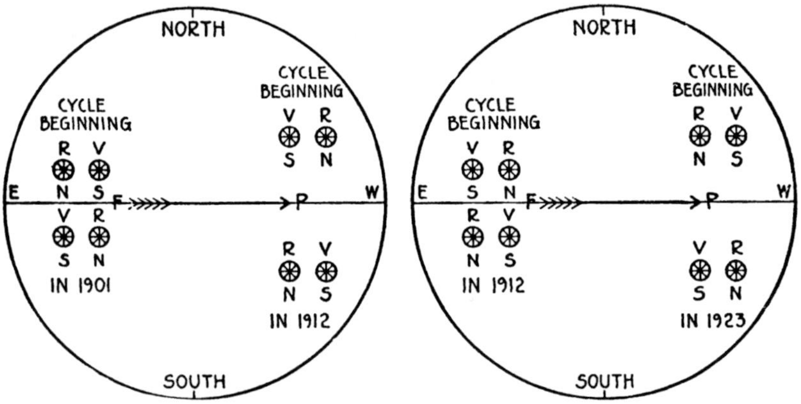

FIG. 8 Sun-Spot Polarities at Minimum of Solar Activity. Two zones in each hemisphere, in which the spots are of opposite magnetic polarity, exist for about two years near spot minimum.

As the cycle progresses towards maximum of spots they draw closer to the sun's equator but the polarity remains always the same. The high-latitude spots of each cycle have opposite polarities from those of the preceding cycle. A strange condition appears near minimum of spots. In the northern (or southern) hemisphere two groups of spots differing a few degrees in latitude may exhibit opposite polarities, one polarity due to a spot in the waning cycle, the other taking its sign from the cycle just beginning. Since the directions of the whirls of the hydrogen flocculi are not reversed, but, on the contrary, there is a change in the sign of polarity at each spot minimum, it is evident that the magnetic period of spots is 22 years, or twice the 11-year period on which the numbers of spots depend.

It is evident that the general magnetic field of the sun must be much weaker than that exhibited in sun-spots on account of the fact that the sun as a whole rotates much more slowly than the vortex whirlpools of the spots. Moreover, in attempting to secure the magnetic field of the sun as free as possible from any local effect of sun-spots, it is obviously necessary to obtain photographs when the sun is entirely free from active spots. Several series of photographs taken under satisfactory conditions have been secured. The difficulty of measuring these photographs, and of securing these measures free from personal and systematic errors, is patent to anyone who has ever engaged in any astronomical measurements requiring the setting on two close components of a double. Moreover, the size of the quantity to be measured is only 0.001 A. As a matter of fact, van Maanen was the only one of the five Mt. Wilson measurers that engaged in the work whose results had a satisfactory degree of consistency. On account of the difficulties in measuring, lines having an intensity greater than 5 in the solar spectrum and weaker than 0 had to be excluded.

The general summary of the results seems to prove conclusively that a general magnetic field exists in the sun, and that in consequence the sun behaves approximately as a uniformly magnetized sphere, with the magnetic axis only slightly inclined to the solar axis of rotation, and with a polarity corresponding to that of the earth. Forty-six spectral lines were investigated, and of these, 30 lines due to *Fe, Cr, Ni, V* and *Ti* show displacements. The strength of the magnetic field determined for each line showed a correlation between field-strength and line intensity, the stronger fields being connected with lines of smallest intensity. Eclipse spectra reveal the information that the weakest solar lines originate at the lowest depths. Since it was possible to measure, for the determination of the strength of field, lines of a solar intensity 5 or less, it is manifest that lines only in a very shallow layer, less than 450 km in depth, can show the influence of the sun's magnetic field large enough to be detected by the present method of attack. The period of rotation of the sun's magnetic axis was found to be 31.52 days. No explanation can be given for this peculiar value of rotation which differs in such marked degree from the equatorial value derived from sun-spots. In fact, no adequate explanation is yet forthcoming to explain the cause of the sun's magnetism.

It is now generally conceded by all scientists that the sun is continually sending off a vast stream of negatively charged particles, or electrons. It is also generally conceded that the presence of spots on the sun is evidence of great solar activity. If the spot is large, and the activity consequently great, the streams of electrons directed toward the earth may reach our upper atmosphere in vast numbers. In the rarefied conditions that must exist in the upper atmosphere, the air becomes ionized. This ionization causes the electromagnetic display known as northern lights, or aurora borealis. As a result of the electrification of the atmosphere, currents are induced in the earth, which may at times be so strong as to seriously interfere with the sending of messages over the telegraph lines or submarine cables. As a further manifestation of the currents flowing through the earth, the navigator's compass may be affected with the result that the north end of the needle may point several degrees from the magnetic north. This leads to a so-called "magnetic storm." The display of a brilliant aurora is very frequently accompanied (and caused) by a spot central on the face of the sun. On May 13, 1921, the spots near the sun's center caused an unusually gorgeous display of aurora visible practically all round the world. The telegraph lines were seriously hampered in the sending of messages. While the aurora was at its height, one of the Atlantic cables connecting Europe and America was burnt out. Whether this was caused by the earth's currents, or was but a strange coincidence, has never been fully determined.

Theoretically, sun-spots should possess an electric as well as a magnetic field. Although observations have been made at Mt. Wilson and elsewhere, no conclusive evidences of a Stark effect have been recorded.

In an important paper presented[1] at the National Academy of Sciences, Adams and Nicholson discuss *The Nature of the Solar Cycle* in the following manner. "The eleven-year period may vary between nine and fourteen years and the amplitude of the cycle by about 50 per cent of its average value. Period and amplitude are apparently unrelated. It seems probable that both the quantity and quality of solar radiation vary during this cycle and many attempts have been made to correlate terrestrial phenomena with sun-spots. Definite correlations have been found with the variations in terrestrial magnetism and its related phenomena. There is evidence of a slight correlation between sun-spots and atmospheric temperature in certain regions on the earth and with other factors of weather and climate for limited regions and for limited time intervals. These correlations are so uncertain that, in the majority of cases at least, predictions on them have very little weight."

[1] *Science*, 75, 594, 1932.

CHAPTER XIX

HEIGHTS AND DISTRIBUTIONS OF VAPORS

NEW information gleaned in the past decade about the structure of the atom has revolutionized the method of attacking solar problems. Practically all of the prominent lines in the spectrum of the sun have been assigned to multiplets with known excitation potentials, the arbitrary intensity scale of Rowland has been submitted to calibration tests which have revealed that the intensities depend on the number of atoms engaged in the formation of the spectral lines. From the weakest lines perceptible in the solar spectrum of intensity -3 (or 0000) to the strongest *Fe* lines at wave-length 3720 and 3735 of intensity 40, the number of atoms involved increases about one million times.

A person having no knowledge of the theory underlying multiplet groups would not advance very far in the practical operation of correlating heights in the chromosphere with intensities either in sun or chromosphere before the fact would be forced upon his attention that generally the lines of greatest intensity reach the greatest heights, and moreover the intensities and heights for any element are greatest for the multiplets of lowest excitation potential. The best element for a study of the explanation of these correlations is unquestionably neutral *Fe*. This element has been assiduously observed in the laboratory and very exact wave-lengths are known. *Fe* is very rich in lines which have been grouped into multiplets with a wide range of excitation potentials. Some of the more important results from the detailed study [1] of this element are given herewith. The lowest atomic level of neutral *Fe* is the a^5D level with average excitation potential of 0.07 volts, the next lowest level is a^5F, with excitation potential of 0.92 volts. The table gives the average heights in the

[1] *Publ. Leander McCormick Obsy.*, *5*, 114; *Astroph. Jour.*, *72*, 146, 1930.

chromosphere attained by lines of different Rowland intensities, with the whole material arranged according to increasing excitation potentials. A total of 789 lines is involved.

EXCITATION POTENTIALS, ROWLAND INTENSITIES AND HEIGHTS*

Excitation Potential	20–40	12–15	10	7–9	6	5	4	3	2	1	0
0.07	2000	1625	900	1100	1075	600	517	428	444	350	.....
	4	4	4	11	4	5	3	7	8	3	.....
0.92	1300	1225	917	767	638	628	504	442	400	333	.....
	7	4	3	15	8	11	12	13	1	3	.....
1.52	1650	1525	1075	875	600	500	435	417	383	.....	.....
	2	4	4	6	2	3	13	9	3	.....	.....
2.18–2.49	.....	.....	775	638	565	524	467	419	403	318	.....
			4	16	24	17	17	27	16	8	.....
2.50–2.99	.....	.....	500	567	537	493	479	433	412	329	300
			1	15	17	31	38	38	24	12	2
3.00–3.49	.....	.....	.....	542	450	447	440	425	364	325	300
				6	9	18	21	26	21	6	1
3.50–3.99	.....	.....	.....	520	400	421	373	387	375	350	300
				5	4	7	15	15	14	5	2
4.00–4.49	.....	.....	.....	400	400	415	390	372	339	321	250
				1	2	10	10	23	32	17	7
4.50–4.99	.....	.....	.....	.....	400	383	400	394	333	300	.....
					4	6	6	9	3	2	.....

* The upper quantity gives the height in kilometers; the lower gives the number of spectral lines involved

The following facts should be noted: (1) The strongest lines of neutral *Fe* in the sun belong to multiplets of lowest excitation potential 0.07 volts. (2) With increase of excitation potential, the maximum intensity of the lines in the multiplets steadily decreases. (3) For multiplets of any given excitation potential, there is a close correlation between intensities and heights. (4) For any given Rowland intensity, such as 6, the heights diminish (vertically in the column) as the excitation potentials are increased. To know the height corresponding to a line of intensity 6, the value of the excitation potential is necessary.

If the material from the flash spectrum on neutral *Fe* is divided at wave-length 4900 A into two groups, we have the information in the following table. The first two horizontal lines of the table give the average heights and average excitation potential of the lines to the violet of 4900 A arranged according to their Rowland intensities. The next two lines give similar data from the lines to the red of 4900 A; and next comes the average for all lines of all wavelengths. The two last lines in the table give the radial motion in the penumbra of sun-spots called the Evershed effect, measured by St. John at Mt. Wilson, the first line giving the value in angstroms and the second line in kilometers per second.

ROWLAND INTENSITIES, HEIGHTS, AND EXCITATION POTENTIALS

0	1	2	3	4	5	6	7	8	9	10	12	15–18	20–40
331	372	390	462	469	501	570	603	720	681	940	971	1325	1593
2.26	2.95	3.01	2.70	2.62	2.54	2.29	2.23	1.82	1.96	1.27	1.50	1.00	0.75
287	308	330	362	405	405	456	514	535	600	633	....	....	....
3.69	3.41	3.64	3.13	3.41	3.16	2.57	2.65	1.82	2.42	2.94	....	....	....
296	332	361	422	450	446	532	576	681	947	898	971	1325	1593
3.33	3.23	3.35	2.90	2.90	2.77	2.39	2.36	1.82	1.98	1.49	1.50	1.00	0.75
.030	.028	.025	.023	.021	.019	.016	.013	.010	.005	.003	.002	.001	.000
2.0	1.7	1.5	1.4	1.2	1.1	1.0	0.8	0.6	0.3	0.2	0.1	0.1	0.0

Early in the year 1909, Evershed announced a remarkable discovery of far-reaching importance in his observations of the displacement of Fraunhofer lines in the penumbra of sun-spots. With the slit of his spectograph placed across the spot, he found that the wave-lengths of lines in the penumbra of the spots were different from the values at the center of the sun. The displacements, which affected practically all of the lines of the reversing layer, were not constant but differed in amount depending on the intensity of the lines investigated. The shift was greater for the weaker lines of the spectrum than it was for the stronger lines. Evershed advanced the hypothesis that the observed displacements are the result of the Doppler effect, and that in consequence, the gases of the reversing layer are in radial motion tangential to the solar surface.[1]

[1] *Kodiakanal Observatory Bulletin,* No. XV; *Kodiakanal Observatory Memoirs, I,* Pt. I.

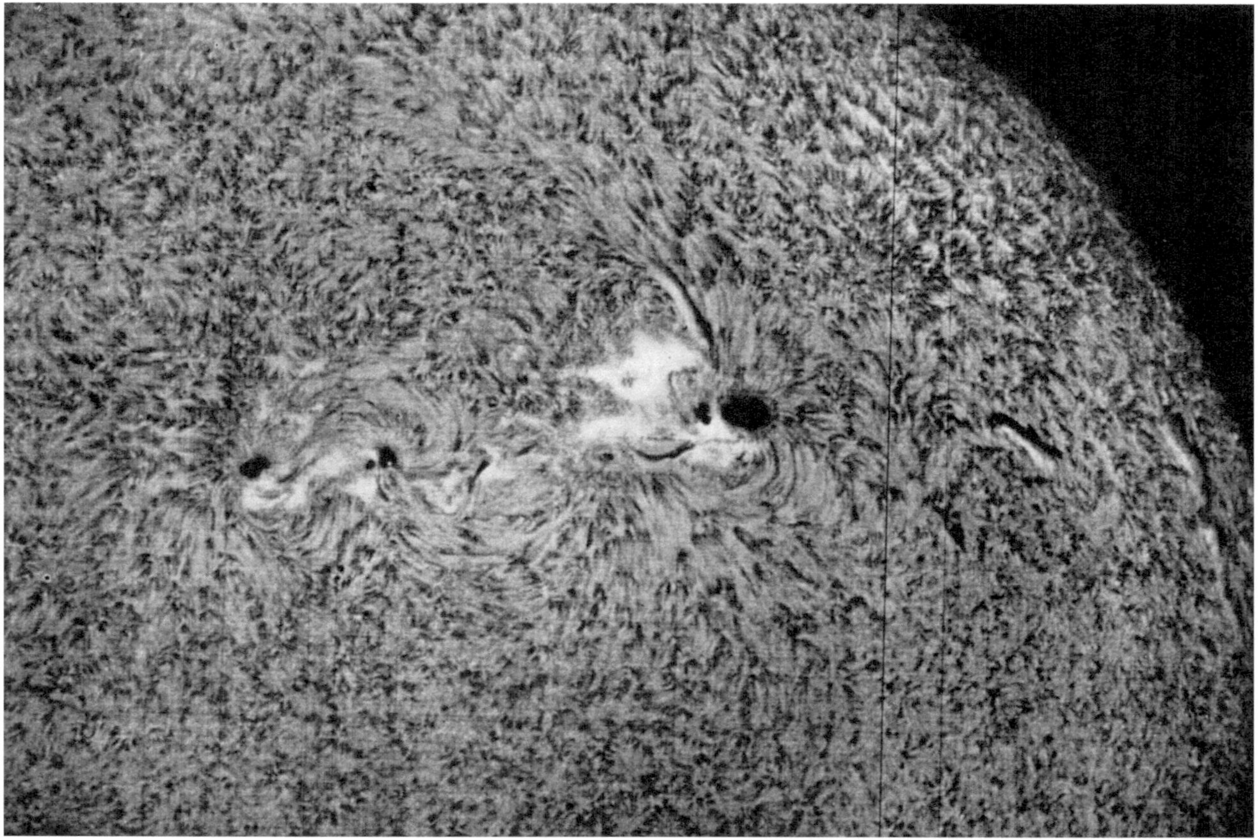

Hydrogen Vortices or Tornadoes above Sun-Spots
Photographed at Mt. Wilson Observatory.

Spectrum of Sun and Spot in the Region of the *b* Lines in the Green

Note the differences in intensities in the spectra. At x is the Fraunhofer spectrum; at y is the sun-spot spectrum; at z are again the two spectra (that of spot in center).

Following the announcement of this important discovery, St. John began an extended series of investigations into the subject, the results of the observations being published in the *Contributions from the Mount Wilson Observatory,* Nos. 69, 74, 88, 348 and 390. The observations were carried out with the 60-foot tower telescope, with the image of the sun 170 mm in diameter, the penumbra of the spots investigated averaging 3.0 mm in diameter. The plates in the violet and green were taken in the third order spectrum, and those in the yellow and red in the second order. For the two cases, the dispersion was 1 mm = 0.56 A and 1 mm = 0.86 A, respectively. Measures were carried out on 506 lines, some of the lines being measured on no less than thirty plates.

The Mt. Wilson measurements, so carefully made by Miss Ware, abundantly verified Evershed's conclusions that the displacements are caused by movements of the solar vapors tangential to the solar surface and radial to the axis of the spot. These motions are none other than the actual flow of the material of the reversing layer *out* of the spots and of the matter forming the chromosphere *into* the spot vortex.

The layers closest to the sun's surface have a motion of translation out of the spot at the rate of two kilometers per second, and this motion becomes less and less at greater and greater elevations until, at a height of about two thousand kilometers, the motion outward of gases from the sun-spot ceases. What happens to the vapors above this level? The information furnished by the investigations of St. John is very definite and apparently admits of no contradiction. From measurements on several elements, it was found that above this level of inversion, the gases of the chromosphere take a motion carrying them *in* to the spot. As greater and greater elevations are reached these movements increase in amount. At the maximum heights reached by lines of the chromosphere, which are attained only by the H and K lines of the element *Ca,* there is a movement of the calcium vapor into the spot at a speed of 3.8 km per second corresponding to a displacement measured by St. John of −0.063 angstroms.

In the following table there is collected the available information from the measures of heights derived from the lines in the flash spectrum and of displacements due to the Evershed effect from 506 lines in the penumbra of sun-spots. The + symbol affixed to an element, as *Ca* +, signifies that the line in the spectrum is enhanced and that it takes its origin from the ionized atom. The conclusions of St. John (*loc. cit.*) are, " In the observation of these velocities we have a methol of sounding the solar atmosphere and of allocating the relative levels of the lines."

MOTIONS OF SOLAR GASES AT VARIOUS LEVELS

Heights in Kms from Flash Spectra	Element	Currents measured in Kms/sec	
		Spots	Disk
14,000	*Ca*+ (H and K)	3.8 inwards	0.5 downwards
12,000	$H\alpha$	3.0 "	
8,500	$H\beta$, $H\epsilon$, $H\zeta$		
8,000	$H\gamma$, $H\delta$	1.1 "	
7,500	*He*		
7,000	*Mg*	0.7 "	
6,000	*Ti*+, *Sc*+, *Sr*+	0.5 "	
5,000	*Ca* (4227)	0.2 "	
2,000	*Al* (15-20)	0.0	
1,500	*Fe* (15-40)	0.1 outwards	0.3 downwards
1,000	*Fe* (10)	0.2 "	
600	*Fe* (8), *Ti*, *Sc*, *V*	0.6 "	0.0
400	*Fe* (4), *Cr*, *Sr*, *Y*	1.2 "	
350	*Fe* (2), *Ni*, *Co*, *Mn*	1.5 "	
300	*Fe* (1)	1.7 "	0.3 upwards
275	*Fe* (00)	2.0 "	

It is evident, from a glance at the above table, that the enhanced lines not only extend to greater elevations than do the unenhanced lines of the same element, but that the Evershed effect for them is more pronounced as well.

Eclipse spectra, taken in connection with the Evershed displacements, can shed some light on the problem of sun-spots, not only regarding the time-honored question of the heights above the photosphere at which the spots themselves originate but also regarding the magnificent work of Hale on spot vortices. After long years of uncertainty regarding the effective temperature of spots, whether they are hotter or cooler than the photosphere, we have finally come to the conclusion that the spots are relatively cooler. This fact first became known through the discovery by Fowler of bands in the red part of the spectrum due to titanium oxide. Bands and flutings of hydride compounds have likewise been found in parts of the spectrum other than the red. There seems only one explanation possible, which is that the compounds can exist only at the cooler temperature of the spots, and that at the higher temperatures of the photosphere the compounds are dissociated and their bands cease to show. All of the researches concerning spots confirm this view of cooler temperatures, the low-temperature furnace lines being strengthened and the high temperature enhanced lines being weakened in spots.

If we are to accept the explanation of the Evershed displacements that they are a Doppler effect which differs for lines of various solar intensities, which in turn depend on the depths at which these lines originate, then we are forced to the conclusion that if there are differences in level between the spots and the photosphere these differences cannot be as much as fifty kilometers.

The motions in the line of sight of the gases in the penumbral regions of sun-spots can be measured with high precision on account of the great dispersion employed at Mt. Wilson. When the measures of the Evershed effect from the individual *Fe* lines are combined into averages, and the spectral lines are then arranged in various manners according to Rowland intensities, heights from the flash spectrum, excitation potentials, and spectral regions, conclusive evidence is found that the size of the Evershed effect does depend primarily on the heights. On the assumption, now thoroughly well founded, that a spot is a vortex on the sun, it appears abundantly verified that the indirect determination of relative heights from the Evershed effect tell a story consistent with the direct measurement of heights from the flash spectrum.

By referring to the above table, these interesting facts are observed: (1) To get the same average height in the red as to the violet of 4900 A, it is necessary to add two units of intensity to the right in the table, or 2 units of greater strength in the red. (2) In quite similar fashion, to reach the same excitation potential in the red as in the violet, it is necessary to go to lines of 2 Rowland units of greater intensity in red than in violet. In addition (3) St. John has found that to get the same measured value of the Evershed effect in the red as in the violet, it is necessary to go to 2 units of greater intensity in the red.

From the calibration of Rowland's intensity scale, it was found that the number of atoms involved in the formation of a line of Rowland intensity n in the violet is approximately the same as are engaged in the production of a line of Rowland intensity $n + 2$ in the red. Hence we see that a spectral line produced by a given number of atoms has the same Evershed effect, the same height in the flash spectrum and the same average excitation potential throughout the spectrum. It is evident that there is one underlying cause, and one only, needed to explain these interesting correlations, namely, the number of atoms involved. But how has this information so vitally necessary for explaining many of the problems connected with solar radiation been determined?

After arranging the lines of different spectra in multiplet groups, it was found by a number of investigators[1] working independently, and almost simultaneously, that simple formulae give the relative intensities of lines belonging to a given multiplet. The formulae are based on the correspondence principle, and give the transition probabilities as a function of the quantum numbers. These probabilities when multiplied by ν^4, ν being the frequency, are proportional to the energy emitted in the separate lines. Measures show that these formulae are only approximate, but on the whole they represent the intensities with fair degree of accuracy. According to Ornstein,[2] the quantities given by the formulae represent the relative numbers of "fictitious resonators," or the relative numbers of atoms involved in the production of the lines of the multiplets.

Let us first see, therefore, how the simple rules provided by the formulae can represent different arbitrary estimates of

[1] Kronig, *Zeits. für Physik, 31*, 885, 1925; Sommerfeld and Honl, *Sitz. der Preuss. Akad. der Wiss.*, p. 141, 1925; Russell, *Proc. Nat. Acad. of Sciences, 11*, 314, 1925.

[2] *Zeits. für Physik, 40*, 412, 1926.

intensities. Take first the estimates by A. S. King, many thousands in numbers of the spectra of the lines in the laboratory. Russell (*loc. cit.*) has found that King's scale, which was intended to represent the actual intensities of the lines, is a remarkably homogeneous scale. These Mt. Wilson intensities in fact are very nearly proportional to the square roots of the relative numbers of atoms predicted by the theory of multiplets. "The agreement is so close that these estimates, especially when averages for several multiplets are available, are clearly almost as valuable as actual measures, when once the significance of the empirical scale has been found."

Another comparison of even greater importance is the calibration of Rowland's scale of intensities for the solar spectrum. Russell, Adams and Miss Moore by means of the intensities of 1288 lines grouped in 288 multiplets have investigated [1] Rowland's scale in the solar spectrum. The relative intensities in multiplets, based on the correspondence principle, can be assumed to be proportional to the numbers of active atoms producing the various lines. These results have most important consequences.

By the observed contours of a number of typical resonance lines of different elements, Unsöld has determined [2] the number of atoms above 1 sq. cm of the sun's surface which are concerned in the production of the lines. Although many theoretical and practical difficulties have still to be surmounted and much more observational work must be accomplished on the contours of lines, nevertheless we have as a result what is perhaps the most important astrophysical publication in recent years. This comes from the skillful hands of Russell [3] entitled "On the Composition of the Sun's Atmosphere." Unfortunately space will not permit a detailed discussion to be given here.

One very important result was the derivation of the energy of binding of an electron in different quantum states by neutral and ionized atoms. These were given for various elements in tabular form together with tables of ionization potentials. These principles were first applied to enhanced

[1] *Astrophysical Journal, 68,* 1, 1926.
[2] *Zeits. für Physik, 46,* 765, 1928.
[3] *Astrophysical Journal, 70,* 11, 1929.

lines. Russell finds that thirteen elements have their most persistent lines in the region of the spectrum accessible to photographic observation. These elements are *Be, Ca, Sc, Ti, V, Sr, Y, Zr, Cb, Ba, La, Hf,* and *Ra;* all but the last, radium, have enhanced lines in the sun. Other important elements like *Cr, Mn, Fe, Co, Ni* and *Mo* have their most persistent lines in the inaccessible region in the sun to the violet of wave-length 3000, but other multiplets arising from low energy-levels are accessible for investigation. Fortunately with few exceptions enhanced lines are found in the solar spectrum for all those elements which have lines of low excitation potential in the accessible region of the spectrum. For reasons stated in foregoing chapters, these elements have their lines strengthened in the chromosphere and hence are prominent in the flash spectrum.

On the contrary, the principal factor which is unfavorable to the appearance of a spectral line in the sun is a high excitation potential. As a matter of fact, there are comparatively few lines in the solar spectrum with excitation potential greater than 5 volts, the only lines of great strength being due to hydrogen. (In the flash spectrum, however, is found the $He+$ line at 4686 A, with excitation potential of 48 volts.)

Hence the excitation potentials for the strongest lines in the visible part of the spectrum give the means of explaining the presence or absence of lines of different elements in the sun. If the most persistent lines are not observable in the solar spectrum, as with the element *Mg*, the method can give only a minimum value of the abundance.

These theoretical considerations depend on a knowledge of temperature and pressure in the sun. According to Eddington,[1] the temperature of the reversing layer doubtless is greatest closest to the photosphere and may be assumed to be 5730° absolute. Russell finds a pressure of 3×10^{-6} atmospheres, in good agreement with the conclusions of other investigators.

The total quantity of the metallic elements in the solar atmosphere seems to be known with some accuracy. The problem of the non-metals is more difficult, mainly on account of

[1] *The Internal Constitution of the Stars,* p. 332, 1926.

less complete observational material. The metals provide all but one-half of one per cent of all the ions and electrons. If the correction for departure from thermodynamic equilibrium should be disregarded, the calculated abundance of hydrogen would be increased thirty-fold.

Assuming the abundance of hydrogen found by Menzel from the flash spectrum (and which the present writer regards to be subject to large systematic errors), then Russell finds the probable constitution of the sun's atmosphere to be:

PROBABLE CONSTITUTION OF THE SUN'S ATMOSPHERE

Element	*By Volume*	*By Weight*
Hydrogen.........	60 parts	60
Helium...........	2?	8?
Oxygen...........	2	32
Metals............	1	32
Free Electrons.....	0.8	0
Total.............	65.8	132

Tables are given for the relative abundance of fifty-six elements and six compounds. Six of the metallic elements, *Na, Mg, Si, K, Ca* and *Fe* contribute 95 per cent of the total mass of all the metals.

Comparisons made with Miss Payne[1] show an excellent agreement in abundances for eighteen of the most important elements, except hydrogen and also the element *K* (probably because Miss Payne's value for the latter depends on only two lines). Comparisons also made with the abundance of elements in the outer ten miles of the earth's crust, including the ocean and atmosphere, and in stony meteorites, show a good agreement throughout (better with meteorites than with sun), except again hydrogen. The difference for *H* is enormous, part of which is probably real and part is due to the uncertainties of the calculation.

Russell calls attention to three puzzles still outstanding: (1) the calculated abundance of hydrogen in the sun's atmosphere is almost incredibly great; (2) the electron pressures calculated from the degree of ionization and from the numbers of metallic atoms and ions are discordant; and (3) the

[1] *Harvard Observatory Bulletin,* No. 835, 1926.

calculated rate of increase of density with depth in the reversing layer is much more rapid than indicated by observations of the flash spectrum.

With the important problems involving numbers of atoms so clearly stated by Russell and many other investigators, we are now in a position to take up anew the question of what may be expected within a multiplet of any element.

THE *Fe* MULTIPLET $a^5F-y^5D^0$

	F_5	F_4	F_3	F_2	F_1
D_4	3820 25 1600 .000 .0102	3887 8 800 .005 .0082	3940 4 500 .016 .0070	....	....
D_3	....	3825 20 1500 .000 .0098	3878 10 1000 .003 .0090	3917 5 600 .013 .0074	....
D_2	....	....	3834 14 1200 .002 .0094	3872 9 1000 .003 .0090	3898 5 600 .013 .0074
D_1	....	....	....	3840 10 1000 .003 .0090	3865 8 800 .005 .0082
D_0	Wave-length Intensity in sun Height in kilometers Evershed effect in sun-spots Sun *minus* vacuum arc in angstroms				3849 7 750 .006 .0078

In important articles, St. John[1] and Burns[2] have discussed their own particular points of view regarding the observational details within a given multiplet. For the purpose of illustrating the "unit nature" of multiplets, both have utilized the same *Fe* multiplet, $a^5F—y^5D^0$, with excitation potential 0.92 volts. After a thorough study of the whole problem, by comparing this multiplet with others, and then

[1] *Mt. Wilson Contributions, 389, 390,* 1930.
[2] *Journal of the Optical Society of America, 20,* 212, 1930.

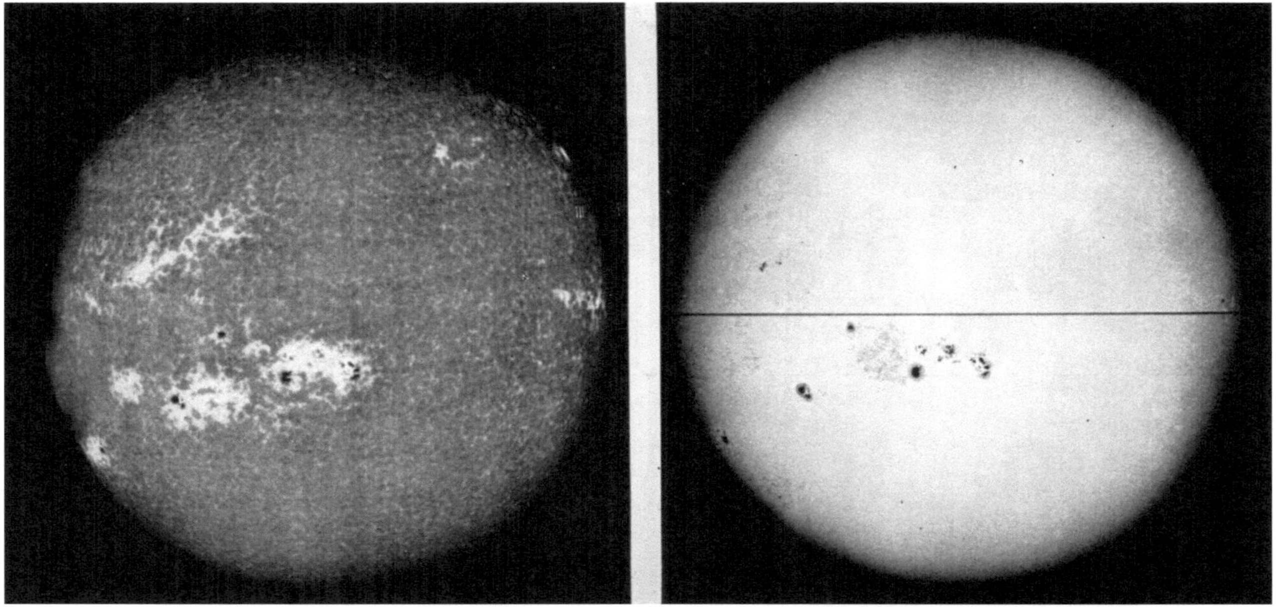

The Sun in the Light of Glowing Calcium Vapor (Left) and Direct Photograph (Right)
Photographed at Yerkes Observatory on February 12 and 13, 1907.

The American Eclipse of 1918
Photographed with the 40-foot camera of the Lick Observatory.

smoothing out accidental errors and getting rid of the effect of blends, the appended table is adopted by the author as his idea of this composite, or ideal, multiplet. The top number in each case is the wave-length, below this is the intensity, then come in order for each line, the height in kilometers from the flash spectrum, the Evershed effect measured in angstroms, and finally, the difference in wave-length between the value in the sun *minus* that of the vacuum arc. The Einstein, or relativity, shift to the red amounts to 0.0082 A, or 21 parts in ten-million in wave-length.

In this multiplet, the numbers of " fictitious resonators " effective in forming the different spectral lines vary greatly. According to Russell (*loc. cit.*, p. 326) more than one hundred times as many atoms are active in producing the strongest line of the multiplet at 3820 A as go to form the weakest line at 3940 A. Hence, within a single multiplet one should expect that the heights found directly from the flash spectrum, or indirectly from sun-spots, would not be constant but would be greatest for the largest numbers of atoms.

St. John and Burns differ radically in the interpretation of the systematic differences between the observations for lines of different intensities in the table. The former takes the ground that the relativity shift to the red in the sun has been confirmed, while the latter is still unconvinced. They agree, however, that the difference sun *minus* vacuum arc *should* be a constant. Burns finds (as shown in the table) that the more intense lines have a greater red displacement than the less intense lines in the same multiplet but this is directly contrary to St. John's deductions. Burns voices his opinion in the following words: " It seems paradoxical to assume as a possibility that a very strong line of a multiplet can originate at a high solar level, where St. John postulates the descending movement, while a weak line of the same multiplet originates only at a lower solar level in the ascending vapor."

The present writer is inclined to agree with St. John and therefore to disagree with Burns in the question of the relativity shift in the sun; he agrees with Burns and disagrees with St. John as to the unequal displacement for different lines in a multiplet; but he disagrees with both St. John and Burns in the interpretation of the data observed within single multiplets.

Two strong multiplets of *Fe* are specially valuable for purposes of comparison, namely, $a^5D - z^5D^o$, with average excitation potential 0.07 volts, and $a^5F - y^5D^o$, with excitation potential 0.92 (the composite multiplet given in the table). These two multiplets have the same maximum Rowland intensity of 20 in each case, and nearly the same minimum intensity. Moreover, the two multiplets cover approximately the same spectral region, from 3820 to 3940 A, and hence no systematic effects depending on wave-length need be considered. Although the Rowland intensities average the same in the two multiplets, there are pronounced differences in the other observed quantities. The multiplet of low excitation potential 0.07 volts compared with that of the higher value 0.92 shows the following: (1) greater intensities in the flash spectrum, (2) greater heights and (3) greater values of the wave-length difference sun minus vacuum arc. Each multiplet separately for (1), (2) and (3) show higher values on the main diagonal than on the side diagonals. For a height of 2000 km from the multiplet with excitation potential 0.07, the difference sun *minus* vacuum arc is + 0.0135 angstroms which, corrected for the relative shift 0.0082 A, gives a residual difference + 0.0053 angstroms corresponding to a downward motion in the sun's atmosphere of 0.4 km per second.

It must be emphasized again and again that the "effective height" above the photosphere at which a line in the Fraunhofer spectrum is formed, the heights from eclipse spectra and the heights from the Evershed effect, do not and cannot refer to the maximum heights to which atoms are ejected by solar activity. All three heights are derived from photographic plates, all depend on the effects produced by the atoms on the photograph. As already stated in the foregoing, the atoms must be in sufficient numbers, or have a concentration adequate to leave a trace of their action on the photographic plate. It is not impossible that under average conditions of solar activity, all *Fe* atoms no matter what are their atomic levels or whether they are neutral or enhanced reach about the same maximum heights in the chromosphere. If the solar activity becomes greater, as in the course of the sun-spot cycle, or locally as in prominences, the heights become greater.

The conditions under which atoms in the chromosphere radiate to form the flash spectrum (or absorb radiation in the dark-line spectrum) are becoming well understood. They depend on temperature, pressure and concentration of atoms. For a single multiplet, it is obviously necessary to assume that all of the spectral lines are formed under nearly identical conditions of temperature and pressure and that all the atoms reach the same maximum heights. The lines differ greatly the one from the other in the number of atoms involved in the emission (or absorption) of radiation. Let us now confine our attention to two lines from a multiplet, the one produced by one hundred times as many atoms as the other. Consider for each line a cylinder of unit cross section. The base of each cylinder we shall call the photosphere, each axis being perpendicular to the solar surface. The top of each cylinder is at an equal, but at a very great distance above the photosphere. The atoms are most concentrated near the photosphere. Looking down into the axis of each cylinder, that is, observing a spectral line at the center of the sun's disk under the ordinary conditions of the Fraunhofer spectrum, it is possible to see down into each cylinder until the atoms become so concentrated that they virtually form a black wall through which vision cannot penetrate. This black wall in each case then becomes the "effective height" above the photosphere at which each spectral line is formed. The stronger line of the two, the one involving the greater number of atoms, has its effective level, or takes its origin, at a greater height above the photosphere than the weaker line of fewer atoms. The above expresses in non-technical language what various authorities for a number of years have been stating in technical terms. This simple scheme seems adequate to explain a great variety of solar phenomena. In particular, it seems easy to understand that within a single multiplet the spectral lines take their origin at different effective heights, the strongest lines at the greatest heights and the weakest at the least heights above the photosphere.

Through a splendid series of researches carried out at Mt. Wilson and elsewhere, a very imposing structure has been built on the assumption that Fraunhofer lines do indeed take their origins at different heights above the photosphere. It seems entirely unnecessary to overthrow this beautiful edifice by requiring that all lines in a multiplet, no matter how much they differ in intensities, must all be found at the identically same heights. It seems equally unnecessary to assume that all lines in a multiplet, involving as they do very different numbers of atoms, must exhibit the same difference in wave-length between sun and vacuum arc. In the composite multiplet, $a^5F - y^5D^0$, the intensities of the strongest and weakest lines of the multiplet are as 25:4 or 6:1, the heights are approximately as 3:1 and the differences sun *minus* vacuum arc are as $1\frac{1}{2}:1$.

From his discussion, St. John finds that the strong solar lines show a greater red shift than do the weaker lines except in the case where strong and weak lines occur in the same multiplet, and then the red shift must perforce be the same for all lines no matter how much they differ in intensities! On the other hand, both Burns and the writer find the greater shift for the strong lines even though the strong and weak lines belong to the same multiplet. Burns attempts, without much success, to explain the greater red shift of the strong lines as due to an "intensity equation" partly due to instrumental causes. Both Burns and St. John find fault with the heights from the flash spectrum in that these heights are not a constant for all of the spectral lines of a multiplet. The problem of the relativity shift in the solar spectrum will be taken up in greater detail in Chapter XXIII.

The theory of the formation of Fraunhofer lines and the problem of determining the relative abundances of elements in the sun's atmosphere is now (1932) receiving much attention by astrophysicists. The puzzles noted by Russell on page 347 show that theory and observation are not yet in satisfactory accord. With the perfecting of the registering microphotometer much observational work has been done on the difficult task of measuring the contours of spectral lines in the stars and in the sun. On account of the vastly greater dispersion possible with the sun, it is evident that the chief advances must come from solar measures. The purpose of these measures is to find the number, N, of atoms involved, particularly in the formation of resonance lines of different elements. In addition to Unsöld, many other investigators have engaged in the problem. Wooley [1] employed a clever method of attack by measuring the widths of the terrestrial atmospheric lines in the B-band of the solar spectrum. With the observations made at different times during the day, the known altitudes gave a ready means of deriving the relative numbers of atoms involved. Wooley found a variation in the widths of lines proportional to $N^{\frac{3}{8}}$ instead of $N^{\frac{1}{2}}$, as expected by Unsöld's theories. He then came to the conclusion that the difference between theory and observation was caused more by Unsöld's method rather than by incorrect calibration by Rowland. H. H. Plaskett [2] measured the magnesium *b*-triplet in the sun's spectrum. He discussed his measures by means of the following assumptions: (1) pure absorption, (2) pure scattering, and (3) combined absorption and scattering. The third assumption seems best to fit both theory and observation. Pannekoek and Minnaert, together and separately, have made valuable contributions to both the theoretical and observational sides. As pointed out in earlier chapters, Milne has done very valuable work on the theoretical side. It is evident that enormous progress has already been made as a result of the new quantum theory, but the goal still seems far distant.

In spite of the imperfect agreement between observation and theory, it will be very important to follow Russell's methods and ascertain the abundances of elements in the sun's atmosphere from the flash spectrum. The best photographs for this purpose are unquestionably those of the Lick Observatory discussed by Menzel and the concave grating spectra taken by Mitchell. The information from

[1] *Astrophysical Journal, 73,* 185, 194, 1931.
[2] *Monthly Notices, R. A. S., 91,* 870, 1931.

the latter should be the more reliable, for the reasons that the grating spectra were photographed almost completely by one spectrograph with constant dispersion, while the Lick spectra were taken with several spectrographs and varying dispersions, with both fixed and moving plates; and in addition they covered only about one-half of the region of wavelengths discussed by Mitchell.

Both Menzel and Mitchell attempted to estimate their intensities on the Rowland scale. In general, each agreed with Rowland and with each other. However, in the foregoing chapters the large systematic differences in intensities between the chromospheric intensities and those of Rowland have been emphasized. These differences may be briefly summarized as follows: (1) The hydrogen and helium lines are much stronger in the chromosphere. (2) The enhanced or ionized lines are likewise much stronger. The great increase in strength has been explained by the fact that the intensified lines are high-level lines. (3) With the neutral lines in the spectra of the metals, similar systematic differences between the chromosphere and Rowland are noted. It was shown that for any element, the strongest lines, or those of low excitation potential, are stronger in the chromosphere than in Rowland; those of medium excitation potential have about equal intensities in the two spectra, while the lines of highest excitation potential are weaker in the chromosphere.

The reasons for these systematic differences for the neutral lines find a ready explanation; namely, lines of low excitation potential involve more atoms than those of the same element of high excitation potential. The greater number of atoms involved permits these lines to be detected to greater heights in the flash spectrum. Apparently, therefore, all high-level lines whether from the neutral or the ionized atom are stronger in the flash spectrum than in Rowland, while on the contrary, those of low levels are weaker than in Rowland.

Remembering, therefore, that Russell's determination of the abundances in the sun's atmosphere depends primarily on the ability to calibrate the Rowland scale, it might be thought that on account of the great differences just noted between the Rowland and the chromospheric intensities, there might be systematic differences in the amounts of the elements found from the Fraunhofer spectrum and from the flash spectrum. In particular it might be thought that *H* and *He*, on account of their great strengths, would be relatively much more abundant from the chromospheric investigations. Manifestly, it is necessary also to calibrate the estimated intensities in the flash spectrum. The reliability of the final results will depend mainly on how accurately this calibration can be carried out.

As Russell has pointed out, the information for each of the elements found in the sun depends chiefly on the strong lines, those of low excitation potential. In spite of the much greater dispersion in the Rowland spectrum than is available in the flash spectrum, the latter has almost as complete information regarding the strong lines. Hence the results for the chromosphere from eclipse spectra should have almost the same accuracy as those of Russell.

Two separate determinations have been made: one from the Lick photographs by Menzel, the other[1] by Mitchell and Miss Williams. The results show that within the limit of errors there appears to be little difference between the relative abundances derived from the Fraunhofer spectrum and those obtained from the flash spectrum. In other words, the two determinations from the chromosphere differ from each other about as much as either differs from Russell's determination. This is true for all of the elements investigated with the exception of *H* and *He*. For obvious reasons, the information from the chromospheric spectra for these two elements should be more reliable than from the Fraunhofer spectrum.

With the results now available, it seems probable that hydrogen is actually more abundant in the chromosphere, but whether the factor is ten, or ten thousand times, we are not now sure. In fact, the lightest elements, hydrogen and helium, still remain puzzles. In spite of the very great number of attempts that have been made to predict from theory

[1] *Publ. Leander McCormick Obsy.*, *5*, Pt. 6, 1932.

the intensities in the hydrogen spectrum, no very great success has yet been attained.

Additional information obtained from eclipse spectra is the distribution of atoms at different heights, or in other words, the density gradients in the chromosphere. Pannekoek and Minnaert, from their 1927 spectra, found density gradients up to 2000 km for the emitting atoms of *H* and *He*. Their measures showed that the density of hydrogen decreased much less rapidly than would be expected in an isothermal atmosphere under equilibrium conditions. The writer and Miss Williams (*loc. cit.*) have derived density gradients for the emitting atoms of neutral *Fe* and for the ionized elements *Ti, Fe* and *Cr* between the heights of 300 and 2500 km. Menzel (*loc. cit.*) has derived a density law up to 2500 km by combining the measures from different elements.

To derive the density gradients, it is necessary to know the heights as accurately as possible. It is also necessary that the atomic origin of the lines be known and that the lines arise from multiplets in which the relative intensities of the lines can be predicted by the theoretical intensity formulae. The inter-system lines are of no use for the purpose for the reason that no intensity formulae are known for them. Further, the lines of the Balmer series of hydrogen cannot be used because no two lines belong to the same multiplet. In the case of an element like *Fe,* there is a great amount of material available. Every multiplet in which two or more lines appear unblended in the flash spectrum furnishes a determination of the intensity gradient at different heights above the photosphere.

Earlier in this chapter, the multiplet $a^5D - y^5D^0$ has been considered. The intensity formulae show that the relative numbers of atoms involved per unit volume are nearly proportional to the squares of the intensities. For purposes of illustration, we shall group together lines of equal heights in the multiplet and round off the predicted number of atoms. From the 12 lines of the multiplet we have the following:

Line	*a*	*b*	*c*	*d*	*e*	*f*
Relative number of atoms	450	300	200	100	60	5
Height in kilometers.....	1600	1500	1200	1000	800	600

With photographs of the flash spectrum taken without slit, measures of the chords to the tips of the cusps give the heights. As explained in the foregoing, a detectable blackening of the photographic plate occurs when there is a given number of emitting atoms along the line of sight, a statement which is true only within a limited region of wavelengths. Hence the number of atoms in line *b*, above the 1500-km level from the photosphere, is exactly equal to the number in line *d* above the 1000-km level, or in line *f* above the 600-km level. Evidently in this multiplet, $1\frac{1}{2}$ times as many atoms are detected by the photographic plate in the act of emitting radiation at the 1200-km level as at the 1500-km level. The number of atoms per unit volume being $1\frac{1}{2}$ times, the density of concentration is therefore 50 per cent greater. Compared with the density at the 1500-km level, the density at the 1200-km level is $1\frac{1}{2}$ times, at the 1000-km level it is 3 times, at 800-km it is 5 times, and at 600-km the density is 60 times. Hence, it is evident that from this one multiplet, the densities of distribution of neutral *Fe* atoms can be obtained at different heights above the photosphere. Other multiplets give similar information, and hence the density gradients may be ascertained by taking the averages for all the multiplets employed. Averages being taken for different heights, it is manifest that the most reliable results are derived from those spectra for which the heights are most accurately known. For any element like *Fe* where many multiplets are available, the material may be divided into different groups of low, medium and high excitation potential. When the number of multiplets are numerous enough, this same process may be carried out with different neutral and ionized elements. When due account has been taken of these matters, it is possible to derive an empirical density law for a number of different elements.

In Menzel's discussion, all elements were grouped together while Mitchell and Miss Williams discussed the elements separately. The results from the two discussions agree, and they show that the density gradients for the various elements are not very different. The density gradients for neutral *Fe* and for ionized *Ti* are not very much greater than for hydrogen; and all values are much less than would be expected in an isothermal atmosphere.

The difference found in the distribution of the atoms of *Fe* of low excitation potential, as compared with that of the atoms of higher excitation potential, might be ascribed to a large temperature gradient in the chromosphere were it not for the fact that the *Ti* + data shows no such effect. The same effect is undoubtedly present in the *Fe* data at much lower levels — in the regions between the effective levels of the reversing layer and the chromosphere. Menzel ascribes this to a temperature gradient, but in view of the fact that the *Ti* + data shows no such effect in this region, even as it showed no effect at greater heights, it is unsatisfactory to assume an explanation which brings the *Fe* data into step but at the same time throws the *Ti* + data out of step.

But why are the density gradients so much smaller than expected? What is the supporting mechanism? Many theories have been proposed. These have been examined in detail by Menzel. Here it will be possible to touch on them very briefly, on account of limitations of space and also for the reason that none of the theories satisfactorily explain the observed facts. Milne (*loc. cit.*) has a beautiful theory in regard to ionized calcium. The atoms fall toward the sun under the influence of gravity until a quantum of energy of the frequency of the *H* or *K* line is absorbed. The momentum associated with the light-quantum tosses the atom high into the chromosphere again, after which once again it begins to fall. This process is repeated indefinitely. This theory may be applied to the other alkaline earths, *Mg*, *Sr* and *Ba*, but it is not applicable to any of the other elements; some other force must support them. As Rosseland [1] was the first to suggest, there is a great turbulence in the sun's

[1] *Monthly Notices, R. A. S., 88,* 377, 1928.

atmosphere in virtue of which the heavy elements will be tossed to great heights in the chromosphere along with the light elements. This would account for the fact that the relative abundances found in the reversing layer, that is, in the lower and denser layers of the chromosphere, are practically the same as found in the higher reaches of the chromosphere from eclipse spectra. Turbulence would also account for the similar density gradients of dissimilar elements.

Gurney's theory [1] is most interesting, especially as it may give a solution for the helium puzzle. Since there are so many *Ca* + atoms in the high chromosphere, a great number of them will become doubly ionized. But the doubly ionized *Ca* ion is incapable of supporting itself by absorbing light-quanta, for the reason that its ultimate lines lie in a spectral region where the number of light-quanta emitted by the sun are relatively few. Hence the *Ca* ++ ion will drop back to the photosphere and in so doing it will attain a tremendous kinetic energy, enough to excite the visible helium lines if it should collide with a *He* atom, or even enough to ionize the *He* and excite the interesting line found in the flash spectrum at wave-length 4686.

[1] *Ibid.*, *89*, 49, 1928.

CHAPTER XX

THE CORONA

THE corona still remains exclusively an eclipse phenomenon. In spite of the amazing achievements of modern science which at times seems to be able almost to accomplish the impossible, no success has attended the efforts made to observe the corona outside of an eclipse.

On account of the dramatic character of the phenomenon and of the great interest in eclipses felt by both the astronomer and the general public, each eclipse as it comes is enthusiastically observed; but the truth must be told that success commensurate with the labor involved is not always forthcoming. "The problems of the corona are many, and few of them can be said to have approached solution. Important facts concerning it have been established, but these facts are more or less isolated, and in general their relations to each other are unknown. The paucity of results obtained thus far is due primarily to the unique condition that the most assiduous of eclipse observers can scarcely hope for more than an hour of totality with clear skies, in his entire lifetime." In this and the following chapter we shall attempt to show how the amazing progress of recent years in every branch of solar research has affected the solution of coronal problems.

Concerning the total eclipse of March 29, 1652, seen in Ireland, Dr. Wyberd writes, "The moon suddenly threw itself within the solar disk with such agility that it seemed to go round like an upper mill-stone. The sun then appeared around the limb, affording a pleasant and remarkable spectacle of rotation." There seemed to be a widespread notion among the early observers of eclipses that during totality very rapid motions took place within the corona and that the corona was some sort of a modern fireworks display with brilliant scintillations and sudden changes. In fact, the early notion seems not to have entirely vanished in the enlightened and scientific age of the twentieth century.

No eclipse expedition worthy of the name will be fully equipped unless it has as part of its program the securing of large scale photographs of the corona. Astronomy of the future needs, as it has in the past, to secure good photographs of every possible eclipse. The Lick Observatory has the most complete series in existence, beginning with the eclipse of 1893. The Lick photographs have always been secured by pointing the camera directly at the sun, the method devised by Schaeberle, and a uniform focal length of forty feet has been employed. This permanent record of the past is always available for purposes of comparison.

The 5-inch aperture, 40-foot focus represents a ratio of aperture to focal length of 1:96. The commercial photographer would look aghast if he were compelled to work with such a slow camera. For investigating the details of the inner corona, a large scale image is necessary, rendered possible by great focal length. Up to the present, the largest scale has been secured at the eclipse of 1900 through the use of a horizontal camera of 12 inches aperture and 135 feet focal length, a ratio $a:f = 1:135$. The brightness of the inner corona permits the employment of short exposures even with the small ratio of aperture to focal length necessitated by cameras of great focal length.

Two methods of mounting cameras of great length are available, either that of pointing the objective directly at the sun, or using the horizontal telescope with coelostat mirror. The former has distinct advantages over the latter, but the erection becomes increasingly difficult as focal lengths are augmented. Miller of Swarthmore College has been very successful with focal lengths of 63 feet. For such a mounting, a double tower is necessary, an inner one to carry the objective, and an outer tower to protect the whole from jars caused by the wind. With this type of mounting the plate is moved by a clock to counteract the diurnal motion. With the Swarthmore tower-telescope on Niuafoou Island, at the most recent eclipse, photographs with superb definition were secured by Marriott.

The horizontal telescope is much easier to construct and erect in the field. The tube may be made of heavy paper and the plate holders carried by a heavy framework surrounded by a dark room. Care should be taken that the tube is not too near the ground and that the tube is protected from the direct rays of the sun by a canvas or paper shelter. The coelostat is the best form of mounting for the plane mirror. Compared with direct mounting, as was effectively shown at the eclipse of 1919, the horizontal telescope has the great disadvantage that the mirror is sensitive to changes of temperature and it may alter its shape and become warped on eclipse day when exposed to the sun's rays. Ordinarily, focus is obtained for eclipse cameras by star trails. If possible a warm night, differing as little as possible from the expected temperature at eclipse time, should be utilized. It is imperatively necessary to keep the mirror protected from the sun's rays on the day of the eclipse until a very few minutes before totality. One should avoid the horizontal telescope if possible, mainly on account of the uncertainties connected with the mirror and its lack of permanent figure. At the Niuafoou Island eclipse, good photographs were secured with the 65-foot horizontal telescope belonging to the U. S. Naval Observatory. The lens is as good a one as the Swarthmore lens of nearly equal focal length used in the tower-telescope and, moreover, Marriott cared for the details of both instruments and the development of the plates. The definition of the tower plates, however, was better than those taken with the horizontal telescope. For the purpose of portraying the structural details of the corona it is evident that the greater the focal length utilized the more valuable will be the results. Cameras of medium and small size have not been superseded by the large instruments and still have useful functions at eclipses. Larger ratios of aperture to focal length are possible than with those of the largest size. Such cameras are more rapid and, consequently, are useful in securing the faint extensions of the outlying corona. They are especially valuable in photometric observations which will be taken up in detail below. Every precaution should be taken to reduce to a minimum the effects of halation and reflections from the glass-side of the plates. All plates used for photographing the corona should be " backed " with some absorbing material on the glass side. Some eclipse observers have had excellent success from using double-coated or triple-coated plates.

In attempting to photograph the corona on small or large scale, it goes without saying that the greatest care should be exercised to secure the most perfect definition, and also that the photographic manipulation should be conducted so as to bring out of the plates as much of the wealth of detail as possible. Development may be carried out in such a way as to attain very different photographic effects. If contrast is needed, as in spectrum work, a " hard " development is required. There are many developers suited for this sort of work with which anyone doing spectrum work is entirely familiar. For developing the corona, however, where it is necessary to bring out as many of the fine details as possible, and where an attempt should be made to minimize the contrast between the bright inner corona and the faint outer corona, an entirely different kind of developer and development is necessary. Old-fashioned " pyro " is probably the best form of developer to use for this purpose, and one would do best to start with a very weak solution and proceed gradually. The proper development of each plate will take at least an hour. Undoubtedly many well exposed coronal photographs have been spoiled through improper care in development. The technique of development of photographs is so well known that no attempt at further details will be given here.

In reviewing the scientific work accomplished at solar eclipses, one is immediately struck by the fact that success has been greater in the direct photography of the corona than in any other branch of eclipse investigations. The self-evident reason is that every well-equipped eclipse expedition, no matter what their other program, attempts to photograph the corona. However, there is another reason not so apparent, which is that perfection of focus and seeing, although desirable, are not absolutely essential in obtaining successful photographs of coronal details. A lack of perfect focus plays havoc with spectroscopic photographs but detracts little from the corona for the reason that its structure is nebulous or filmy in detail. These remarks should not be interpreted to mean that one should be satisfied with anything short of absolute perfection in the determination of the best focus.

The general form of the corona can be predicted in advance of the eclipse. At sun-spot minimum are found the long equatorial streamers and the short plume-like polar brushes which were well seen at the eclipse of 1900 (p. 164), or in 1922 (p. 365). At sun-spot maximum the corona is nearly circular in shape, thus resembling a gigantic dahlia. How closely are these typical shapes connected with the sun-spot curve? It is rather curious to find that in the thirty years following 1893, when large-scale photographs of the corona were first successful, sixteen total eclipses were observed either near maximum or minimum of spots, or were distributed along the branch of the curve descending from maximum. The eclipse of 1914 was the only corona seen on the ascending branch of the sun-spot curve, the corona of 1916 having been obscured by clouds. Fortunately for our knowledge of this subject, the eclipses of 1925, 1926 and 1927 were all on the ascending part of the curve. With the eclipses of 1929 and 1930 on the descending branch of the sun-spot curve near maximum, our information regarding the shape of the corona can now be subjected to more exact scrutiny. We ask ourselves the question whether the typical coronal shapes take place exactly at maximum and minimum, or whether they start before or after the times of maximum and minimum. Take one particular eclipse. Sun-spot maximum occurred in August, 1917. The eclipse of June, 1918, did not show the typical corona of sun-spot maximum, for the polar streamers were shorter than those near the equator, and the corona departed more from the circular shape than was actually anticipated. Generally when the maximum of spots is past, the streamers draw away from the poles, and the longest rays are found in the sun-spot zones, making the corona rectangular in appearance.

Although it may be said with truth that the shapes of the

1905
Spain

1908
Flint
Island
Pacific

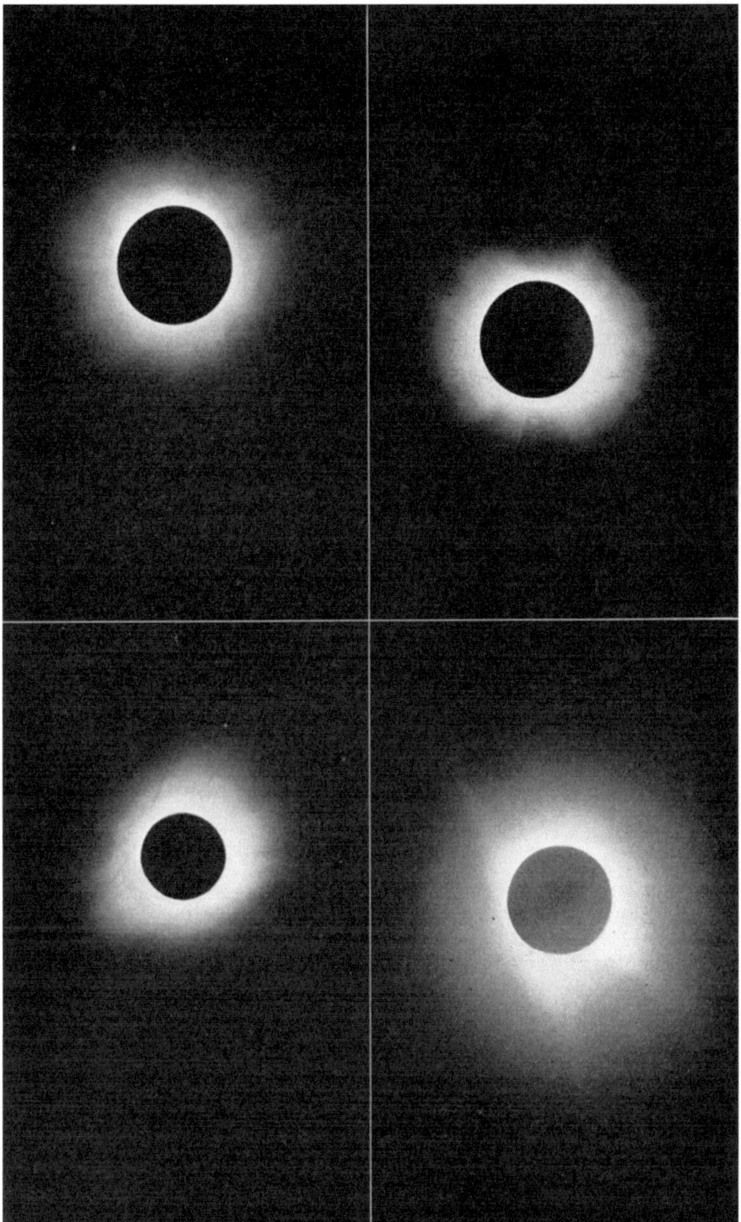

1918
America

1922
Australia

Three Coronas Photographed by Lick Astronomers and One (1922) by Canadian Observers

Solar Corona, September 21, 1922
Exposure 2 seconds. Lick Observatory Expedition to Australia.

corona at minimum of sun-spots all resemble each other in having long equatorial streamers and pronounced polar brushes, yet each corona has its own particular features, its own peculiar structure. The coronas of 1878, 1889, 1900 and 1922, all taken at minimum phase of sun-spots, could never be mistaken one for the other. In fact there were two eclipses in the year 1889, on January 1 and December 22, with very pronounced alterations in shape in the coronas. These variations in form, as pointed out by Wesley,[1] were precisely in accordance with the change to be expected with the sun returning toward a condition of spot activity. The shapes of the coronas from 1922 to 1927, inclusive, furnish interesting comparisons. The eclipse of September 14, 1922 represented the typical minimum corona. The eclipse of September 10, 1923 exhibited the same features. Even on January 24, 1925, there was still the typical minimum shape, of pronounced polar brushes, but the equatorial streamers were not so long as those of 1922. To the right of the vertical, however, there was a long pointed shaft of light giving mute evidence that the time of minimum was well past. On January 14, 1926, the corona had lost all of the characteristics of the minimum type and closely resembled that associated with maximum of spots, though the time was two and one half years before the expected maximum of spots. The corona of June 29, 1927 was distinctly that of maximum type.

Corona times near sun-spot maxima resemble each other even less than do the different aureoles at sun-spot minima. The coronas corresponding to maximum of spots are, it is true, approximately circular in outline, but prominent rays and streamers shoot out at various angles. We are forced therefore to the conclusion that changes are continuously going on in the corona. How rapid are these changes? Can they be detected from photographs taken with the same camera, at the beginning and ending of a total eclipse? It is possible that the few short minutes of totality may afford too short an interval to permit these changes to be detected with certainty, from photographs taken at any one location, but mo-

[1] *Observatory, 13,* 105, 1890.

tions might possibly be detected from photographs secured at widely separated stations at the same eclipse.

Any theories dealing with coronal structures must depend on the measured changes within the corona itself, and it is therefore imperatively necessary that reliable information be secured regarding these motions. The importance of this problem has been fully recognized by eclipse observers for many a long year. Manifestly, no real progress was possible as long as it was necessary to depend for information on drawings of the corona. In fact, sketches made during the time of an eclipse have been a continued disappointment, but the non-success was no greater than what should have been expected. Even a skillful draughtsman subjected to the excitement and unfamiliarity of a total eclipse could not be expected to see and draw, in the hurried interval of a couple of minutes, more than a few details of coronal structure. These details of necessity must be exaggerated in the drawing. A sketch made by a second person, as well trained as the first draughtsman and secured at the same time and location, probably would stress other details regarded as important. But to make matters worse, most of the drawings of the corona made in the past were secured by men who were observing their first eclipse, and frequently these very men had little experience or skill in the making of sketches. With the advent of the photographic plate, the wholesale drawing of the corona during totality has been pushed more and more into the background.

To detect motions in the corona with the greatest certainty, several prerequisites are necessary: First, the photographs to be measured should be on as large a scale as possible; second, the interval in time should be as great as possible; and third, the photographs to be compared should be secured with cameras of nearly the same focal length and they should resemble each other in general appearance as much as possible. Plates developed with different effects of contrast and taken under different conditions of seeing cannot furnish motions with the highest degree of precision.

Attempts were made as early as 1889 to secure the necessary information by the comparison of photographs at the eclipse of December 22. An interval of two and a half hours elapsed between the time of totality at Cayenne and Cape Ledo in West Africa. Unfortunately, clouds were experienced at the latter station. As a result of the long duration of totality at the eclipse of 1901, an exceptionally favorable opportunity existed for determining motions from observations at a single station. Perrine[1] secured successful photographs with the 40-foot camera. Measures of short exposure negatives, taken near the beginning and end of totality, showed no displacements of coronal masses in the interval of a little more than five minutes. On account of the accuracy of the measurements, it was possible to have measured with certainty a velocity of 20 miles per second across the line of sight. Motions should have been suspected if they had been as great as 12 or 15 miles per second. The 1901 photographs were near sun-spot minimum. At the eclipse of 1905, Hansky secured photographs on the same scale as Perrine's photographs.[2] The corona was one of typical maximum form and presented few conditions for the best results. The corona was very intense, especially near the solar surface, and its rays, being projected from all sides of the disk of the sun, so superposed themselves the one upon the other and became so entangled that it was almost impossible to distinguish the same details on successive photographs. By taking the original negatives and making glass positives, by proper shading and by local development, much detail was secured in the corona. From the measures of 43 separate rays, information was obtained concerning the motions of the coronal material within the period of totality lasting a little more than three minutes. The velocities determined were little greater than the errors of observation. None of the velocities investigated exceeded 25 km per second.

In accordance with the plans of the Lick Observatory of always attacking eclipse problems of the greatest importance, an attempt was made at the eclipse of 1905 to detect changes in the coronal structure by establishing three stations at widely separated localities and securing large-scale photographs at each station. Parties were sent to Labrador, to

[1] *Lick Observatory Bulletin, 1,* 151, 1902. [2] *Mitt. Pulk., 2,* No. 19, 107.

Spain and to Egypt. Unfortunately for the success of the scheme, cloudy weather prevailed in Labrador where were stationed Curtis and Stebbins. Thin clouds were met in Spain, but they did not greatly interfere with the photographs secured by Campbell and Perrine. Clear skies greeted Hussey in Egypt where totality occurred 70 minutes later than in Spain. A careful comparison was made of the photographs [1] secured at the two stations. A number of fairly well-defined nuclei were found, both east and west of the sun. Details of structure within the nuclei appeared to change, but the nuclei as a whole remained in the same position. Measures of the greatest accuracy were impossible on account of the poor definition of the Egyptian plates caused by poor seeing. The conclusion was that " the masses in question could not have moved so much as one mile per second during the interval of 4200 seconds. Greater speeds might well have occurred within the principal coronal streamers, or within some of the arched forms which enclose prominences, without our having detected them; for their structure is quite uniform, and well-defined nuclei are absent."

The American eclipse of 1918 afforded an additional opportunity by comparing plates taken at three separate stations where were located expeditions from Lick, Lowell and Sproul Observatories. At Goldendale, Washington, the focal length was 40 feet, at Syracuse, Kansas, the camera was 38.7 feet in length, while at Brandon, Colorado, the camera was 63 feet long. One plate secured by each of the first two cameras and two plates with the 63-foot instrument were compared. Each of the three cameras pointed directly at the sun. There were no definite nuclei or other distinctive features present in the streamers that were sufficiently well defined to be used as points of measurement by Miller.[2] There were, however, three arches surrounding three prominences and attention was confined to these. The first arch was around the " Pyramid " or " Eagle " prominence in the northeastern quadrant. Towards the pole side of this prominence there were four well-defined arches. The second arch was around a prominence in the SE quadrant, while the third

[1] *Lick Observatory Bulletin*, 4, 121, 1907. [2] *Publ. A. S. P.*, 32, 207, 1920.

arch was near the "Heliosaurus" prominence. Each of the three prominences displayed four separate arches, and the positions of one arch only of each were measured. These measures gave fairly accordant results and seemed to show that the arches had changed in the twenty-six minutes' interval between the Lick and the Sproul photographs. The average rate of speed at which these arches receded from the sun was about ten miles per second.

At the eclipse of 1926, Horn-D'Arturo compared photographs taken by the Italian expedition in East Africa with those taken by the British expedition in Sumatra two and a half hours later. He measured the coronal domes at position angle 335° and found that if they moved at all, their motions were not greater than 2 km per second. Von Klüber[1] reached a similar conclusion from measures on the photographs of the German expedition in Sumatra compared with Horn-D'Arturo's measures.

Horn-D'Arturo found that on the photographs taken in Africa there were well-defined streamers at position angles 350° and 320°, both traceable to heights greater than one solar radius, whereas on the Sumatra photographs the streamer at 350° had disappeared, the streamer at 320° was intensified and a new, well-defined, very straight streamer had appeared at 325°, traceable to a height of $1\frac{1}{2}$ solar radii. If the new streamer be interpreted as matter just ejected from the sun, then the matter must have been moving at least 50 km per second.

From the photographs compared by Miller, it was evident that the polar rays east of the axis of symmetry are curved away much less than the rays to the west, the difference not being due to the effect of projection. It is therefore evident that the velocities with which material is ejected from the sun to form the coronal streamers must be very small compared with the motions with which we are familiar in the prominences. Any theory dealing with the formation of the coronal structure must take cognizance of these moderate velocities of ejection.

The corona owes its entrancing beauty to the far-flung,

[1] *Zeits. für Astrophysik*, 4, 1, 1932.

pearly-white streamers, so conspicous in each corona observed. The contrast with the rosy-red prominences close to the edge of the black moon makes a never-to-be-forgotten spectacle. In addition to the streamers, there are other features of the corona that deserve attention, and it is necessary to find out their relations to solar phenomena such as prominences, sun-spots and faculae. "Arches," "hoods" or "striated cones" have been observed since Cleveland Abbe first saw them at the eclipse of 1869. They are specially conspicuous at the time of sun-spot maximum. The first good photographs obtained of them were by Schaeberle at the eclipse of April, 1893. The coronal arches were noted by him[1] to be associated with prominences. Miss Clerke[2] draws the conclusion that "Each pearly pavilion is erected over a red flame. Coincidences of the kind are of perpetual occurrence." These hoods were specially marked in the coronas of 1896 and 1898, but were practically missing from the minimum type of corona of the year 1900.

At the eclipse of 1901, a different kind of "disturbance" was noted in the photographs secured by Perrine, which consisted of a conspicuous center, apparently at the sun's limb, from which strong streamers stretched out to great distances from the edge of the sun. This disturbance was all the more interesting for the reason that Perrine found it to take its origin over the region of a prominent sun-spot. In fact, this spot was the only one known to exist on the sun at the time. Similar disturbances were seen in the photographs of the eclipse of 1918, but Campbell and Moore could find no relation between them and any sun-spot or chromospheric phenomena known to exist on the sun. As already stated, the arches were specially prominent at this recent eclipse and they extended to much greater distances from the sun's limb than they did in 1893, although both eclipses were near the sun-spot maximum. In 1923, Miller photographed a disturbed area in the corona intimately associated with a spot near the sun's limb, shown at page 209. As

[1] *Contributions from the Lick Observatory, 4,* 94, 1895.
[2] *Problems in Astrophysics,* 129, 1903.

a whole, disturbances are much more conspicuous in the coronas seen at maximum of sun-spots for the reason that prominences are much more numerous and active and are found in all heliographic latitudes, while at sun-spot minimum the prominences are feeble and are confined to zones near the solar equator.

Much valuable information regarding coronal disturbances was obtained [1] at Niuafoou Island in 1930. This was the first time in the history of eclipses that so much detailed structure was visible in the spectral lines of the corona, this structure being made possible by the fine definition of the concave grating spectra taken without slit. A comparison of all the 1930 grating spectrograms, but more particularly a study of the H and K lines, reveals the interesting fact that the sun, two years after sun-spot maximum, was in a condition of great activity on the day of the eclipse. The spectra and the direct photographs taken by the 63-foot tower telescope by Marriott show prominences completely encircling the sun. Comparisons were also made with spectroheliograms on four successive days, taken at Mt. Wilson or Kodaikanal. The superb definition of the eclipse photographs shows both the prominences and the inner corona in great beauty.

A comparison of all the photographs, the spectroheliograms with the eclipse plates both direct and spectrographic, shows that the sun was very active not only at eclipse time but throughout the whole period of nearly four days covered by the plates. At the time of totality there was a prominence almost exactly at the south point of the sun extending to a height of 25,000 km. At this time the axis of rotation passed 26° to the west of the south point. The photographs, direct and spectrographic, show prominences even at the north and south poles of the sun's axis.

Marriott's photographs and the eclipse spectra exhibit that the whole southeast quadrant was a tremendously stormy region on the sun. The center of the longest streamers of the whole corona was situated 30° to the east of the south point of the sun, or at position angle 150°. Intertwining the coronal streamers is a beautiful series of coronal

[1] *Publ. Leander McCormick Obsy.*, *5*, 155; *Astroph. Jour.*, *75*, 1, 1932.

domes. At totality, at the base of the long streamers and domes is an extended group of prominences centered at 150° of position angle. These prominences, though apparently very active as shown by the eclipse photographs, did not reach the greatest heights on the sun. These greatest heights were actually attained by prominences also in the southeast quadrant but at 115° of position angle. This is the location of the conspicious feature called the " strawberry dome," with its magnificent series of arches and with its delicate intertwined structure of filamentous details. Facing page 373, at A and B are found drawings from the eastern edge of the sun from the region where the " strawberry-shaped " disturbance is found on Marriott's photographs. At A are drawings from the K and Hα lines of the spectra of the chromosphere, and at B is the same region from the direct photographs and from the lines 5303 and 6374 of coronium. At C on the western edge of the sun, the detail in the 5303 line resembled the cocoanut trees which grew in great profusion on " Tin-Can " Island.

For many long years all eclipse observers have called attention to the connection between coronal streamers and prominences. This dependence of streamers and prominence activity is abundantly verified in 1930, but this eclipse demonstrates the fact that the longest coronal streamers, on which the shape of the corona more or less depends are not necessarily connected with the prominences which at the time of the eclipse are of greatest height.

The long streamers of the corona are best seen near the time of minimum of sun-spots. The splendid series of photographs secured by Lick Observatory expeditions have been utilized by Moore and Baker to measure the direction of the axis of symmetry of the polar brushes, photographs at five separate eclipses exhibiting the polar rays sufficiently well-determined for the purpose. The direction of the sun's equator is inclined about 7° to the ecliptic and 26° 16′ to the terrestrial equator. The measures given in the following table show that the axis of symmetry of the polar rays coincides, within errors of measurement, with the rotation axis of the sun.

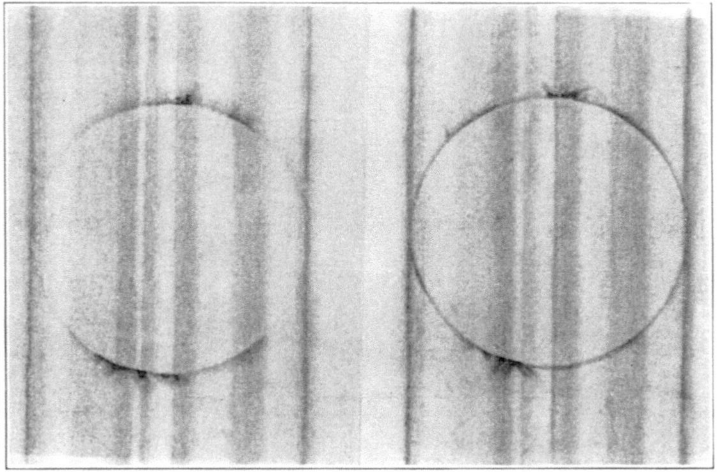

Left: Photograph of 1930 corona by Marriott. *Right:* Drawings from spectra by concave grating, showing (*A*) the ring for 6374 and (*B*) the green ring for 5303 of "coronium."

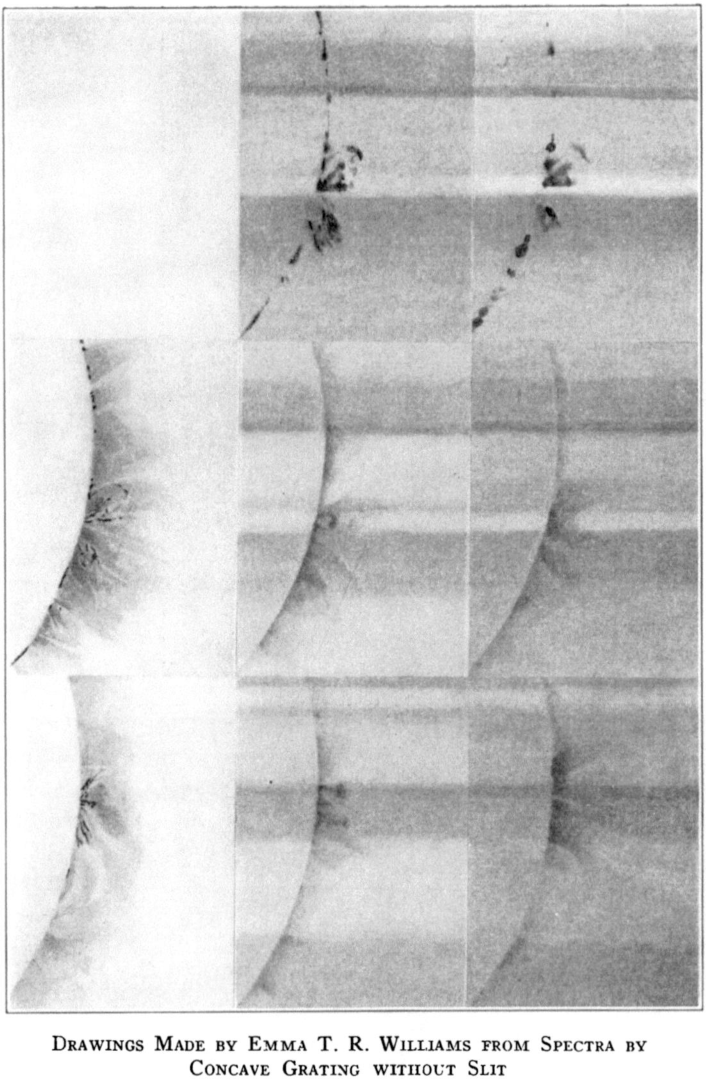

A East

B East

C West

DRAWINGS MADE BY EMMA T. R. WILLIAMS FROM SPECTRA BY CONCAVE GRATING WITHOUT SLIT

A and *B* refer to the eastern section of the sun shown in Plates I and II, and *C* to the western section. (Please note that all positions in *C* are inverted.)

In *A* are given the K line of $Ca+$ and the $H\alpha$ line of the chromosphere from spectra taken immediately following the first flash. In *B* and *C* are given (from left to right) details from Marriott's photographs near the beginning (*B*) and end (*C*) of totality, and the structure in the coronal rings of spectra at λ 5303 and λ 6374, respectively.

Axis of Symmetry of Polar Streamers in Solar Corona
(Position angles measured from North point of Sun)

Date of Eclipse	Axis of Polar Streamers			Rotation Axis	Polar Streamers minus Rotation Axis
	Moore	Baker	Mean		
	°			°	°
1898 Jan. 21	9.4 W	9.3 W	9.4 W	8.2 W	1.2 W
1900 May 28	17.1 W	17.6 W	17.4 W	16.9 W	0.5 W
1901 May 18	21.5 W	20.7 W	21.1 W	20.3 W	0.8 W
1908 Jan. 3	0.9 W	0.8 E	0.0	1.2 E	1.2 W
1922 Sept. 21	24.4 E	24.8 E	24.6 E	25.0 E	0.4 W
				Average	0.8 W

How vacillating and uncertain have been the steps of progress in acquiring information of the spectrum of the corona! And how unsatisfactory that knowledge still is! The reasons are very patent: the coronal spectrum is very weak and the opportunities for investigating it are very few. All observers up to the eclipse of 1882 took it for granted that every spectral line seen at mid-totality, when to the eye there was no visible trace of the chromosphere, must perforce take its origin in the corona. It was with a great shock of surprise that H and K of calcium were seen in 1882 projected on the black face of the moon, where presumably there is no light at all. About thirty bright lines were photographed by Schuster and by Abney at the eclipses of 1882, 1883 and 1886. Even as late as 1893, the H and K lines were assumed to belong to the corona.

Knowledge of the spectrum of the corona may almost be said to begin with the eclipse of 1893, when Fowler with a prismatic camera photographed nine rings, all of which agree in position with lines reported for the 1886 eclipse. At the same eclipse, Deslandres photographed three coronal lines at 3987, 4086 and 4231 A, and also three others at 3164, 3170 and 3237 A, in the ultra-violet. These last three have not been observed since by others, nor have two of Fowler's nine. It was not until 1898 that coronal spectra had acquired sufficient precision to distinguish between the chromospheric line 1474 K at wave-length 5317 and the "coronium" line fourteen angstroms farther to the violet, at 5303. Photographs were secured at the eclipse by Fowler, by Campbell, by Naegamwala, and by Newall and Hills. Additional photographs were secured by Frost in 1900, by Dyson in 1900, 1901 and 1905, by the U. S. Naval Observatory parties in 1900, 1901 and 1905, by Fowler and Lockyer in 1900, and by Lockyer and by Campbell in 1905. More recently the Lick expeditions have secured good photographs, by Lewis, and by Campbell and Albrecht in 1908, and by Lewis, by Campbell and Moore in 1918 and again in 1922. More recently excellent spectra have been secured, in 1925 by Curtis and Burns, in 1926 by Davidson and Stratton, in 1929 by Grotrian, and in 1930 by Mitchell.

At the eclipse of 1918, Slipher found on his photographs the green coronium line extending across the space occupied by the moon's image — due to scattering of light in the earth's atmosphere. At the eclipse of August, 1914, Carrasco, and Bosler and Block, independently discovered a new coronal line in the red, at wave-length 6374 A. In Lick *Observatory Bulletin, 10*, 1, 1918, Campbell and Moore assemble all the reliable observations of coronal wave-lengths determined since the eclipse of 1893. These values for forty lines are given in the second edition of this book on page 329. As an illustration of the precision of knowledge at that time regarding coronal wave-lengths may be cited the measures for the strongest line, that in the green. Before the recent eclipse, the most accurate results came from the 1918 observations, by Campbell and Moore, and by Adams, St. John and Miss Ware. Uncorrected for rotation (0.035 A), and corrected (assuming the corona rotates at the same rate as the photosphere), the values in international angstroms are given:

	Uncorrected	*Corrected*
Campbell and Moore	5302.83	5302.80
Adams, St. John and Ware	5303.02	5303.06

The difference in the corrected values is 0.26 A, while the mean of the two measures is 5302.93. The mean value is in excellent agreement with the 1930 result of 5302.91 A from the measures by Mitchell of eight spectra. Six spectra at the 1930 eclipse give 6374.28 A, for the line discovered at the 1914 eclipse.

It might be said that the corona exhibits three separate spectra: first, the bright-line spectrum of " coronium " existing only in the inner corona and extending on the average to about 5′ or 200,000 km from the sun's edge; second, the continuous spectrum of the middle corona; and, third, the Fraunhofer lines showing feebly in the outer corona.

In the *Revised Rowland Tables,* page 226, 1928, is given the list of bright lines attributed to the corona by Campbell and Moore but revised to include[1] the 1926 eclipse. In *Zeitschrift für Astrophysik, 2,* 106, 1931, Grotrian gives the results of the Potsdam expedition to Sumatra of May 9, 1929. In the following table, the 1930 eclipse is also included. After eliminating the lines due to the high chromosphere, there are only eighteen lines out of forty which are now believed to be truly coronal in origin, although a line or two may still be suspicious. The first column gives the wavelengths in international units. Those at the top of the table, given to two places of decimals, are from the 1926 eclipse by Davidson and Stratton. The values of 5302.91 and 6374.28 are from the 1930 spectra. The line at 6704 was discovered at the 1929 eclipse and verified by the 1926 eclipse photographs. The line at 6776 announced by Mitchell at the 1930 eclipse needs to be verified. The intensities given in the remaining columns are estimates where the strongest lines, 3388 in the violet and 5303 in the green, are usually assumed to be of intensity 20. The intensities given in the table are those assigned by the observers except for the 1918 eclipse where Moore and Campbell's estimates are each multiplied by the factor 2 so that the line 5303 may have an intensity 20. For the 1922 eclipse, the intensities have been furnished by letter through the kindness of Moore. In the second column an average intensity is assigned from the mean of all the eclipses.

The intensities of the coronal lines in the table must be considered to be on a *relative* scale and not *absolute* for the

[1] Davidson and Stratton, *Memoirs of the Royal Astronomical Society, 64,* 105, 1927.

reason that 5303 varies from eclipse to eclipse and is not constant as was assumed in forming the tabular values.

Before taking up the striking differences in the intensities from eclipse to eclipse shown by the table, one must consider the factors upon which the coronal intensities depend. They may be grouped under four heads: (1) color sensitivity of the photographic plate, (2) the dispersion, (3) prism or grating, and (4) slit or slitless. The improvements in plates in recent years, making them more sensitive at the red end, are mainly responsible for the discovery of the lines in the red. Prisms of glass absorb at the violet end of the spectrum;

WAVE-LENGTHS AND INTENSITIES OF CORONIUM

Wave-Length I.A.	Mean Intensity	Hills and Newall, 1896	Fowler, Shackleton, and Lockyer, 1898	Dyson Mean 1900–1901–1905	Lewis, 1908	Moore and Campbell, 1918	Moore, 1922	Davidson and Stratton, 1926	Grotrian, 1929	Mitchell		
										1905	1925	1930
3328....	8	......	....	......	......	......	......	......	8	....	....	....
3387.96..	20	......	....	12	30	......	......	20	20	20	20	20
3454.13..	8	......	....	9	8	......	......	5	8	10	5	6
3600.97..	10	......	....	......	15	10	6	9	12	....	15	15
3642.87..	3	......	....	2	5	4	......	3	......	4	?	4
3800.77..	3	3	3	3	3	2	......	4	1	3	3?	3
3986.88..	8	5	5	3	4	2	4	10	5	8	4	8
4086.29..	6	......	....	......	6	2	2	8	6	....	4	6
4231.4...	8	10	5	5	8	6	8	10	8	6	6	6
4311....	2	......	....	2	1	......	......	2	1	....	....	....
4359....	4	3	3	4	......	......	......	4	1	2	....	2
4567....	4	8	3	6	......	2	......	6	3	....	....	....
4586....	2	......	1	4	......	......	......	4	......	....	....	....
5117....	2	......	....	2	1	......	......	2	5	....	....	....
5302.91..	20	8	10	20	10	20	20	20	20	20	20	20
5536....	1	......	....	2	......	......	......	2	......	....	....	....
6374.28..	12	......	....	......	......	......	......	5	6	....	15	15
6704....	2	......	....	......	......	......	......	......	4	....	....	....

gratings (of speculum metal) and the silver-on-glass reflectors absorb in the ultra-violet. Increase of dispersion spreads the monochromatic images farther apart and weakens the continuous spectrum, with the result that the emission lines by contrast are easier to see on the photographs taken with higher dispersion.

At the 1930 eclipse on Niuafoou there was a striking difference between the coronal spectra taken with the concave gratings and with the prismatic camera of Dr. C. E. Adams of the New Zealand expedition. The prismatic camera gave a very brilliant spectrum — in fact, more brilliant than that of either concave grating. Both forms of spectrographs were without slits. The dispersion of the gratings, however, was ten times the average dispersion of the prismatic spectrum. Many coronal rings were seen on the grating photographs, the lines 5303 and 6374 being well visible even on the brief exposure for the second flash. On the prismatic spectra, on the contrary, the only coronal line visible on any of the photographs was 5303, and even this line was seen with great difficulty. It is readily seen therefore that the dispersion employed has a very important bearing on the intensities, and hence any conclusions drawn from variations in intensities between the lines of one eclipse and another have very little meaning unless the instrumental conditions are fairly constant.

The specially interesting features of the coronal spectra of 1930 are the large dispersion and absence of slit. On the photographs, the image of the sun is 14.5 mm in diameter and the dispersion 10.9 angstroms per mm. All of the coronal rings show much interesting detailed structure, but this is specially true for the strong 5303 and 6374 rings. On the cuts, facing page 372, the vertical line coincides approximately with the axis of rotation of the sun. One can see at a glance that if a slit had been used in 1930 passing radially through the center of the sun, the intensity of 5303 (or 6374) would have depended on the position angle of the slit. If placed near the solar axis, the coronal lines would have been feeble while if it had happened to pass over a more brilliant portion (which could not have been told in advance) the intensity would have been much greater. Hence, for discussing the distribution of coronium about the sun, slitless spectra (if obtained with large dispersion and good definition) are vastly superior to spectra taken with a slit.

With these instrumental sources of variations in intensities clearly in mind, let us look at the estimates of intensities in the foregoing table. There are several striking peculiarities. Many of these, and others, have already been noted by Campbell and Moore. The most remarkable involves the line at 3601 A. It was not observed by any one until the eclipse of 1908, in spite of the fact that it is one of the strongest lines of the coronal spectrum and the quartz spectrographs were adequate to secure it. The photographs by Mitchell at the three eclipses of 1905, 1925 and 1930 were all taken with the same spectrograph. The plates of the three eclipses have about the same sensitivity in the violet. No trace of 3601 is found in 1905 although 3388 and 3454 farther to the violet are strong lines. By way of contrast, in 1925 and 1930 the line 3601 was very strong. (Other peculiarities of this line will be referred to later.) Likewise, 4086 was not observed before 1908. At the 1929 eclipse Grotrian reports three lines missing, 3643, 4586 and 5536. The last line of the three is the weakest of the whole coronium spectrum; it is, moreover, near a *Fe* + line in the chromosphere of 5534.85 and intensity 15 in the flash spectrum. Davidson and Stratton ascribe this line to the corona. The first of the three lines at 3643 is on the average a fairly strong line. Mitchell could not be sure of it from his 1925 spectra, however, though it was well visible in 1905 and 1930.

One's first thought presented for an explanation of these peculiarities is that the intensities vary, and, having settled this to one's satisfaction, the next thing is to try and connect the variation with the sun-spot cycle. There we strike a snag. The years 1900 and 1901 had coronas of minimum type while the 1905 eclipse was of maximum type, and yet 3601 was invisible in all three years and, moreover, it has been a strong line in every eclipse without exception since 1908.

The first step towards finding the origin of the bright lines in the corona evidently will be to explain the peculiarities of intensities just noted by finding relationships between the different lines. To this end, Stratton and Davidson [1] summarize the similarities of coronal structure noted previously by others. Lockyer and Fowler place 4086, 4231, 4400, 4586 and 5303 in one group; 3801, 3987 and 4567 in a second; and 4359 in a third. Campbell and Moore have three groups: 3388, 3601 and 5303; 3801, 3987 and 4567; and 4231 by itself in the third group. Davidson and Stratton grouped 3388 with 3987; 3454 with 3643; and 3601 with 4086 and 5303. They further suggest that " 3643 should go

[1] *Observatory*, *54*, 197, 1931.

with 3801 and 3987," thus making one group of 3388, 3454, 3643, 3801 and 3987.

The 1930 slitless spectra make it very plain that coronium has a very uneven distribution at different position angles surrounding the sun. The limitations of the spectra taken with a slit are at once apparent. If the slit happens to fall on an intense part of the coronal ring, then all lines belonging to a group physically connected should be intensified together; but if the slit crosses a fainter stretch of the corona, all lines of the group should be weakened together. Evidently the comparisons of intensities of lines at one eclipse with those at another (e.g., 1926 and 1929) may mean comparatively little, so much depends on the position angle of the slit. Unfortunately, for the purpose of discussing groups, most of the coronal spectra up to date have been taken with a slit, and, moreover, it must not be overlooked that the slitless spectra obtained at early eclipses have left much to be desired in the matter of definition.

As already noted, the line 3601 is remarkably peculiar, missing in 1905 and very strong in both 1925 and 1930. In the last two years, the lines are very intense on both limbs of the sun but in a very restricted region near the sun's equator. The line 4086 resembles 3601 very closely, and undoubtedly these form a pair. With either one or the other of these lines the following have been grouped: 3388, 4231, 4586 and 5303. No confirmation of this grouping is given by the grating spectra.

Chief interest naturally centers in the green line at 5303. The 1930 spectra, showing details that in shapes resemble eruptive prominences, exhibit the identical details in the line 3388 but in much weaker intensity on account of absorption by the speculum of the grating and the silver of the coelostat. The 1905 coronal spectra likewise show structural details similar in character in the two lines. It is interesting to find the two strongest lines, 3388 and 5303, unquestionably forming a pair. As already stated, 3601 and 4086 cannot be grouped with this pair.

The lines 3454 and 3643 certainly do not form a pair, the latter being missing on Grotrian's 1929 spectra while the former line is strong. Mitchell's 1925 spectra likewise show 3643 missing or doubtful. There seems little reason for grouping 3643 with 3801 and 3987, as suggested by Stratton and Davidson. If the three lines form a group, their relative intensities must increase and decrease together in different eclipses. The grating spectra and also those of Dyson and of Lewis do not confirm this grouping.

All previous authorities have made a group of 3801, 3987 and 4567. The relative intensities of the three with prismatic spectra have been observed as follows: (1896) 3, 5, 8; (1898) 3, 5, 3; (1900, 1901, 1905, Dyson) 3, 5, 6; (1908) 3, 4, invisible; (1926) 1, 5, 3. The relative intensities vary so enormously from eclipse to eclipse that it does not look reasonable to combine all these or to place any combination of these three lines into pairs. The conclusion is confirmed by the slitless grating spectra. After examining all lines of the corona there seem to be only two pairs of lines which appear to show similarities in form, namely, 3388 and 5303, and 3601 and 4087.

All observers who have discussed slitless coronal spectra have come to the conclusion that the green ring at 5303 is weakest near the sun's axis and strongest in the sun-spot zones. A summary of the information[1] including the 1930 eclipse shows that the maximum intensity is always found near prominences but does not necessarily coincide with the prominences.

In comparing the images of 5303 and 6374 in the 1930 spectra, it was surprising to find how little they resembled each other in their structural details of greatest strength and how little either resembled the high-level lines of K and Hα of the chromosphere. Some similarities can be detected between the faint streamers in the lines 5303 and 6374 and those in Marriott's direct photographs, but in the stronger details few likenesses can be found. It was further found that the radiation of 6374 always sticks close to the sun's edge and is more concentrated and more uniform than 5303. The details referred to above are found in the drawings reproduced at pages 372 and 373. From these comparisons it

[1] *Astrophysical Journal*, 75, 21, 1932.

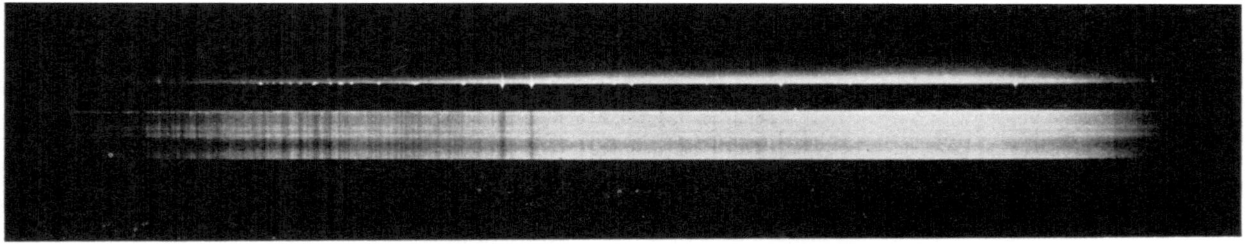

Lick Photograph, June 8, 1918; Spectrum of Corona, West of Sun (above) and Solar Spectrum (below)

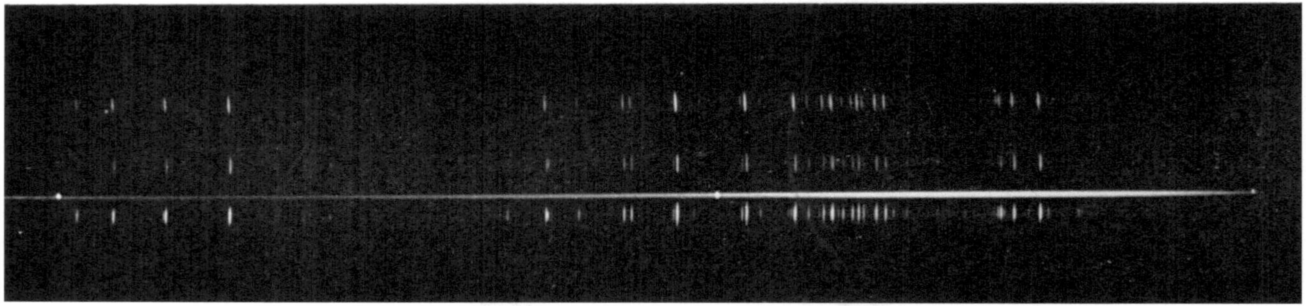

Lick Photograph, 1918, Spectrum for Obtaining Exact Wave-length of Green Coronium Line with Iron Comparison Spectrum

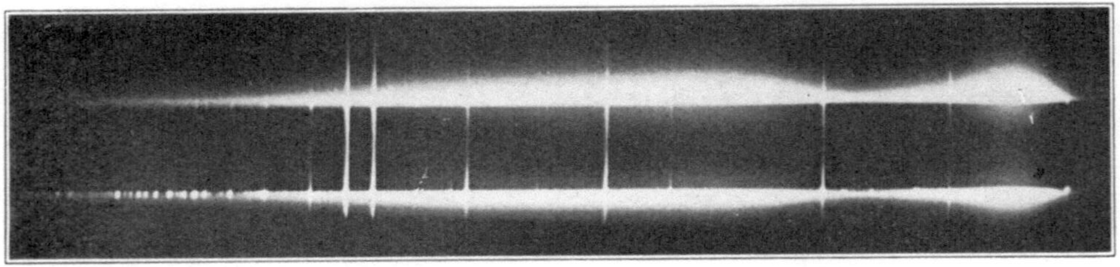

GENERAL SPECTRUM OF THE CORONA, JUNE 8, 1918

Single prism, Lowell Observatory Expedition. The strongest lines are due to the chromosphere and do not originate in the corona. Note that the bright lines stretch across the dark moon. These are caused by diffusion of the light of the chromosphere by clouds in the earth's atmosphere.

is evident that 5303 and 6374 cannot take their origin in the same atom, or at least not in the same atom in the same state of ionization.

On account of the apparent simplicity of the spectrum of coronium, Nicholson, in 1911, began a number of investigations of great interest, the results of which have appeared in the *Monthly Notices of the Royal Astronomical Society.* An attempt was made by him to connect the wave-lengths by series relationships, and furthermore to find an explanation for these series on the hypothesis that the spectral lines took their origin in an atom consisting of a heavy nucleus surrounded by negatively charged electrons. Nicholson's work was the first attempt to explain spectral series by means of Planck's quantum theory of radiation, according to which interchange of energy between systems of a periodic kind can take place only in certain definite amounts, or quanta, determined by the frequencies of the systems. With the enormous increase in knowledge of atomic structure in the past twenty years, Nicholson's investigations have for us now merely a historical or academic interest.

Pannekoek[1] has assumed that the coronal lines may be due to Ca^{++}. At present we know so little of the spectral lines which take their origin in the singly-enhanced, doubly- or triply-enhanced elements, that it is quite futile to guess which of the many possible elements will give the particular series of lines seen in the visible portion of the coronal spectrum. One guess is just about as good as another, and therefore we shall refrain from hazarding a conjecture. The only possible method of securing real progress is to improve the observational data as much as possible.

Freeman[2] believes that argon is the cause of the coronal lines. He finds coincidences in wave-lengths for two-thirds of the forty lines in Campbell and Moore's list (1918). Many of the lines identified by him are no longer regarded as coronal in origin. Russell and Bowen[3] come to the conclusion "that the attribution of the coronal lines to argon is without foundation." This is unquestionably another

[1] *B. A. N.*, *14*, 1922.
[2] *Astrophysical Journal*, *69*, 179, 1929.
[3] *Astrophysical Journal*, *68*, 177, 1928.

case, so frequent in the history of astrophysics, of the identification of origin from mere coincidences in wave-length.

Hylleraas[1] criticizes the suggestion by Rosenthal that the coronal spectrum is due to helium, and finds there is no quantitative agreement to support this theory.

Hopfield[2] finds in the laboratory spectrum of neutral oxygen a line with wave-length 6374.29 and also two lines at 6300 and 6364 which agree in wave-length, within limits of error, with unidentified lines in nebulae. Several workers have produced the green auroral line at 5577 in the laboratory. There is a remarkable agreement in wave-length between the laboratory line and the coronal line at 6374.28.

More recently, de Bruin[3] calculated three additional lines in the neutral oxygen spectrum with wave-lengths 5302.70, 6704.07 and 6775.90. He assumes that these coincide in wave-length with coronal lines and hence concludes that "the mysterious coronium turns out to be neutral oxygen." For the 5303 line, the best value for the coronal wave-length comes from taking the mean of the 1918 and 1930 eclipse results, which gives 5302.92; a difference too large to permit identifying the oxygen wave-length with the coronal. Still more recently, Frerichs and Dingle in *Nature* (*129*, 901, 1932) independently discuss de Bruin's work. The former measures the wave-length 6374.292 for the oxygen line. Both of these physicists come to the conclusion that de Bruin's coincidences are merely accidental and that "the great mystery of the coronal lines remains unsolved."

In addition to the bright line spectrum, the corona shows a continuous spectrum in the inner corona, 8′ or 10′ deep, with Fraunhofer absorption lines visible in the middle and outer corona. The early observations seemed to indicate that at sun-spot minimum the green coronium line was weak and the Fraunhofer spectrum strong, while at sun-spot maximum the emission lines were stronger, and the dark lines weaker. It is only recently that observers have recognized that the presence of thin clouds at the time of the eclipse may greatly affect the visibility of the Fraunhofer lines ob-

1 *Zeit. für Physik*, *69*, 361, 1931.
2 *Physical Review*, *37*, 160, 1931.
3 *Nature*, *129*, 468, 1932.

served in the corona. At the eclipse of 1901, photographed by Perrine in Sumatra through thin clouds, the coronal spectrum showed Fraunhofer lines in the region corresponding to the invisible moon! It is important, for testing various theories of the corona, that the distance out from the sun at which the absorption lines become invisible in the corona should be determined by adequate photographs taken under clear skies.

A great increase in our knowledge of the coronal spectrum came as the result of the observations of the Lick Observatory[1] at the Australian eclipse of September 21, 1922. On account of the perfect weather conditions and long duration of totality an opportunity was afforded for carrying out an extensive program with exposures exceeding 5 minutes.

The spectra taken with each of the instruments of one-prism dispersion exhibited the continuous spectrum and most of the bright lines confined to a region 4′ to 6′ from the sun's limb. The maximum was found for the green line with an extension of 8′ from the edge of the sun. The spectrum of the outer corona seemed unquestionably to show the Fraunhofer lines which were specially visible to the violet of $H\gamma$ where the continuous spectrum was less intense. No trace of the sky spectrum was found to exist beyond the limits of the coronal spectrum. Since the sky was remarkably clear, without the slightest evidence of haze or clouds, it is manifest that the Fraunhofer lines of the 1922 eclipse did not take their origin by reflection in the earth's atmosphere. Moore's observations in thus proving the Fraunhofer lines of the outer corona are caused by the scattering of the sun's light by some means in the corona are of the utmost importance in advancing our knowledge of the perplexing solar aureole. Comparisons of the measures of the lines of the coronal and sky spectra at two points 20′ east and 20′ west of the sun's limb showed that the coronal lines on both sides of the sun were displaced to the red of the corresponding sky lines. The amount of displacement corresponds to a velocity in the line of sight away from the observer of 26 km per second. Since the light observed was reflected sunlight, the measured radial

[1] Moore, *Publ. A. S. P.*, 35, 59, 1923.

velocities supply evidence that the particles of the corona at 20′ from the sun's edge are moving away from the sun with a speed of the order of 20 to 30 km per second.

It is highly probable that the relative intensities of the emission lines of the corona vary with the sun-spot period, though information on this point is very meager. It is difficult to compare spectra secured by different observers at different eclipses using instruments of vastly differing resolving powers, especially since the coronal lines are so weak and are seen projected on a background of continuous spectrum. Under these conditions, the intensities of the bright line spectrum secured with low dispersion instruments will obviously be less intense than for instruments of high dispersion.

The spectra secured with the slit spectrographs and also by means of an instrument without slit seemed to prove conclusively that the coronal emission lines in 1922 were much fainter than those of the eclipse of 1918, also obtained by the Lick-Crocker expedition. The corona of 1922 was of the sun-spot minimum type. Hence, the suspicions of previous observers that the bright lines of the corona are fainter at sun-spot minimum than they are at sun-spot maximum seem to be completely confirmed.

Various observers have investigated the rotation of the corona by measuring from spectra the Doppler effect of motion in the line of sight. It has been generally assumed that the corona rotates with the sun, which is at a rate of 2.0 km per second at the limb of the sun. A rotational speed of this size corresponds to a shift in wave-length of the coronium lines amounting to 0.035 A. And yet, the wave-lengths for this line in 1918 by the Lick and Mt. Wilson observers, each using an efficient instrument, differ by 0.2 A, or six times the rotational shift! It is probably no very great exaggeration to say that at the present time we know absolutely nothing regarding the rotation of the corona. It is not impossible that the wave-lengths of the lines 3388 and 5303 may be determined with greatly increased accuracy, but these lines at best are faint, and the exposures available are very short. The best method of determining the rotational speed will be unquestionably by securing as accurate wave-lengths of these lines as possible, with the slit of the spectograph stretching across the sun's equator. If different values of wave-length are obtained for east and west limbs, thus indicating rotation, the observer should be very careful that he has eliminated all possible instrumental sources. In place of using the solar spectrum as a comparison, it is probable that more accurate wave-lengths will be obtained by using an artificial source, in the manner employed with stellar spectra. The comparison spectrum could readily be superposed on the black moon.

Wright and Curtis describe [1] in detail their unsuccessful attempts, made at the eclipses of 1923, 1925, 1926 and 1929, to determine the rotation of the corona by the method applied to the Orion nebula by Buisson, Fabry and Bourget.

In spite of the pitifully small amount of time available for the investigation of the spectrum of the corona, much knowledge has been acquired; but, unfortunately, suspicion has also been cast on some of the information we thought was fully secured. A brief summary of what we now *think* we know may not be out of place. Three types of spectra must be distinguished: the continuous spectrum close to the edge of the sun, the emission spectrum, and the Fraunhofer spectrum. The coronal spectrum is now fully distinguished from that of the chromosphere. The lines of calcium and hydrogen, frequently photographed at mid-totality, do not belong to the corona, but are caused by the light of the chromosphere being diffused in the earth's atmosphere, usually by thin clouds.

In order to ascertain the atomic origin of coronium, coronal spectroscopic work needs more information about (1) wave-lengths of greater precision, and (2) detection of similarities in structure. To attain the highest accuracy in wave-lengths a slit is desirable. The most useful form is a three-prism spectrograph designed to gather as much light as possible. To secure a wide range of wave-lengths, optical parts of quartz are necessary. To detect similarities in the structure of coronal lines slitless spectra are needed. All

[1] *Sproul Observatory Publications*, No. 11, 1930.

coronal investigations require large dispersion in order to increase the contrast between the bright-line coronal spectrum and the continuous spectrum. The great difficulty is the pitifully small amount of light available together with the short exposures permitted. Unfortunately, a slit cannot be used with a concave grating — there is not enough light. Without a slit there is little hope of finding detailed structure in the coronal lines unless the sun is in an active condition.

Out of eight eclipses observed, the author has actually seen seven coronas, the 1927 phenomenon being blotted out by clouds. The corona that exhibited the most spectacular beauty and made the most lasting impression was not the first eclipse witnessed, that of 1900, but the eclipse of 1918. This was the eclipse of pronounced color, the prominences were large and brilliant. The contrast between the warm, rosy color of the prominences and the pearly-white filmy structure of the corona left a never-to-be-forgotten impression on the mind, that time can never efface. On account of the greater number of prominences visible at sun-spot maximum, the eclipse near maximum provides a more wonderful picture to the unaided eye than does the minimum type of corona. The eclipse of 1927 therefore presented to those fortunate enough to witness it, as gorgeous a spectacle as that of the eclipse of 1842 (see p. 132) which aroused both the populace and the astronomers to such a high pitch of enthusiasm that every eclipse from that day to this has been assiduously observed.

To observe and enjoy the beauty of the corona one needs little but his own good eyes. A good pair of field glasses or a small telescope will permit the study of the prominences or of the details of the inner corona. Baily's beads, at the beginning and at the end of totality can be more fully enjoyed by the use of telescopic power. It should hardly be necessary to add that one should protect the eyes from the glare of the sun while the crescent sun is diminishing before the advent of totality.

CHAPTER XXI

CORONAL THEORIES

AS THE Fraunhofer lines undoubtedly exist in the spectrum of the outer corona, they must be caused by the reflection of chromospheric light by matter existing in the corona in a finely-divided state. Fortunately, there are methods available for testing the question of scattering in the corona, namely, by observations for the determination of the polarization of light. Since the eclipse of 1860,[1] when Secchi and Prazmowski first took up the subject, polarization observations have found their place at almost every total eclipse. At the beginning of the investigations, they were carried out entirely by visual methods; but in a manner similar to what has happened in other branches of astronomical work, photographic observations have gradually displaced visual ones, and as a consequence greater and greater precision has been attained. The accurate determination of the percentage of polarized light has an important bearing on the study of the distribution of matter in the corona. To be of the greatest value, the polarization should be known for different distances from the sun's limb.

In general, there are two different methods of analyzing polarization; one is by the use of double-image prisms, the other by means of plane mirrors. There are many variations of these methods possible. In 1900, Wood employed a direct vision prism before the object-glass of a telescope, the eye-piece containing a Savart plate and a Nicol prism. This combination gave a continuous spectrum crossed by very distinct diagonal interference bands, manifesting fairly strong polarization estimated to equal between 10 and 15 percent. The character of the interference bands indicated that the bright-line spectrum was not polarized, or in other

[1] *Comptes Rendus, 51,* 195, 1860.

words, that the light causing these lines was not reflected sunlight. The appearance presented in the telescope however differed so materially [1] from what had been expected that it took many of the precious seconds of totality for the observer to readjust his ideas; and all the while he could not help but feel that something radically wrong must have happened to the apparatus. At the same eclipse, Dorsey [2] photographed the corona through a double-image prism in the manner utilized by A. W. Wright in 1878. The method gives two photographs of the corona on each plate; one having cut out of it all the light polarized along the line joining the two images, and the other all that polarized at right angles to this direction. Hence, if the corona is polarized radially or tangentially, one image will be deficient in light along the diameter perpendicular to this direction. Which image is deficient along the line joining the centers of the two photographic images depends on the kind of double-image prism used and whether the polarization is radial or tangential. Dorsey also examined the corona visually by a polarimeter consisting of a telescope in the focal plane of which was placed a biquartz half an inch square. The eye-piece contained a Nicol prism, and between the objective and the biquartz was a double pile of plates. The conclusions were that the corona is polarized radially, the visual observations giving the amount of eleven per cent at a distance of 8′ from the moon's limb, a value agreeing well with Wright's 11.2 per cent found at 7′ from the moon's limb at the eclipse of 1878.

No attempt can be made to give here a complete account of the numerous observations made at various eclipses to determine the amount of polarization. Special mention, however, should be made of the excellent work by Newall and by Turner, each of whom has observed polarization effects at several eclipses. In *Lick Observatory Bulletin, 6,* 166, 1911, R. K. Young discusses the measures of photographs secured by Perrine of the Lick Observatory at eclipses of 1901, 1905 and 1908. At the Sumatra eclipse,

[1] *Publications of the U. S. Naval Observatory, 4,* D 116, 1905.
[2] *Publications of the U. S. Naval Observatory, 4,* D 117, 1905.

THE CORONA PHOTOGRAPHED WITH POLARIZED LIGHT. LICK OBSERVATORY EXPEDITION, JANUARY 3, 1908

Note the difference in the distribution of coronal light in the two photographs.

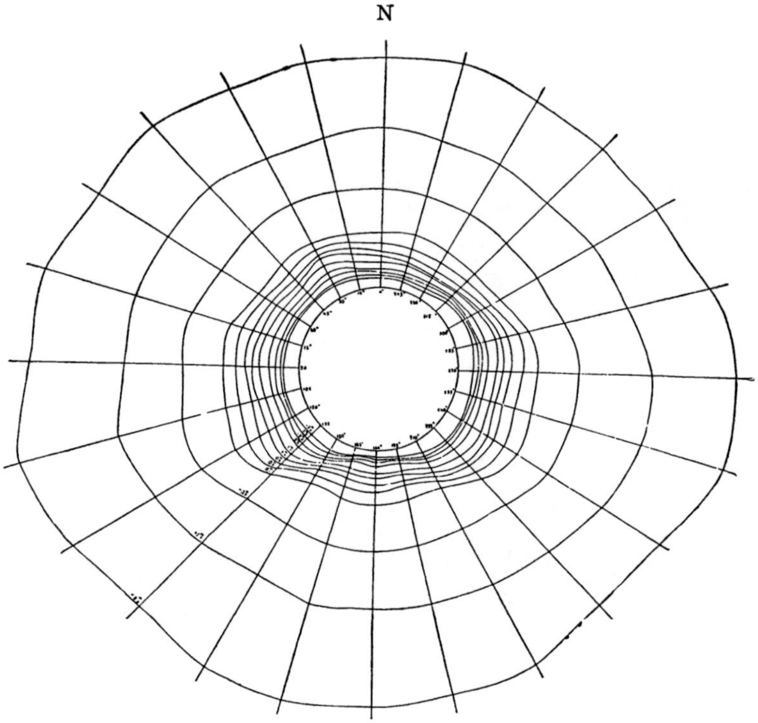

Curves Showing Intensity of Light in Corona, August 21, 1914
Measures by Bergstrand in units of stellar magnitude.

the plates were secured by a double-image camera with the prism set in succession at five different positions separated by angles of a quarter of a right angle. At the two succeeding eclipses, photographs were secured by the same double-image camera and also by a reflecting polarigraph. This latter consisted of three cameras with lenses of three inches aperture and fifty inches focal length. In front of each of two of the lenses was placed a glass reflector, so inclined that the light from the corona was incident at the polarizing angle, the planes of polarization of the two reflectors being perpendicular to each other. The third camera was used merely as a check. The measures showed that the polarization was radial and that the percentage of observed light increased rapidly from the limb, reaching a maximum of thirty-seven percent at 5′ distance, and then diminished slowly, being thirty-five percent at 9′ from the limb. Assuming the well-known law of the reflection and scattering of light that it varies inversely as the fourth power of the wave-length, the value of 11 per cent in the visual region 5600 A would correspond to 33 per cent in the photographic region 4270 A. A close accord is thus seen to exist between the visual values obtained from the eclipses before 1901, and the photographic results from the three eclipses of 1901, 1905 and 1908. In view of the very great difference between the amounts of polarization in the visual and photographic regions, it is highly desirable that values in the visual region be obtained by photographic methods by the use of a color filter and isochromatic plates. In view of the experience of Wood at the eclipse of 1900, it is urged that all observations in the future for polarization effects be made photographically. Observations might possibly still be made visually by experienced observers, but such values should be looked upon merely as checks on the more accurate photographic results. Observers should be most careful to know the amount of polarization caused by the apparatus itself so as to eliminate these effects from the total polarization observed during the progress of the total eclipse. As is well known,[1] every

[1] Wood, *Astrophysical Journal, 12,* 283, 1900.

form of apparatus that disperses light, at the same time polarizes it. A Rowland grating gives strongly polarized spectra; and with prismatic spectra, as the dispersion is increased by additional prisms, the polarization is likewise increased by the new surfaces added.

At the eclipse of 1918, Lewis of the Lick Observatory party, by means of two separate double-image cameras, secured successful photographs in two different regions of the spectrum by using blue and green color filters. The effect for the blue was found to be greater than for the green. Quantitative values for the amount of polarization could not be furnished however for two reasons: first, the law of diminution of the intensity of coronal radiation at different distances out from the moon's limb is unknown; and second, the effect on the corona, of polarization of the light of the sky surrounding the corona, has not been fully investigated. At the eclipse of 1905, Newall[1] found that at a distance of three-quarters of a degree from the center of the corona, the strength of the Savart bands from the sky neutralized those from the corona. This signifies that from the veil of the illuminated sky between the observer and the corona there came as much polarized light as from the corona three-quarters of a degree from the center. Unquestionably, the character and intensity of the atmospheric polarization vary considerably at different eclipses, which of necessity are observed under different conditions of clouds and moisture in the terrestrial atmosphere. On account of the great intensity of the corona close to the sun, for instance, at 1′ from the limb, it is difficult to measure the intensity of the darkening of the photographs and hence to evaluate the amount of polarization so close to the edge of the sun. Newall in 1901 obtained "quite marked polarization" at 1′ from the limb. The Savart photographs for testing polarization seem to possess some advantages over the double-image or reflection methods (see Newall, *loc. cit.*).

The only means of unravelling some of these puzzles seems to lie in determining the law of change in the intensity of the corona at different distances out from the limb of

[1] *Monthly Notices, R. A. S., 66,* 475, 1906.

the sun. The law best known is that of Turner,[1] as the result of photographs obtained in 1898, that the intensity of the corona varies from the edge of the sun outwards inversely as the sixth power of the distance measured from the center of the sun. At the eclipse of 1905, Schwarzschild[2] confirmed Turner's law, and this same eclipse, Graff[3] assumed the correctness of this law to determine the law of blackening of his photographic plates. But at this same eclipse of 1905, Becker[4] found the intensity of the corona subject to a different law, that it varied inversely as the fourth power of the distance counted from a point one-seventh of a solar radius inside the edge of the sun. At the eclipse of 1908 by means of measures carried out by the bolometer, Abbot[5] confirmed Becker's law rather than that of Turner, while R. K. Young (*loc. cit.*) found an intensity depending on the inverse sixth and eighth powers of the distance measured from the center of the sun.

A very different law was found by Bergstrand in a very important publication entitled "Études sur la distribution de la lumière dans la couronne solaire," Upsala, 1919. From photographs secured at the eclipse of August 21, 1914, an attempt was made to determine the relative intensity of light distributed within the corona. The measurement of the absolute intensity and the estimation of the total light of the corona, compared for instance with that of the full moon, did not form part of the program. The problem is one of photometry, and for its solution can be brought the vast experience gained by many years of investigation in determining the magnitudes of the stars. Of the several methods available, Bergstrand adopted the plan of employing twin photographic objectives, mounted equatorially in such a manner that the two solar images could be impressed upon one and the same photographic plate. On the day of the eclipse the times of exposure of the two objectives were

1 *Popular Astronomy, 14*, 548, 1906.
2 *Astron. Mitteil. zu Gottingen, 13*, 1906.
3 *Astron. Abhandl. der Hamburger Sternw. in Bergedorf, 3*, 1, 1913.
4 *Memoirs, R. A. S.*, 57 and *Phil. Trans. Roy. Soc. 207 A*, 1908.
5 *The Sun*, 133, 1911.

made identical, but the aperture of one of the objectives was reduced by means of a suitable diaphragm to one-third that of the other.

The intensity of the silver deposit measured on the plates is the summation of two separate effects, one of which is due to the corona itself while the other comes from the diffuse light of the sky. Added to these two, there is in reality a third effect found close to the moon's limb, that of a halo caused by reflection from the glass-side of the plate of the strong illumination of the inner corona. Fortunately, the intensity of the corona could be separated from the two other effects. The values thus secured do not in any manner confirm Turner's law of the inverse sixth power nor yet the law of Becker according to which the intensity varies inversely proportional to the fourth power. In fact, Bergstrand finds that the intensities near the solar equator differ greatly from those near the poles, the equatorial rays having an intensity three times as great as the polar rays. The equatorial and polar intensities, however, can be brought into relationship with each other in a very simple manner by supposing that the corona is composed of two phenomena. One of them, the "interior corona" exists exclusively in the equatorial zone. In both of these phenomena, the intensity of the light is inversely proportional to the square of the distance measured from the edge of the sun, the intensity of the "equatorial corona" being, however, about double that of the "interior corona." On page 389 is given a curve representing Bergstrand's values. Measures were carried out on solar radii separated from each other by 15°. Position angles are designated from the north towards the east. The intensity at a distance of one radius from the edge of the sun is taken as unity, and values are represented in terms of stellar magnitudes, where a difference of five magnitudes represents a change of a hundred-fold intensity. The strongest coronal rays at times depart sensibly from the direction of the solar radius. This is shown by a jet which leaves the sun at position angle 25° but which does not go out radially and is found in the external curves between 30° and 45°. Some of the most intense rays apparently do not take their origin from the edge of the sun but rather from the front or back side of the solar disk. Moreover, the structure of the corona is highly complicated, since the distribution of jets is not uniformly distributed in all longitudes and since they frequently depart sensibly from the radial direction. On account of the greater strength of the equatorial rays, it was possible for Bergstrand to observe these rays on the photographic plates to a distance of ten radii, or five solar diameters from the edge of the sun, before they diminish in intensity to that of the diffuse sky light. In the polar direction, equality was attained at a distance of three and a half solar diameters.

Valuable observations were secured by photography at the 1925 eclipse [1] by Pettit and Nicholson. They used a Ross 6-inch doublet of 15-foot focus fed by a coelostat. Exposures were made in photographic light and also at wave-lengths greater than 6100 A. Standard photometric squares were impressed on each plate. The plates were mounted on a turntable and measures were made by the Koch microphotometer. Two corrections were applied to the measures, one on account of general scattered sky light, when it was assumed that the mean density of the plate at the four corners gave a measure of the sky illumination. The other correction was for halation and scattered coronal light. No correction was applied for the polarization at the coelostat mirror. Pettit and Nicholson find that their measures in the photographic region confirm the inverse sixth power from the center of the sun, while those in the visual region obey the inverse seventh power. However, the measured values both in the photographic and visual regions more closely conform [2] to the inverse fourth power measured from a point not coinciding with the center of the sun. The 1925 photographs were measured to a distance of three radii from the sun's center while Bergstrand's 1914 photographs were measured to ten radii. Also at the 1925 eclipse, King and Miss Harwood [3] found that the intensity of coronal radiation at dis-

[1] *Astrophysical Journal, 62,* 202, 1925.
[2] *Handbuch der Astrophysik, 4,* 333, 1929.
[3] *Harvard Circulars,* 312, 1927.

tances from the sun's limb greater than 10′.5 follows the law of inverse squares.

Following the eclipse of 1926, Stetson and Andrews[1] discuss the measures made by different observers (including themselves) and come to the conclusion that no single law can express the brightness of the solar corona as a simple function of the distances either from the limb or from the sun's center. Within $2\frac{1}{2}$ radii of the center, the intensity appears to decrease as the seventh power, from $2\frac{1}{2}$ to 3 radii the inverse fourth power fits better, while further out the law is the inverse square.

At the 1927 eclipse, Belanovsky and Perepelkin[2] secured photographs through thin clouds which were measured by the Hartmann microphotometer. From the measures, lines of equal intensities, or isophotes, were drawn which showed that the corona decreased in intensity according to power 2.7 with respect to distances taken near the edge of the sun.

At the 1929 eclipse, the German expedition secured[3] excellent photographs with a horizontal camera of 28-foot focus and with an astrographic telescope of 11-foot focus. A thorough discussion has been made by von Klüber. The photographs of both cameras were measured by the Hartmann microphotometer and isophotes were drawn giving lines of equal intensity in the corona. The astrographic plates show that the corona decreases in intensity according to power 2.5 in units of distances from the edge of the sun. With the horizontal camera, at the sun's poles the plates show that the decrease in coronal intensity is according to the power 2.0, while at the equator the exponent 2.3 more correctly represents the observations. All of the 1929 measures together are best represented by the inverse power 2.4 times the distance from the sun's edge.

It is evident that our knowledge regarding the distribution of light within the corona is in a very unsatisfactory state, since the law of the intensity has been found to be inversely as the second, fourth, sixth or even eighth power

[1] *Astrophysical Journal, 69,* 227, 1925.
[2] *Monthly Notices, R. A. S., 88,* 740, 1928.
[3] *Zeit. für Astrophysik, 2,* 289; *3,* 142, 1931.

of the distance from the sun. It is consequently of great importance that a well-devised form of apparatus be constructed for use at eclipses, and that photographs be secured both in the violet and visual regions on a carefully prepared plan at several future eclipses. The apparatus used by Bergstrand seems to leave little room for improvement. If the photographic plates secured at the eclipse could be impressed by light from a standard source, and if in addition photographs of the full moon were obtained, we should then be in a position of having information additional to that acquired during the progress of an eclipse. We need to know whether the intensity of the distribution of light within the corona follows the same law at every eclipse, or whether this law varies according to the sun-spot period, and we need to know the law both in the blue and yellow regions. Many of the coronas have been observed through clouds or haze, or with varying conditions of transparency. Unfortunately it has not been possible to make proper allowances for these varying factors with the result that the observations are undoubtedly affected by systematic errors. Not until we secure more and better standardized observations can we expect to advance much in the solution of coronal problems.

In considering the measurements made for determining the total light of the corona, we see similar evidences of large systematic errors depending on transparency conditions at the time of the eclipse. The following table is taken from *Handbuch der Astrophysik, 4*, 336, 1929, and brought up to date. The total light of the corona is expressed in terms of the total light from the full moon.

The photographic determinations at the eclipse of 1886 and the two eclipses in 1889 give values that are apparently too small, while one value in 1898 is too large. Giving half weight to each of these four determinations, the weighted mean shows the total light of the corona to possess 55 per cent of the light of the full moon. Results from one expedition may be given here. Kunz and Stebbins observed the 1918 eclipse with a potassium photo-electric cell. They compared the light of the corona with a standard candle,

Total Light of the Corona

Eclipse	Method	Observer	In terms of full moon
1886	Photographic	W. H. Pickering	0.025
1889, January	"	Holden	0.04
1889, December	"	Holden	0.02
1893	"	Turner	0.6
1898	"	Turner	1.1
1898	"	Bacon and Gare	2.7
1905	"	Graff	0.26
1905	"	Schwarzschild	0.17
1908	"	Perrine	0.11
1886	Visual	Abney and Thorpe	0.8
1889, January	"	Leuschner	0.4
1893	"	Abney and Thorpe	1.1
1905	"	Fabry	0.75
1905	"	Knopf	0.85
1908	Bolometer	Abbot	0.20
1918	Photo-Electric	Kunz and Stebbins	0.50
1922	"	Briggs	0.41
1925	Microphotometer	Parkhurst	0.27
1925	Thermocouple	Pettit and Nicholson	0.52
1926	Thermopile	Stetson and Coblentz	0.52

with two electric lamps, with the full moon, and with an area of the sky during totality and during full sunshine. Their numerical results are as follows:

Observed total light of the corona	0.60	candle-meters
Same, corrected to outside of atmosphere	1.07	candle-meters
Observed ratio of corona to full moon	0.6	
Same, corrected to outside of atmosphere	0.50	
Observed ratio of corona to sky circle of diameter one-half degree and 8° from uneclipsed sun	0.105	
Same during totality	640.	

The photo-electric cell with which these measures were secured had a maximum sensibility at wave-length 4500 A in the blue.

Further details about the earlier eclipses may be obtained from *Handbuch der Astrophysik.* In order to illustrate the inconsistencies in the best work along these lines, a few words may be said about the more recent eclipses, beginning with 1925. At this eclipse, Harvard Observatory occupied four stations in order to measure the total light of the corona by a photometer of "pinhole" type designed by

King. It was found that the integrated brightness within a circle 3° in diameter, given in stellar magnitudes, is: photographic −10.96, photovisual −11.61; within a circle 6° in diameter the values are −11.40 and −11.71, respectively. After eliminating the effect of the illuminated sky, the stellar magnitude of the corona is: photographic −10.76, photovisual −11.57. With one of these photometers at the 1926 eclipse, Stetson and his co-workers obtained conflicting results in the circles 3° and 6° in diameter; but the measures seemed to show that the corona was 40 per cent brighter in 1926 than in 1925.

It is interesting also to compare the measures of the total illumination of corona *plus* sky. In 1925 the horizontal illumination was 0.24 foot-candles, or equal to the total intensity 30 minutes after sunset; in 1926, measures by the Macbeth illuminometer gave 0.14 foot-candles; in 1929, a value 0.15 and 1930 (Niuafoou Island) 0.38 foot-candles. The measures on the three last eclipses were carried out by Stetson's instruments. The above measures seem to show that although the 1926 corona was 40 per cent brighter than that of 1925, the total illumination of corona and sky together was 40 per cent fainter.

Another instance of the uncertainties underlying measures of coronal radiation was shown at the 1926 eclipse. With the photographic photometer it was found that the total brightness of the corona was but one-tenth that of corona plus sky as measured with the illuminometer.

At the eclipse of 1918, Aldrich[1] found that the total brightness of the sky during totality was less than that of twilight one hour after sunset of the same day. At the Australian eclipse of 1922, Ross directed a camera, from which the lenses had been removed, toward the south celestial pole and exposed a photographic plate during totality. Other plates from the same box were exposed on the evening of the same day for equal intervals of time at 6:14, 6:17, 6:20, 6:23 and 6:26 by the clock. The central portions of the plate were cut out and all six plates were developed together. The plates showed a regular gradation.

[1] *Smithsonian Miscellaneous Collections, 69,* No. 9, 1919.

It was found that the illumination at the south celestial pole corresponded with that when the sun's center was 97° 29′ from the zenith.

It is evident from the above summaries of measures that our present knowledge regarding the total amount of light received from the corona is in a very unsatisfactory state. Photometric methods present many complications. When results at one eclipse are compared with those of another, the instruments must be thoroughly calibrated and methods properly standardized before we can have any confidence in the conclusions. To obtain the total light of the corona, it is necessary to extrapolate to the edge of the moon and even to the edge of the sun. Unfortunately we seem to have little sound information about the law of intensity in the corona close to the sun's edge. The greatest difficulty of all is that the corona perforce must be observed through clouds and haze at half the eclipses where the corona is at all visible and through varying transparencies at the other half. Up to the present we know of no adequate methods of making proper allowance for varying atmospheric conditions. Taking everything into consideration, it seems highly probable that the accidental and systematic errors existing in measures of coronal radiation certainly amount to 20 per cent, and possibly 50 per cent, of the final values. Roughly speaking, the total light of the corona is one-half that of the full moon and one-millionth that of the noon-day sun.

Hence we must not take too seriously the attempts to correlate intensity of coronal radiation with sun-spot activity. It will be necessary to await more accurate observations of the future. It seems entirely probable that the inner corona at sun-spot maximum must be brighter than at sun-spot minimum. Moreover, as the inner corona contributes the most energy to the total coronal radiation we would therefore logically expect that the total energy at maximum of spots is greater than at minimum. But to confirm this from observations already secured is another story. When we consider the total amount of time available for coronal investigations, we must not be too discouraged with the results obtained to date.

In former chapters it has been shown that for half a century it has been recognized that there is a close connection between the shape of the corona and the sun-spot cycle. At minimum of spots the corona shows the long equatorial extensions and the strong polar brushes, while at spot maximum the corona is more nearly circular in outline. The first to call attention to this was Ranyard in 1879 in the " Eclipse Volume " of the Royal Astronomical Society. In 1897, Hansky made the connection more certain by publishing a series of reproductions of the corona arranged according to the sun-spot curve.

In *Handbuch der Astrophysik, 4,* 317, 1929, there is stated the general problem. For the past forty years, since accurate photographs became available, the dates of maxima and minima of spots have been as follows:

Maximum	*Minimum*
1893.6	
	1901.7
1905.6	
	1913.6
1917.6	
	1923.6
1928.5	

The eclipses of the past decade have been beautifully situated with respect to the spot cycle. The eclipse of 1922 took place shortly before minimum of spots, that of 1923 almost exactly at minimum. A year and a half after minimum came the 1925 eclipse, while those of 1926 and 1927 were just before and 1929 and 1930 just after maximum.

The information on the activity of the sun, as recorded in the sun-spot cycle, comes, as the name signifies, from observations on spots, their Zürich relative numbers, the Greenwich means areas, the latitude of spots, etc. Similarly observations made on prominences, give also the number, mean areas, mean latitude, etc. The sun-spot and prominence curves closely parallel each other. In a sense, the information from prominences supplements that from spots in that the spots are phenomena observed on the face of the sun while the prominences are photographed only at the sun's limb.

To observe spots a moderate equipment, merely a telescope, is all that is necessary. The activity of the sun manifesting itself in spots can be observed as the spot moves across the face of the sun; measurements of polarity, etc., continued from day to day, give a fairly faithful indication of the relative activity of the sun. In comparison, the prominences are transitory phenomena. Even if individual prominences were more permanent than they are, the rotation of the sun would carry them quickly out of sight. Occasionally when a total eclipse comes, the direct photographs of prominences and the eclipse spectra can be compared with spectroheliograph photographs. Then we see the limitations of the latter method. The eclipse photographs on October 21, 1930, showed the sun with a very stormy region in the southeast quadrant, while the Mt. Wilson spectroheliogram taken only a few hours earlier showed nothing particularly remarkable about the activity in this quadrant. Unless the prominence exhibits both height and contrast, through masses of gases in eruption, it is usually not a conspicuous object on the spectroheliograph plates. Hence, all things considered, it has been generally felt by the average astronomer that the information from spots gives a more reliable indication of the activity of the sun than is obtainable from observations of prominences. As already stated, however, the two types of observation supplement each other.

When attempts are made to find correlations between coronal disturbances and either spots or prominences, it is evident that many more connections must always be found between prominences and coronal structure than between spots and corona. The spots are seen on the *face* of the sun while both the corona and prominences stretch out from the *edge* of the sun. It must not be thought that there must be a more intimate connection between prominences and coronal activity than there is between sun-spots and coronal disturbances. The eclipse photographs show prominences and disturbed regions in the corona but from the nature of things cannot show spots. Unless the eclipse astronomer observes the spots on the final days before the eclipse—and he is

THE INNER CORONA IN 1926

Photographed with the 63-foot tower telescope by the Swarthmore College Expedition.

The Lion-Like Prominence at the 1929 Total Eclipse
Photographed with the 63-foot camera by the Swarthmore Expedition.

usually too busy with a thousand and one preparations that must be made — or unless he looks up the literature afterwards (and this is rarely done), the possible connections between individual spots and coronal disturbances pass unnoticed.

Ludendorff has done a valuable piece of work in providing a simple method of measuring the shape of the corona. He had the happy inspiration of utilizing the published photographs and half-tone reproductions to draw roughly by eye lines of equal intensities (isophotes) in the corona. In some instances, isophotes were already available in the original sources, especially in the corona of 1905 from the investigations of Graff and in that of 1914 from the hands of Bergstrand. Comparisons of the results from Ludendorff's tracings with the more accurate methods of Graff and Bergstrand showed no systematic differences. From all of the tracings of each eclipse, two quantities a and b, were determined by the method of least squares. The value a is the ellipticity of the corona at the edge of the sun, while $a + b$ is the ellipticity at a distance of one radius out from the sun's edge. From the examination of thirteen eclipses between 1893 and 1927, Ludendorff found that the quantity a was nearly constant for all eclipses. On the contrary, the value b varied. For the coronas near sun-spot maximum, the value of b is very nearly zero, the corona being approximately circular in outline. Near minimum of spots, however, b has increased in value, the corona being more elliptical.

Mitchell has found a closer connection between the coronal shape and sun-spot numbers than there is with the phase in the spot cycle measured from the time elapsed from minimum or maximum of spots. The table in *Handbuch der Astrophysik*, *4*, 338, 1929, is here extended to include the eclipses of 1922, 1929, and 1930. The values from the 1929 eclipse are from isophotes discussed by von Klüber (*loc. cit.*). Photographs of the 1922 eclipse were kindly sent me by Moore, and of the October, 1930 eclipse by Marriott. The measures of both eclipses (following Ludendorff's methods) were made by Miss Williams.

Spots Decreasing				Spots Increasing			
Eclipse	Spot Numbers	*a*	*b*	Eclipse	Spot Numbers	*a*	*b*
				1893.3	88	0.03	*0.03**
				1926.0	77	0.06	0.01
1918.4	69	0.13	0.10	1927.5	67	0.04	0.00
1929.3	59	0.10	0.02	1905.7	59	0.01	0.00
1908.0	45	0.06	0.02				
1930.8	35	0.04	0.23				
1898.1	27	0.06	0.12				
1896.6	22	0.03	0.23				
1901.4	11	0.04	0.26	1923.7	11	0.06	0.18
1900.4	9	0.03	0.29	1914.6	9	0.05	0.16
				1925.1	8	0.05	0.10
1922.7	6	0.04	0.24				

* All values of *a* and *b* are positive except *b* of the 1893 eclipse.

On the left side of the table are the eclipses between maximum and minimum of spots, or on the descending branch of the sun-spot curve, while on the right hand are the eclipses after minimum with spots increasing in numbers. One should read down on the left and then up on the right side with spots increasing. Instead of arranging the material according to the phase in the spot cycle, as was done by Ludendorff, it has been arranged according to the mean of the spot numbers in a synodic month of the sun's rotation, i.e., the mean of 13 days before and after the eclipse, also the day of the eclipse itself. It might be added that von Klüber, from isophotes drawn from microphotometer measures of the 1926 eclipse, obtained + 0.05 and + 0.02 for *a* and *b* respectively, in excellent agreement with Ludendorff's rougher measures.

Bergstrand [1] also takes Ludendorff's figures and applies a correction to the values of the isophotes in polar regions to allow for the effect of equatorial and mid-latitude streamers being superposed over polar rifts. He finds a correlation with sun-spot numbers but a closer connection with prominences in high latitude zones.

Lockyer [2] makes a very valuable compilation by collecting

[1] *Arkiv. Mat. Astr. och Physik. A, 22,* No. 1, 1930.
[2] *Monthly Notices, R. A. S. 91,* 797, 1931.

into one diagram, beginning with the year 1860, information regarding sun-spots, prominences and coronal shapes.

On account of the connection between coronal and other solar disturbances, already alluded to, and on account of the fact that prominences are much more widely distributed in heliographic latitudes than are spots, it is not surprising that Lockyer finds a close connection between the shape of the corona and prominences, the maximum type of corona occurring when prominences are near the sun's poles.

The quantities *a* and *b* in the table are exceedingly interesting. The mean of all 16 values of *a* is 0.05. The only eclipse with outstanding values, that of 1918, is explained in *Handbuch der Astrophysik*.

The three eclipses 1900, 1901 and 1922 took place on the average 0.9 years *before* the time of minimum of spots. Each of these eclipses showed the typical " minimum " corona; the mean value of *b* is 0.26. On the other hand, the three eclipses of 1914, 1923 and 1925 took place the same interval (0.9 years) *after* spot minimum, and yet the coronas had lost their minimum characteristics, *b* having the mean value 0.15. The three eclipses 1896, 1898 and 1930 took place about four years before minimum of spots. The mean value of *b* amounting to 0.19 is greater in size than that at a time less than one year after minimum. Hence, it is evident that two years before spots and prominences are at a minimum, the corona takes on the " minimum " shape.

With the recent spot maximum occurring at 1928.5, the eclipses of 1926, 1927 and 1929 were all "maximum-type" or circular coronas. The 1926 eclipse took place 2½ years before spot maximum. It would seem also that the "maximum " type of corona takes place two years before maximum of spots. The greatest surprise in the whole of the tabular values is for the eclipse of October, 1930. Taking place a year and a half after the 1929 eclipse and only a little more than two years after maximum of spots and probably four years before the next minimum, the value of *b* was approximately equal to that of a minimum-type of eclipse.

Bernheimer [1] agrees with the writer that the minimum

[1] *Meddel. Lunds Astron. Obs.*, Ser. I, No. 126, 1931.

type of corona takes place before minimum of spots. He also finds that the coronal shapes are more closely connected with sun-spot numbers than with the phase in the spot cycle.

After observations of sun-spots during three hundred years we are forced to the conclusion that the 11-year curve must be regarded as quite erratic. The solar activity from numbers and areas of spots, or from frequency and areas of prominences, gives us curves which are similar but are not identical. Other closely allied curves, such as terrestrial magnetic phenomena, at times depart from the general run of the spot and prominence curves. In spite of numberless investigations, it is still unsafe (see p. 120) to predict, even a few years in advance, what the sun-spot cycle is going to do. Recognizing these limitations, or in other words, realizing that the sun has sporadic and unexpected bursts of activity, it is well not to make too great claims for the dependence of anything on the sun-spot curve.

The corona takes its shape primarily from the lengths and position angles of the longest streamers. The eclipse of 1930 showed that the longest streamer was connected with solar activity that had persisted for several days. Hence it appears certain that the corona can have no constant shape and that no doubt it varies from day to day depending on whether the active solar areas are near the sun's limb when only they could give an effect projected on the sky background. The corona we happen to see or to photograph during the few fleeting seconds of totality is a temporary phenomenon which will subject itself to exact analysis with great reluctance.

In spite of all that is said above, the author is going to be rash enough to predict that the 1932 corona, two years before the expected minimum of spots, will show the minimum type of corona with long equatorial streamers and strong polar brushes.

When we attempt to solve the enigma of the corona, we are face to face with one of the most difficult problems in the whole realm of astronomy.

First of all, the corona is not an atmosphere of the sun consisting of atoms and molecules attracted by gravity. But gravity on the sun is 27 times more powerful than the value on the earth, while the corona has been observed to the enormous distance of ten million miles from the surface of the sun. Thus it is easy to see that if the corona were truly atmospheric in its nature, the resulting pressure would be colossal. The chromosphere is indeed the solar atmosphere, not one of oxygen and nitrogen as on the earth, but of the gases found to exist in the flash spectrum, each heated to high temperatures due to their proximity to the sun, and yet the pressure at the base of the chromosphere is much less than one-thousandth of the earth's pressure at sea-level.

In the forceful words of Simon Newcomb[1] we may remind the reader that "the great comet of 1843 passed within three or four minutes of the surface of the sun, and therefore directly through the midst of the corona. At the time of nearest approach its velocity was 350 miles per second, and it went with nearly this velocity through at least 300,000 miles of corona, coming out without having suffered any visible damage or retardation. To form an idea of what would have become of it had it encountered the rarest conceivable atmosphere, we have only to reflect that shooting stars are instantly and completely vaporized by the heat caused by their encounter with our atmosphere at heights of from 50 to 100 miles; that is, at a height where the atmosphere entirely ceases to reflect the light of the sun. The velocity of shooting stars is from 20 to 40 miles per second. Remembering, now, that resistance and heat increase at least as the square of the velocity, what would be the fate of a body, or a collection of bodies like a comet, passing through several hundred thousand miles of the rarest atmosphere at a rate of over 300 miles a second? And how rare must such an atmosphere be, when the comet passes not only without destruction, but without losing any sensible velocity? Certainly so rare as to be entirely invisible, and incapable of producing any physical effect." Other comets, the great comet of 1882 for instance, have almost grazed the sun's surface.

[1] *Popular Astronomy,* 265, sixth edition, 1887.

Any adequate theory of the corona must be capable of explaining[1] the following facts:

1. The total brightness of the corona is very small, being about one-half that of full moon and one-millionth that of the sun. Half of the total light comes from a zone extending only 3′ from the limb of the sun.

2. Its spectrum shows the bright emission lines of coronium extending to a maximum distance of 8′, and also a continuous spectrum in the inner corona and Fraunhofer lines in the middle and outer corona.

3. The emission lines of the spectrum are fainter at minimum of spots than at maximum.

4. The emission spectrum due to "coronium" contributes only a small fraction of the total energy of coronal light. The predominant part lies in the continuous spectrum. The distribution of energy in this continuous spectrum differs little from that of the sun.

5. Polarization is a maximum at a distance of about 5′ from the sun's limb, and it diminishes more rapidly towards the sun than away from it.

6. Matter of any kind so close to the sun must be very hot and must reflect and scatter the solar rays.

7. According to the observations of Abbot, the coronal materials are deficient in heat rays.

8. If the sizes of the coronal particles change in diameter at different distances from the sun's limb, a corresponding change in color of the corona would result. No change in color is noticeable.

9. The internal motions in the corona are very small.

10. The sun exhibits a magnetic field.

11. According to Bergstrand, the intensity of the corona varies inversely as the square of the distances from the sun's surface. Other authorities derive the inverse fourth, sixth, seventh or eighth powers.

12. It is necessary to explain the changing form of the corona with variation in the sun-spot period.

In the bright inner corona is found the emission spectrum

[1] See also Abbot, *Smithsonian Misc. Collections, 52,* 31, 1908, and *Lick Observatory Bulletin, 5,* 15, 1908.

of coronium and a strong continuous spectrum. In the less intense outer corona, the Fraunhofer lines are seen. In the inner corona, due to proximity to the sun's surface, are found temperatures approximating those in the photosphere. On account of minute pressures existing in the corona, the atoms may readily lose one, two, or more external electrons. The long free paths permitted by these small pressures cause conditions that cannot be approximated in our terrestrial laboratories. Unfortunately, due to the paucity of light and the small amount of time available for investigations, coronal wave-lengths are much less accurately determined than those of the nebulae. The source of the coronium lines will unquestionably be found among the elements in the top part of the periodic table. The "coronium" problem will be more difficult than that of "nebulium."

As the coronium emission constitutes a very small part of the total energy found in the corona, it may practically be left out of consideration in attempting to find a theory to explain the radiation of the corona. Any adequate theory must give solutions to the following problems: (1) What is the cause of the strong continuous spectrum of the inner corona, and (2) What is the explanation of the Fraunhofer spectrum in the outer corona?

The first theory of the corona to be propounded was the one, so very popular at the time and thought to be capable of explaining away most of the astronomical problems — the meteoric hypothesis. In virtue of this theory, the corona is nothing more nor less than the trails of myriads of meteors as they fall into the sun. Even at the time[1] a great authority on meteors, Newton of Yale, pointed out that the details observed in the corona were "inconsistent with any conceivable arrangement of meteoroids in the vicinity of the sun." The hypothesis was such an artificial one and had so little to commend itself that it is surprising that it found such a large place in astronomical literature — but then we must not forget that very little was known concerning the corona.

Schaeberle's mechanical theory of the corona has much

[1] *Nature*, Sept. 30, 1866.

more to recommend it — but it too does not seem to have been able to bear the test of time, at least not in its original form. By virtue of this theory, the corona is caused by light emitted and reflected from streams of matter ejected from the sun by forces acting along lines normal to the surface of the sun. The forces are most active near the center of the sun-spot zones, and consequently, are confined almost wholly to the equatorial regions. Hence, as a result of this theory, the rays seen around the poles of the sun can have no existence, except that of streamers from the equatorial regions seen projected by perspective above and below the poles. In order that the force of ejection may be sufficiently great to overcome the attraction of gravitation, it was necessary to ascribe to the materials forming the longest rays initial velocities as large as 400 miles per second. While velocities of this size are not impossible on the sun, the truth of the matter is that no such motions have ever been discovered in the corona. Hence it is necessary to discard the theory or to modify it in some essential details. Moreover, "according as the observer is above, below, or in the plane of the sun's equator, the perspective overlapping and interlacing of the streamers cause the apparent variations in the type of the corona." This explanation might satisfy an annual variation (which is not known to exist), but fails to account for the change in form coincident with the eleven-year sun-spot period.

There are, unquestionably, many forces acting on the coronal materials repelling the particles away from the sun in opposition to gravitation, and as these forces have been considered one by one, different coronal theories have been propounded. In 1885 the "electrical theory" was announced by Huggins.[1] The eclipse of 1889, of the minimum sun-spot type, having exhibited strong polar rays much resembling the lines of force about a magnet, Bigelow[2] brought forward the "magnetic theory," and Ebert[3] the "electro-magnetic theory." Bigelow's theory is very successful in explain-

[1] *Proc. Royal Society, 39,* 108, 1885.
[2] *The Solar Corona discussed by Spherical Harmonics,* 1889.
[3] *Astronomy and Astrophysics, 12,* 804, 1893.

ing the details of the minimum type, but not so fortunate with the other forms of corona. The recent investigations at Mt. Wilson have shown that a magnetic field does exist around the sun, as demanded by Bigelow's theory.

Great have been the claims of the exponents of the " radiation-pressure theory " not only for explaining the details of the corona, but also for furnishing a rational elucidation of why comets' tails always point away from the sun, and what causes the aurora borealis, zodiacal light, etc. That a ray of light exerts a pressure on any surface on which it impinges comes as a direct result of the Electro-magnetic Theory of Light published in 1873 by Clerk Maxwell [1] and, as was shown by Bartoli [2] in 1876, can be deduced from the second law of thermodynamics.

Since the light-pressure depends directly on the intensity of the sun's radiation, which decreases inversely as the square of the distance, as is also the case with gravity, the ratio of pressure to weight is therefore a constant independent of the distance from the sun. The manner in which this ratio is found was first shown by Bessel [3] in 1836, who computed the magnitude of the repulsive force from the curvature of the tail of the comet in 1811. Bredichin,[4] more recently, from measures of many comets' tails, has found them to be of four different types, in which the repulsive forces are respectively 18.5, 3.2, 2.0, and 1.5 times the attraction of gravity; the straight tail, according to his ideas, consisting of hydrogen, the plume-like tail of hydrocarbons, and the short stubby one of metallic vapors, chief among which are iron and sodium. The electrical force, on which Bredichin explains his repulsions, has been shown by Lebedew [5] not to have a sound physical basis.

This objection cannot be raised to the principle of Arrhenius. That light actually exerts a pressure has been shown by Lebedew,[6] and Nichols and Hull [7]; the latter, in-

[1] *Electricity and Magnetism,* 792.
[2] *Il Nuovo Cimento, 15,* 195, 1883.
[3] *A. N., 13,* 185, 1836.
[4] *Annales de l'Observatoire de Moscou,* (2), *1,* 45, 1886.
[5] *Astrophysical Journal, 16,* 155, 1902.
[6] *Annalen der Physik,* (4), 6, 433, 1901; *Astrophysical Journal, 15,* 60, 1902.
[7] *Physical Review, 13,* 307, 1901; *Astrophysical Journal, 15,* 62, 1902.

deed, have succeeded[1] in producing a laboratory comet's tail, although, as pointed out by them, other forces than light-pressure probably helped to give the repulsion.

However, a rigid application of the theory of Maxwell is possible only when the body acted upon is large compared with the vibrations of light itself. When the body is of a size approximating the wave-length of light, Schwarzschild[2] has shown that the maximum value of the repulsive force is about twenty times the attraction of gravity.

Undoubtedly, the radiation-pressure theory has been of the very greatest assistance in dealing with the corona, for the reason that it provides us with a knowledge of an additional force acting in a direction in opposition to gravitation. Miller has published a series of excellent papers[3] enquiring into the question whether the coronal streamers exist in accordance with a modified Schaeberle mechanical theory, that their motions are produced by ejection, by the rotation of the sun, by the attraction of the sun and by the radiant pressure of the sun. For the purpose of the investigation, Miller examined the excellent series of photographs of the corona obtained by the Lick Observatory expeditions from 1893 to 1918, inclusive, also plates secured by himself in 1905 and 1918, and Lowell photographs taken in 1918. The conclusions drawn are that the force of repulsion is surprisingly large, being almost equal to the attraction of gravitation, and as a result, it is unnecessary to assume the very large velocities of Schaeberle's original theory. The facts accumulated seem to be in fairly satisfactory accord with the theory of Arrhenius expressed as follows:[4] " It is very probable that those drops, for which gravitation is just compensated by the pressure of radiation, will be the chief material of the inner corona. For drops of other sizes are selected out, the heavier ones by falling back to the sun, the lighter ones by being drawn away by the pressure

[1] *Astrophysical Journal, 17,* 352, 1903.

[2] *Sitzungsberichte der math.-phys. Classe der k. b. Akademie der Wissenschaften zu München, 31,* 293, 1901.

[3] *Astrophysical Journal, 27,* 286, 1908; and *33,* 303, 1911; and also *Publ. A. S. P., 32,* 207, 1920.

[4] *Lick Observatory Bulletin, 1,* 152, 1902.

of radiation, so that just those drops which, so to say, swim under the equal influence of gravitation and pressure of radiation will accumulate in the corona."

By properly choosing his mathematical constants, it was possible for Miller to reproduce each and every coronal streamer so far discovered on eclipse photographs, even those streamers found near the sun's poles. Hence, the mechanical theory supplemented by that of radiation pressure seems to have solved many of the perplexing questions regarding the corona, but there are still many difficulties to be surmounted before we can feel ourselves on thoroughly sure ground. It is unquestionably necessary to take into consideration magnetic and electric forces. Like radiation pressure and gravitation, these forces modify the other effects in a quantitative manner.[1]

That matter can exist in the finely divided state required by the theory of Arrhenius was shown[2] by "the eruption of Krakatoa, which drove the fine ashes up to an elevation of 30 km (18 miles). The finest particles of these ashes were slowly carried by the winds to all parts of the earth, where they caused, during the following two years, the magnificent sunrises and sunsets which were spoken of as 'the red glows.' This glow was also observed in Europe after the eruption of Mount Pelée. The dust of Krakatoa further supplied the material for the so-called 'luminous clouds of the night,' which were seen in the years 1883 to 1892 floating at an elevation of 80 km (50 miles), and hence illuminated by the light of the sun long after sunset."

If the temperature of the inner corona is approximately 5000° C, it is quite difficult to see, as was pointed out by Abbot,[3] how matter can exist in the solid or liquid state or how the dust-particles of the inner corona can be "drops of liquid metal."[4] Another criticism of almost insurmountable character is that voiced by Eddington.[5] Owing to the fact that conditions in close proximity of the sun cannot

[1] See Pringsheim, *Physik der Sonne*, 330, 1910.
[2] *Worlds in the Making*, 7.
[3] *Lick Observatory Bulletin*, *5*, 20, 1908.
[4] *Ibid.*, *2*, 188, 1904.
[5] *Monthly Notices, R. A. S.*, *80*, 723, 1920.

be duplicated in the laboratory, we are ignorant of the true laws of radiation-pressure, which may have "encouraged quite exaggerated ideas of the possible effects of radiation-pressure." The "upper limit" to its power of supporting or driving out matter has been calculated by Eddington, and has been found to be equivalent to a pressure of 2 dynes per sq. cm. This can be likened to a wind of this strength, and the exact effect of any material will depend on its power of absorption — of stopping the wind instead of letting it blow through. Allowing an ample margin for uncertainties of observation, Eddington calculates that the "pressures of radiation cannot carry a total weight of more than a milligram per sq. cm." Applied to the chromosphere, the density is found to be of the approximate size of 10^{-12}, a quantity which indeed appears so absurdly small that it seems to contradict all our ideas concerning prominences. The density of the corona and in comets' tails on the same hypothesis is a thousand times smaller, or 10^{-15}.

Apparently therefore, the radiation-pressure theory is not entirely free from perplexities, but these difficulties may not be entirely insurmountable. Before further progress is made we must ascertain, by observational means, the size and the number of particles per unit volume forming the corona at different distances from the sun, and also the temperatures according to the Stefan and Wien-Planck formulas. The luminosity of the bright inner corona is caused by particles which are heated to incandescence by solar radiation and which scatter sunlight, the particles being subject to gravitation, to radiation pressure and to electromagnetic forces largely unknown. Owing to their proximity to the sun, these particles cannot be in the solid or liquid condition and must therefore exist in the gaseous state. The molecules of the coronal gases strongly illuminated by sunlight probably act like the fine particles of a fog in scattering light. According to Fabry,[1] the part of the luminosity of the corona which gives the continuous spectrum may be due to this diffusion and not to reflection by small particles. As the result of some experiments on the diffusion of light by air,

[1] *Observatory*, *41*, 211, 1918.

Fabry estimated that a truly gaseous corona having a density only one-thousandth-millionth that of air at atmospheric pressure would scatter sufficient sunlight to account for the luminosity of the corona, and the polarization effects which have been observed. The explanation by Fabry has very much to commend it, but it is very much to be doubted whether it is based on sufficient experimental evidence. Moreover, it is difficult to see why the light of the corona, if caused by molecular scattering, is white in color and not blue like the sky.

Since the heavenly bodies, the sun, the stars, nebulae, etc., are bodies at high temperatures radiating energy, a knowledge of radiation laws is of vital importance to the solution of astronomical problems. Unfortunately until recently, little was known of these laws, but with the advent of the Bohr atom and Saha's theory of ionization, but specially on account of the enormous activity in the past decade in investigating problems of radiation, we have already made many discoveries and many more are to follow.

Saha's theory of ionization (see Chapter XVII) has already been very successful in interpreting the importance of enhanced lines in the flash spectrum and in furnishing an adequate explanation of the differences between the Fraunhofer spectrum, the sun-spot spectrum and the chromospheric spectrum. Cannot this same theory be expanded so as to furnish an adequate explanation of the cause of radiation in the corona?

It has been very difficult to understand the process whereby the coronal particles are enabled to emit light at the very great distances of ten million miles from the surface of the sun, under conditions of very low pressure and very long mean free path. But the corona is an appendage of the sun, and somehow or other (we have never realized just how) it may be possible for the corona to borrow some of its radiation from the sun. But how about a body like the Orion nebula? It gives a bright-line spectrum. Are astronomers to keep on saying, as they have for a number of years, that the Orion nebula emits light because the kinetic energy causes the particles to be heated and that the luminosity is wholly the result of heat? It is excessively difficult to imagine a body of such vast dimensions heated to luminosity but yet surrounded by the intense cold of inter-stellar space. Apparently some cause other than that of temperature must be sought.

In studying the problem of galactic nebulae, Hubble[1] comes to the conclusion that their luminosity is derived from the radiation from stars in the vicinity. If this were a simple case of reflected starlight, the spectra of the nebula would agree exactly with the spectra of the associated stars; but such is found not to be always the case. Hubble remarks, "It is doubtful whether or not a mass of diffuse nebulosity isolated in space and with no stars involved could hold together and at the same time shine by light generated by collisions of molecules. At temperatures corresponding to intensity-distribution or width of lines in nebular spectra, the average speeds of the molecules would be so high compared with the velocities of escape that the nebulosities would probably dissipate rapidly. On the other hand, if molecular speeds were sufficiently small to admit of cohesion in the mass, the nebulosity would probably be too cold to radiate light. This argument suggests that diffuse nebulosity is not intrinsically luminous, but is rendered so by external causes."

If it is possible to discover the mechanism whereby the nebulae are rendered luminous, the same mechanism will probably explain the cause of the coronal radiation. The fact must not be overlooked that each and every star has a corona surrounding it, but these coronas can never be rendered visible to astronomers since the total light of each corona must be comparatively feeble compared with the luminosity of the sun which it surrounds. Milne[2] has shown that the light of the corona can have no sensible effect in modifying the spectrum of the sun.

The mechanism for explaining the radiation of corona and nebulae unquestionably is found in the electron. The sun and the stars are vast radiating spheres at very high

[1] *Mt. Wilson Contributions*, No. 241, 1922.
[2] *Phil. Trans. Roy. Soc'y, A, 223*, 201, 1922.

temperatures, and from each of these suns, billions of electrons are shot off every second of time. The number of the electrons emitted depends on the intensity of the radiation while the energy of the electrons depends primarily on the temperature. The energy of the electrons will carry them to the greatest distances from the radiating body in the case of the very hottest stars.

Schwarzschild proposes a theory with attractive possibilities, namely, that the corona consists of "electron-gas," that is, a gas of very long mean free path which is capable of reflecting and polarizing light. But as Schwarzschild has pointed out, the electrons, each of which carries a unit negative electric charge, must require an equally large positive charge on the surface of the sun; this causing such a strong electrical field that even the fastest moving electrons would be stopped at distances less than a millimeter. In consequence, the corona could have but little extension. Quite independently, Mitchell (2d ed., p. 356) comes to the conclusion that the radiation of the outer corona is caused by electrons. Ludendorff attempts to get rid of the great electrostatic charges required by the electron-gas theory by assuming that atoms carrying positive charges must be mixed with the free electrons.

In a series of excellent articles,[1] Wilhelm Anderson reviews the various coronal theories. He accepts the electron-gas theory. After making certain plausible assumptions concerning conditions that exist in the corona, he calculates, on the basis of convective equilibrium, the effective molecular weight of the coronal material, and he finds a good agreement with the atomic weight of the electron. The thermal radiation of the inner corona is naturally greatest near the solar surface, so, likewise, is the intensity of the light of the photosphere reflected by the electrons. Anderson takes the three laws of intensities of radiation in the corona, the inverse sixth, fourth and square, and he finds a good agreement between the two latter at a distance of 0.28 solar radii. On the assumption that at this distance the diffuse reflected light has 38 per cent of the intensity of the total illumina-

[1] *Zeits. für Physik,* (4), *33,* 1925.

tion, he calculates the results from theory and finds a good agreement with the fourth-power law close to the sun, but at greater distances out a better agreement with the inverse-square law.

It therefore seems to be quite probable that the coronal radiation takes its origin in the electron, the energy coming from two separate causes: (1) from thermal radiation resulting from collisions, and (2) by reflection and scattering of the photospheric light. The first effect is visible only in the inner corona, while the second manifests itself both in the inner and the outer corona.

In quite similar manner, Zanstra [1] has made an application of the quantum theory to explain the luminosity of diffuse nebulae. The nebula appears to have little radiant energy of its own, but borrows or takes energy from the star (or stars), the electron being the medium for the transfer of energy. Bowen [2] explains the spectrum of nebulium as the result of the long mean free paths of the electrons, the jumps being " forbidden " under laboratory conditions.

Hence, it seems to be definitely proven that both the solar corona and the nebulae are rendered visible as the result of the action of the electrons. Whatever the exact mechanism may be, both the number of electrons and their intensities of emission depend on the temperatures of the stars. According to the investigations of Eddington and Milne, radiation pressure greatly assists in the discharge of electrons.

It is unnecessary to assume with Schwarzschild that the corona consists almost exclusively of electron-gas. Zanstra made no such assumption in explaining the radiation in nebulae. In the inner corona is found the emission spectrum due to coronium. This radiation can be caused only by the collisions of electrons. But colliding with what? Evidently with protons and with atoms which have lost one, two, or more external electrons. Above the 14,000-km level reached by the highest lines of the chromosphere, there are unquestionably other ions reaching far greater heights but whose radiations give no light in the visible spectrum but are found

[1] *Astrophysical Journal, 65,* 50, 1927.
[2] *Ibid., 67,* 1, 1928.

in the ultra-violet beyond the reach of spectroscopic investigations. Coronium atoms have been detected to distances twenty times greater than the highest in the chromosphere. Hence, in the inner corona it is necessary to assume ionized atoms and protons. These are most numerous near the solar surface. They thin out rapidly in the inner corona and few of them are found in the outer corona. In the outer corona there can be relatively few ionized atoms, and hence the outer corona is visible mainly as the result of scattering of the photospheric light by the electron.

Regarding the shape of the corona, we must recall that near sun-spot minimum the spots in the dying cycle are near the solar equator while those awakening into life are at higher latitudes. Hence when the sun is quiescent, the effect will be that of very long streamers going out in straight lines, the maximum lengths being attained not at the equator but at the higher latitudes of the awakening spot zone, a fact already noticed by Campbell. A spot on the sun, or an active prominence, may be a local center of activity on the sun, with the result that coronal streamers or hoods surrounding the prominences may result. Since the field strength in spots is far greater than that of the sun as a whole, the comparative inactivity of the sun at its poles is exhibited by the short polar brushes. Even before passing the epoch of minimum of spots, the sun takes on new energy, and hence electrons are discharged with increasing strength in regions not necessarily limited to the solar equator. Consequently, the corona takes on first a rectangular shape, and then a contour more and more circular as the time of sun-spot maximum is approached. Owing to the greater average vigor of electronic discharge at sun-spot maximum, there should be a greater intensity of the emission lines of coronium than at minimum of spots; and this is found actually to be the case. After passing through the minimum of spots, the awakened solar activity shows itself in three different portions of the sun: (1) In the photosphere, the increased radiation causes spots to appear. (2) In the chromosphere, the increased radiation carries the elements of medium height to greater average elevations. (3) In the corona, the increased radiation causes an increase in strength of the emission lines of coronium and also makes the corona lose the shape associated with minimum of spots, of long equatorial extensions and short polar brushes.

According to the recent work of many investigators, there is some doubt existing as to whether the atmospheres of stars are in thermodynamic equilibrium. If they are not, then the result will be that the theoretical conclusions now (1932) believed to be true must probably be modified in essential details. Furthermore, as conditions existing in the sun cannot be approximated in laboratory experiments, little is actually known of the exact physical laws involving the discharge of electrons from the sun or those underlying radiation pressure at the solar surface. It seems almost useless to test any theory from laboratory experience.

It is now more than fifty years since the first attempt was made by Huggins to photograph the corona in full sunshine. The authority of his name, great in the annals of spectroscopy, gave a degree of plausibility to the problem. The task to be overcome is to separate the light of the corona from the strong illumination of the sky. The chief names connected with this work are those of Hale, Riccò, Deslandres, Wood and Hansky. It was natural that the methods so successful in photographing the prominences should be first tried, and, in order that the atmospheric glare might be reduced as much as possible, the observations were made from mountain tops, Pike's Peak and Mt. Etna being occupied for this purpose. No success being secured, a series of attempts were made by heat-measuring instruments like bolometers or thermopiles. Photographic methods, of using color filters and plates sensitive to different parts of the spectrum, have been thoroughly tested. Since the Great War, when such noted success was attained in airplane photography by using plates sensitive to the deep red, attempts were revived on the corona. Lindemann had found it possible to photograph stars of the first magnitude near the sun. Each and every one of the plans, however, at times carried out with great skill and ingenuity, have always ended in the same manner — failure to photograph the corona in full daylight.

The observations by Abbot in 1908 and those by Kunz and Stebbins in 1918 have shown the cause of the failures, namely, the intrinsic feebleness of the corona. Even in the brightest parts, the inner corona is no brighter than the surface of the full moon, which has a brilliancy six-hundred-thousand times less than the sun. The corona is about equal to the intensity of the illuminated sky at eight or ten degrees distant from the sun, but close to the sun's edge the light of our central luminary is so overpowering that it appears indeed well-nigh impossible to photograph the corona in full daylight.

In view of the above conclusions, it is interesting to read of the recent work [1] of Lyot of the Meudon Observatory. At the summit of the Pic du Midi, at an altitude of 2800 meters, he used a telescope stopped down, and with a disk in the focal plane extending 30″ beyond the sun's edge. Observing directly with an eyepiece, he saw rose-colored prominences. Setting the slit of a spectrograph close to the limb, he was able to see and to photograph the 5303 and 6374 coronal lines, superposed on the Fraunhofer spectrum. Polariscopic observations were made and direct photographs were obtained which resembled in appearance the inner corona. It is surprising to find success where others had failed. The work is beset with such great observational difficulties that future researches will be followed with the greatest of interest.

The past decades have been golden years for the progress of astronomy, particularly on account of the attack on atomic structure by the astronomer, physicist and chemist, combined. The importance of the electron has thus been recognized. The theory of ionization which has already been so successful in furnishing an explanation for many of the difficulties connected with the flash spectrum, the chromosphere and sun-spots is but a branch of a larger theory of photo-electricity dealing with the production of light by the passage of electricity through gases. Photo-electric action involves both ionization and radiation. When an electron strikes an atom and a transfer of energy takes place, there may be complete ionization, as shown by the production of

[1] *Comptes Rendus, 191,* 834, 1930; and *193,* 1169, 1931.

positive and negative ions, or there may be partial ionization, that is, a disturbance of an atom which is not detectable as ionization but is shown by the production of radiation. Both radiation and ionization are caused by the action of electrons in their bombardment of atoms. The fundamental basis assumed for Saha's remarkable theory is that the electrons obey the same laws as gases, or, in other words, that they have the same physical properties as the atoms.

Moreover, recent work in astrophysics has demonstrated conclusively that the source of energy is found in the electron, that even mass itself has its origin in the electron. Eddington has shown that the stars are all slowly losing mass, for the reason that radiation is energy, the dissipation of energy means the discharge of electrons, which is synonymous with saying a diminution of mass. Consequently the sun and stars are continually losing mass by the discharge of electrons. Hence it is inevitable, as a result of this theory, that unless energy is being created — possibly by cosmic rays — the end of the universe may be foreseen.

It is unfortunate that the corona can be investigated only at the rare occasions of total eclipses and that an individual, no matter what his enthusiasm or skill, will have during his whole lifetime less than one hour within which to make all of his observations of the corona. In the immediate future, total eclipse expeditions mean long trips from home to locations where there is little reason to expect good atmospheric conditions; hence progress in knowledge concerning the corona promises to be very slow.

CHAPTER XXII

THE EINSTEIN THEORY OF RELATIVITY

EINSTEIN'S theory of gravitation has been justly regarded as the greatest triumph of mathematical reasoning that has taken place since the time of Newton. It is safe to say that no scientific achievement of recent years has aroused so much popular interest and enthusiasm as that evoked by the verification of the Einstein prediction from observations made at the total eclipse of May 29, 1919.

The first step in any scientific investigation is to get at the facts, derived from observation usually by a series of measurements. Great precision, patience and care are necessary to enable the observer to record the true facts devoid of any inferences or illusions of the mind. Lord Kelvin has said that, "Nearly all of the greatest discoveries of science have been but the rewards of accurate measurement and patient, long-continued labor in the minute sifting of numerical results." After the observations have been secured, they are classified and analyzed and tested to see whether they will conform to some known law. The laws of nature differ greatly from human or civil laws since the former must be universally true and must apply under all conditions. If a law of nature is found to be deficient, even in some minor detail, it must be revised to satisfy the conditions, or, if this is impossible, then the law must be discarded.

The grandest and best known natural law is that of gravitation discovered by the great Newton. Its importance lies primarily in the fact that it applies to all bodies, it is universal. In the two and a half centuries that have elapsed since the falling of the apple, the law of gravitation has experienced one grand triumph after another, with the inevitable result that we have come to look upon this law as practically infallible. Other laws may come and go and be revised into changed form, and go on again to be again and again revised, but the law of gravitation has stood firm without change, with the enunciation of its principle in the same form as when handed to us by Newton. Every body in the universe attracts every other body in the universe with a force that is proportional to the product of the masses and inversely to the square of the distances apart.

And now we are told that the law of gravitation must be discarded, that space is curved and that we live in a world of four dimensions. The mind reels, and refuses to be convinced. Common sense tells us, we say, and the experience of the ages has proven, that space has but three dimensions, length, breadth and thickness, so why say such an apparently foolish thing that time, which we *know* has nothing to do with space, must be considered as a fourth dimension? The average man of intelligence immediately calls to mind some of the properties of the mathematician's space of four dimensions, — a man enclosed in a steel-proof vault, if living in a four-dimensional space, could get outside of the vault without passing through any of the walls. One smiles incredulously, and thinks of a remark by Bertrand Russell that " mathematics may be defined as the subject in which we never know what we are talking about, nor whether what we are saying is true."

Is the theory of relativity of Einstein an unreality in the brain of the mathematician, or must we accept it as true and believe in a four-dimensional space? If the Einstein theory is accepted, then as an important consequence the law of gravitation *must* be revised, there is no middle course. The new theory states that there is no " force " of gravitation, for this " force " is but one of the inherent properties of four-dimensional space. To decide which of the two conflicting hypotheses best represents the law of gravitation, Newton's or Einstein's, we shall have to pass in review the salient facts. Science desires the simplest explanation. Einstein's own conception of the law of gravitation, as expressed in the New York *Times* is, " Please imagine the earth removed, and in its place suspended a box as big as a room or a whole house, and inside a man naturally floating in the center, there being no force whatever pulling him. Imagine further, the box being, by a rope or other contrivance, suddenly jerked to one side, which is scientifically termed '*difform motion*,' as opposed to '*uniform motion*.' The person would then naturally reach the bottom on the opposite side. The result would consequently be the same as if he obeyed Newton's law of gravitation, while, in fact, there is no gravitation exerted whatever, which proves that difform motion will in every case produce the same effects as gravitation.

" I have applied this new idea to every kind of difform motion and have thus developed mathematical formulas which I am convinced give more precise results than those based on Newton's theory. Newton's formulas, however, are such close approximations that it was difficult to find by observation any obvious disagreement with experience." Does this theory of Einstein concerning a man falling from the top of a step-ladder to the floor of a room really furnish a *simpler* explanation than the classical law of falling bodies of Newton? If it does, then a very radical change in our modes of thought must take place.

History shows that the present is not the first time that the perfect science of astronomy has been forced to change its point of view. In the early days of civilization it was believed that the earth was flat, — and this notion is still not quite eradicated from human thought. Later was evolved the theory that the earth was the center of the universe and about it revolved the sun, moon, planets and stars. The mechanism required by the Ptolemaic system to explain the motions of the heavenly bodies by cycles and epicycles was such a cumbrous one, that it led to a remark by a former king of Spain that if he had been present at Creation he could have furnished the Creator with some very valuable suggestions. The work of Copernicus, Tycho, Kepler, Galileo and Newton revolutionized scientific thought, and as a result the earth was displaced from her important position as the center of the universe. Although the geocentric point of view is still retained in some of our scientific explanations, we realize that the sun is the center of our system, with the earth occupying no more exalted a position than that of being but one of the planets. The Newtonian mechanism reduced celestial motions to the greatest simplicity. In obedience to the law of gravitation, every planet describes a perfect ellipse about the sun as focus, and these elliptical orbits would repeat themselves indefinitely were it not for the gravitational forces arising from the other planets. After allowing for all of the planetary disturbances, or "perturbations," Newcomb found that the orbit of the innermost planet Mercury was rotating in its own plane at the rate of 42 seconds of arc per century in excess of that required by the traditional theory. Various attempts were made to explain away this discrepancy in the motion of Mercury — by assuming a planet or planets inside the orbit of Mercury, or outside of its orbit, by postulating a belt of matter in a flattened disk or series of ellipses surrounding the sun, or by presupposing that the sun has an oblateness of shape and distribution of mass different from that usually taken for granted — but all to no avail. In every case, the assumption of mass required to produce the observed effect on Mercury would have caused disturbances not observed in the other planets. The solution of the problem came only with the theory of relativity which furnished a motion of the perihelion of Mercury of 43″ per century, in almost exact agreement with the observed discrepancy. And so, when in addition, the observations of the 1919 total eclipse gave deflections in accordance with the Einstein prediction, the confirmation of the theory of relativity seemed almost complete.

Newton's law, however, is not invalidated by the latest discoveries but rather is supplemented by them. It is possible in very few instances, and as the result only of the most refined measurements, to be able to detect the difference in observed effect between the Einstein and the Newtonian laws. In addition to his great work on gravitation, Newton proposed an erroneous theory of light. According to this theory, light consists of minute corpuscles expelled

Einstein Cameras Used by the British in Brazil, May 29, 1919
Arranged horizontally with coelostat mirrors.

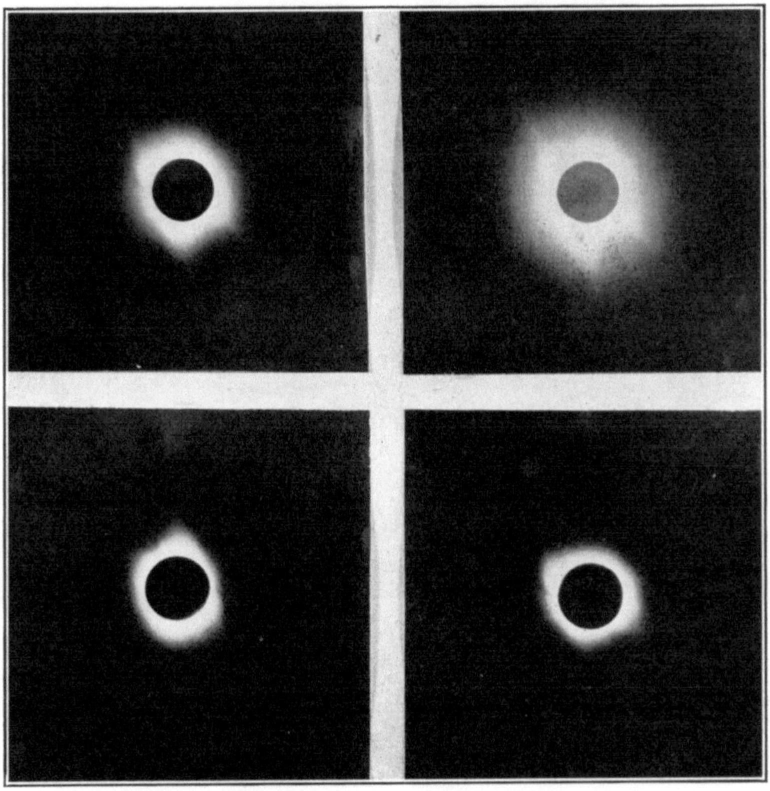

CANADIAN PHOTOGRAPHS, AUSTRALIA, SEPTEMBER 21, 1922

Upper pair: Exposures 10 sec. (left) and 20 sec. (right).

Lower pair: Through Nicol prism turned 90° between exposures, each of 5 sec.

from a luminous source with high velocities. In the estimation of Kayser,[1] "Newton's errors impeded science even to the middle of the nineteenth century." It was not until the observations of Hooke, Huyghens, Young and Fresnel showed the corpuscular theory to be untenable that the wave theory was finally established. But how do the waves of light travel from the sun to the earth? It was necessary for the physicist to invent some material medium for their propagation, and hence the existence of an imponderable and universal "ether" was devised. With the discovery of the laws of electromagnetism, it was shown by Maxwell that electromagnetic disturbances also were propagated with the speed of light, and the conclusion was drawn that light itself must be electromagnetic in character. Additional functions were imposed by physicists upon the universal medium with the result that Maxwell and Kelvin were able to make estimates of the density and rigidity of the ether with almost as much assurance as if they had been dealing with perceptible matter. (More recent researches have shown that not only is light electrical in nature but so also is mass and energy.)

Light and all electromagnetic manifestations were believed to be actions taking place in the all-pervading ether. Light from the distant stars reaches us in the form of wave motions in the ether, and it has been necessary to assume that the sea of ether between us and the distant stars is unbroken. Moreover, in order to account for astronomical aberration and other phenomena, it was proven beyond doubt that the ether must necessarily be at rest, that it pervades everything but that it is not dragged about by ponderable material. Since motions of all kinds are relative in their nature, it was of the utmost importance to theories of physics that a state of absolute rest should exist in nature and this appeared to be found in the condition of the ether. Accordingly, attempts were made to detect the drift of the earth through the motionless ether. In 1887, the first experimental determination was carried out by Michelson. The same experiment was later repeated by Michelson and

[1] *Handbuch der Spectroscopie*, I, 5, 1900.

Morley, by Morley and Miller, and quite recently by Miller alone. This experiment has been described so frequently that it will be unnecessary here to go into details. The velocity of light was known to be, in round numbers, 186,000 miles per second, and this was supposed to represent the velocity of the wave motion through the ether. The velocity of the earth in its orbit is 18 miles per second, and hence if the ether is stationary with respect to the sun, the drift of the earth through the ether should be at this speed. Accordingly, if a beam of light reaches the terrestrial observer in the direction of the earth's motion, it will be traveling with respect to and overtaking the earth at a speed of 185,982 miles per second, while if the beam of light comes in the opposite direction its velocity with respect to the observer will be 186,018 miles per second. Although it is impossible to measure the velocity of light along a single straight course with sufficient precision to compare the two velocities above, nevertheless these velocities can be compared by a differential method. This was accomplished by the Michelson-Morley experiment. This consisted in dividing a beam of light by means of an unsilvered mirror, allowing each half of the beam to travel the *same* distance in directions perpendicular to each other, and then reflecting the two beams back to the same point at which they were originally separated, and then recombining them. It is evident that on account of the motion of the earth, the path through the ether of the half beam which is traveling parallel to the direction of the earth's orbital motion must be slightly longer than the other half of the beam. Consequently, when the two halves of the beam are reunited, the waves of light should be a little out of phase. By turning the apparatus through a right angle, a check is obtained on the equality of the distances traversed by the two beams. The precision of the experiment was so great that it should have been possible to detect the drift of the earth through the ether equal to one-tenth of its orbital velocity about the sun. But no effect whatever was detected and there was no indication that the earth was moving at all through the ether.

The most obvious explanation of the failure of the Michelson-Morley experiment to detect the drift of the earth is that the ether is stationary with respect to the earth, or is carried along by it. But if we admit this possibility, we are in a quandary to explain the aberration of light from the stars. Everyone is familiar with the fact that on a rainy day when there is no wind the rain falls vertically, and to keep dry, when standing still, it is necessary to hold an umbrella over one's head. If the holder of the umbrella starts to walk he tilts the umbrella forward in the direction of his motion, for experience has told him that only by so doing can he keep dry. The faster he walks the more the umbrella is tilted forward. This is a simple case of relative motion. So when a telescope is pointed in the direction of a star, the telescope tube must be tilted forward by an appreciable angle so that the rays from the star may pass down the tube to the eye of the observer. The angle through which the telescope has to be inclined is known as the angle of aberration. This angle was discovered by Bradley in 1728. Its value is $20''.47$. On account of the great precision of modern astronomical measurements this angle is considered large.

But how explain the Michelson-Morley paradox that the earth must be drifting through the ether while at the same time the effect of this drift has never been detected? Apparently there is no way out of the dilemma but that our ideas must in some manner be revised. In 1893, a satisfactory explanation was offered by Fitzgerald, and in 1895 independently by Lorentz, that the negative result of this experiment becomes quite intelligible if it is assumed that when a body is in motion through the ether its dimensions in the direction of motion become slightly shorter than when it is at rest. This would indeed seem to be a strange and a highly arbitrary hypothesis if unsupported by other scientific evidence. The work of Lorentz and Larmor on the electromagnetic theory showed that it was necessary to assume that a body when in motion does actually contract by just the amount demanded by Fitzgerald's explanation. If this hypothesis is true, then as a result we are inevitably forced to the conclusion that time also must be measured in a different manner for an observer at rest and for one in motion. The Fitzgerald contraction is not merely an idle arbitrary speculation, but after repeated experiments it has been shown to be true and mathematically exact in the well-known laws of electromagnetic forces.

The amount of contraction depends on the velocities, and in the majority of the cases of our experience, or for average velocities, is excessively minute. In the case of the earth the contraction is only one part in 200,000,000, which corresponds to a contraction of $2\frac{1}{2}$ inches in the earth's diameter in the direction of its motion. The Michelson-Morley experiment therefore failed to detect the motion through the ether for the simple reason that the dimensions of the apparatus were actually automatically contracted by an amount just sufficient to compensate for the effect sought. Other ingenious experiments, electrical and optical, were tried for the purpose of detecting the drift through the ether, but always with the same result. There thus appears to be a "conspiracy" among the various agencies at work to prevent man from measuring his motion through space.

In 1905, Einstein published his restricted Principle of Relativity that "it is of necessity impossible to determine absolute motion by any experiment whatever." According to Einstein, all motion is relative, and no experiment can be possibly devised so as to decide which of two systems is at rest and which in motion. It is therefore impossible by any experiment to detect uniform motion with respect to the ether. Hence the assumption of the ether, brought into physical science for no other purpose than to explain the propagation of light, becomes entirely unnecessary. If the ether has no position whatever in space, the statement that it exists has no meaning. Hence, if the ether is retained by the physicist he must assign a new set of properties. A second consequence is that the velocity of light with respect to two observers moving relatively to each other must always be the same, no matter in what direction the light is traveling, as the velocity must be equivalent to that when determined by an observer at rest, of 186,000 miles per second. These radical changes can exist only under the condition that certain relations, known as relativity transformations, exist between the space and time measurements made by the two observers.

The principle of relativity thus entails changes in our mode of thought of the most revolutionary kind, and our preconceived notions in consequence are turned topsy-turvy. Let us see a few examples. If an observer is traveling at the rate of 161,000 miles a second in a vertical direction his arm, 30 inches long, is of its natural length when extended horizontally, but is contracted to 15 inches when hanging by his side or raised vertically above his head. This you say is foolishness. Well, take a yardstick and measure the length of the arm. Horizontally the arm measures 30 inches on the scale, and vertically exactly the same 30 inches by the scale. Yes, but when you hold the scale vertically it too has been contracted and each inch on the scale is in reality only half an inch, so that the arm measures 30 *half-inches* or 15 inches. "Yes," you will say, "but whoever heard of anyone moving with such a colossal speed as 161,000 miles per second?" Why not? It is impossible to measure absolute velocities in space, and for all we know to the contrary we may be speeding along at this terrific rate. The β particles, referred to in Chapter XIV, move with velocities of 100,000 miles per second.

A conclusion perhaps even stranger than the foregoing must be drawn from the principle of relativity. Suppose observer A is moving through space at a speed of 161,000 miles per second relative to B, and allow each observer to hold an identical yardstick. On account of his fast flight, A's stick is contracted to half of the length it has when at rest, and hence B's stick appears to be twice the length of his own. But since motion is purely relative in character, B must be moving with respect to A with the speed of 161,000 miles per second, and hence B's yardstick is contracted; and it appears to be twice as long, half as long, or equal to the length of an identical stick, depending on which stick is the moving one or whether they both are at rest.

If the speed of the observer is increased beyond 161,000

miles per second, the contraction becomes greater and still greater; and at the velocity of light all lengths dwindle to zero and vanish. The observer changes to an object of two dimensions, length and breadth, but without thickness, but he himself is blissfully unconscious of his sorry plight. Distance is annihilated and so also is time. If traveling at the velocity of light, the observer could go to the most distant star without becoming a day or even a second older, for the reason that no time had elapsed. These paradoxes make the theory of relativity very difficult of comprehension to anyone who is not a mathematician and they seem moreover to controvert our " common-sense " view of things.

Everyone will agree that if one measuring rod identical in length with another can be twice as long, half as long or equal in length to the other, there must be something wrong somewhere with the method of measurement. Einstein inquired into our methods of measuring length and time and pointed out the lack of definiteness in the concepts of space and time as ordinarily used. What do we do when we measure the length of an object like a book? We take a foot-rule or meter stick and determine its " true " length. We measure the size of the book, however, only by comparing it with a scale whose length we suppose to be known. But what do we actually know of the length of the foot-rule? We take it for granted that it has been compared with some standard foot-rule or yardstick. We know the lengths of our measuring rods only by comparing their lengths with other rods, and so on and on through an almost endless series of comparisons, until we come to the Standard Yard preserved in London, or the Standard Meter kept in Paris. The latter is a rod of definite shape, of a certain particular metallic composition, and at a given temperature, the length between two specified lines is exactly a meter. The length of the meter is approximately one ten-millionth part of the earth's quadrant. What would happen if the earth and all it contains were suddenly contracted to half its present dimensions? All objects including ourselves and all measuring appliances would dwindle to half size. We would never have any idea that the earth had contracted if we kept our attention on terrestrial objects. But lo, and behold, the sun has suddenly doubled its size! If the sun and the external universe could have contracted at the same time as the earth and the observer, we could never have become aware of the difference. The " true " length of a book or a rod thus depends on the definitions and postulates which we adopt. Manifestly there is no such thing as " absolute " space, everything is merely relative to the observer.

You will insist, however, that if there is no " absolute " space, there must certainly be " absolute " time, for time seems to go on quite independently of any observer. Each of us unconsciously feels that when we die, the world will carry on just about as before. Time and space thus seem on a very different basis. But how do we measure time? Ordinarily by a clock. We assume that the pendulum swings in a perfectly uniform manner and that the clock hands move equal distances in equal intervals of time. But how *prove* our assumption? We are forced for our definitions of time intervals to go back to the earth, and to assume that the earth is a colossal top set spinning on its axis, and its mass is so great that like a gigantic fly-wheel its motion *must* certainly be uniform. The assumption seems sound and all of the accurate measurements of astronomy are based on it. What would happen to the vaunted predictions of astronomy if the earth changed its speed of rotation? In fact, it was shown in Chapter IV that some suspicion has lately been cast on the uniformity of rotation of the earth, — and our time-keeper may after all be a faulty one. Apparently therefore, we seem forced to the conclusion, since time-measurements depend on measures of distances, that there is no absolute time any more than there is absolute space. The determination of both space and time consist essentially in the measurement of certain lengths, and such measures have no meaning except in relation to ourselves and our every-day experiences.

Let us return again to the measurements of the book. We recognize the object because we have seen other books. We glance inside the pages, and find a treatise on higher mathematics or some ebullition of modern fiction, both unreal and devoid of meaning to our feeble intellect. We likewise recognize the foot-rule and we accept some one's word for it that this rule has been compared with other rods, etc., etc., and that we know its length. Applying the measuring rule to the book, and spending a few seconds of time in the operation, we find a length of eight inches. A Martian dropped to the earth would not recognize a book, for in the advanced civilization that is supposed to exist on the ruddy planet, books do not exist. Neither would he recognize the foot-rule, and certainly the marks that designate the inches would be utterly unintelligible. The simple process of measuring the length of a book would be completely impossible to the gigantic brain of the Martian. He has not been taught from his childhood up to measure with our particular rods.

The measurement of a length, simple though it seems, is after all a very complicated process. We see the book, and note that it occupies a definite position on a table in a room furnished as a library. Having two eyes, we have learned unconsciously to estimate the distance to the book, and so we telegraph the thought to our brain that we must reach forth the hand and pick up the book. (It is not necessary to trace further the train of thought.) But it is easy to see that in measuring the length of the book we are putting into practice the training of hundreds and even thousands of years, acquired at times with great difficulty by ourselves and our ancestors.

Knowledge thus comes to us only as a series of experiences or events. Each experience is perhaps an instantaneous photograph on the brain. But every photograph when taken requires a definite exposure-time, sometimes longer, sometimes shorter in length, depending on the intensity of the light and the speed of the photographic plate. And so there are brains which work slowly and those which grasp more quickly. We see therefore that throughout all of our existence we have tacitly assumed, without realizing it, that measures of space are not at all independent of time, and that durations of time are not devoid of their inherent reference to space. There are no durationless events, and each and every one of our experiences involves a consciousness of both space and time. Thus neither space nor time is absolute, and every observation is relative to the individual observer. But each observer being confined to the earth, by common consent we have come to the conclusion that Truth is attained only when our observations agree with that of the average individual and that we are in error when we differ from the general average.

Throughout the whole of our scientific consciousness we have assumed that in agreeing with the average observer we have thereby derived the " true " length or have determined the " true " duration of time. The great triumph of Einstein consisted in making clear that the " true " measures of each observer, and hence of the average observer, are only relative in their nature and not absolute in value. Consequently, all of the observations with which we are familiar are relative and depend only on the individual as represented by the average observer. The method of advance was manifestly to divorce the series of events from the observer, and look upon them from the point of view of the objects themselves. This was Einstein's method. He assigned space and time solely to the observer, but gave to nature an unfamiliar combination of space and time of four dimensions. However, this is not the same four-dimensional space that has lately been much discussed by the mathematician, a space of length, breadth and thickness and a fourth dimension at right angles to the other three. The relativity world of four dimensions we have in reality been quite familiar with throughout our whole lives, as is signified by the words right-and-left, forward-and-backward, up-and-down, and sooner-and-later. The first three dimensions give the familiar world of *objects,* the four dimensions taken together furnish the world of *events.* In this four-dimensional world a particle occupies not one point, as we are accustomed to think of in space of three-dimensions, but a series of points representing its positions at successive instants of time. Its history then is represented by a line, called its " world-line." If an observer is admitted, he has his own world-line and immediately he imagines that his world-line, of all lines, is the most important in the universe.

He immediately proceeds to divide up the four-dimensional entity into a space of three dimensions with time as the fourth, and we have the world familiar to us all. But the theory of relativity rules out the observer from any consideration whatever, and there now results a space-time nature in which is scarcely recognized space and time. There is thus no " shape " to anything, no difference between straight and crooked, it is impossible to measure an angle, which is in fact the difference in direction between two straight lines. This four-dimensional world which was invented by Minkowski, naturally has no reality, it does not exist and its properties cannot be interpreted by our ordinary experiences. The fourth dimension at right angles to the other three is not even the time plotted as a quantity directly, but the product of time into the square root of minus one. This quantity, $\sqrt{-1}$, is known to every intelligent boy as the symbol of an imaginary quantity. Hence it is futile to attempt to assign to this unfamiliar and unreal world any of the attributes derived from the common experience of life. Many writers in their attempts to popularize have used such catch-phrases as " time is curved," " there is no *now*," " a phenomenon may be seen before it happens," and so forth and so on.

" Relegating space and time to their proper source — the observer — Einstein bids us contemplate the residuum of what we observe. This residuum is the true world. It is shapeless, because we have abstracted shape; yet it is metrical and has quantitative properties which can be expressed in mathematical terms. Clearly we cannot describe this true world in terms of familiar things, because the whole point of Einstein's theory is that we must subtract the ideas which we ourselves have added in order to form familiar things. Mathematics is the only language in which the inherent qualities of this unfamiliar world can be described. But though it seems unfamiliar, nature is left simpler by this purification. A closer unity is perceived in the bases of phenomena apparently diverse; and, for example, the effect of gravitation on light is clearly foreseen. Further, the laws of nature must relate to this four-dimensional residuum and the space and time we ourselves introduce cannot be relevant. This led Einstein to the conclusion that Newton's law of gravitation, which refers to one particular separation of space and time, cannot be the exact law; and he proposed a new law applicable to the four-dimensional world."[1]

Although there is no distinction between straight and crooked in the four-dimensional entity, it is however possible to join two points or events by a certain unique track which plays a part corresponding to the straight line in our familiar life. This unique track is called a *geodesic,* and instead of being the shortest distance between two points — the important property of the straight line in ordinary space — the geodesic is of the maximum length. We see, therefore, what unfamiliar conceptions are involved in the theory of relativity, and how radically changed must be the point of view that makes the distance between two points a maximum and not a minimum!

Let us see some of the necessary consequences of the principle of relativity in its relation to the propagation of light. As already stated, this principle enunciates that the velocity of light is independent of the observer, or, stated in other words, that no matter what is the velocity of the observer the wave-front of an emitted beam of light is always a sphere with the observer as the center. According to the older views, a definite point of the ether, the source of light, was the center of the sphere, but according to the newer ideas it is the observer himself who is the center of the sphere. The observer is thus in a sense the center of the universe, he is at rest and the universe is carried by for his inspection. If an observer A is the center of a sphere, the wave-front will have advanced after time t so that the square of the radius of the sphere will be represented by

$$x^2 + y^2 + z^2 = c^2t^2$$
$$\text{or} \quad x^2 + y^2 + z^2 - c^2t^2 = 0$$

where the observer is the origin of coordinates and c is the velocity of light. For any other observer B, the wave-front

[1] Eddington, *Contemporary Review, 116,* 643, 1919.

will again be a sphere but with B as its center, and the mathematical equation will be

$$x'^2 + y'^2 + z'^2 - c^2t'^2 = 0$$

According to the relativity assumption, the sphere observed by A is identical with that observed by B. Hence a linear transformation between x, y, z, t and x', y', z', t' must transform either of the above equations into the other. If now we put $ict = w$ (where i is the imaginary quantity, $\sqrt{-1}$), then the equations reduce to $x^2 + y^2 + z^2 + w^2 = 0$ and $x'^2 + y'^2 + z'^2 + w'^2 = 0$, each of which represents a sphere in four-dimensional space. x, y, z, w form a set of orthogonal coordinates, and x', y', z', w' must represent a second set of orthogonal axes in this same space. In space of three dimensions we are familiar with such sets of axes. An observer at any one place chooses three axes, one north and south, the second east and west and the third vertically up and down. Another observer at a different place on the surface of the earth takes a similar set of axes with reference to his directions of north, east and vertical. But what are up and down, and north and south to one observer are not the same directions that appear to the other observer. A rotation of axes and a transfer of the origin will transform the set of axes of one observer into those of the other. Similarly in the four-dimensional space-time continuum the coordinates of A can be transformed into those of B by a rotation of axes. But the sphere observed by A being the same as that observed by B, or by any other observer, it therefore must appear as an objective reality. But A and B divide up the space-time continuum differently. Space and time depend only on the consciousness of the individual observer; hence the coordinates that A calls pure space and pure time will not be space and time to observer B but will be a mixture of the two. It is only the combination of space and time, not space and time separately, that represents reality independent of the peculiarities of the individual observer. The mathematical expressions by means of which x, y, z, w are rotated into x', y', z', w' are the famous transformations of Lorentz which were derived a decade before Einstein's work was published.

Starting with these transformations and assuming that light follows a geodesic and not a straight line (which is now a meaningless term), Einstein developed his famous theory of gravitation. The only constant introduced into the discussion is c the velocity of light, the maximum velocity of which we have any observational experience. Virtually the only assumption made is the one of fundamental importance, that the velocity of light is identical to all observers. This is at best an assumption, and if it can be shown by any experiments that this premise is not true, then the whole Einstein structure will fall into ruins. By means of a little-used and very difficult branch of mathematics, a theory was developed which explains gravitation, the motion of Mercury and the gravitational attraction of the sun on light rays. As the author does not claim to be one of the original dozen who could understand Einstein's equations he will not introduce any further mathematical formulas. These newer conclusions will herald an advance over the old law of gravitation only on the condition that Einstein's theory represents the observed facts more closely than the theory of Newton.

We may perhaps obtain an inkling of the meaning of four-dimensional space by starting out with the properties of three-dimensional space, with which from childhood we have been familiar. In the world of *events* all four coordinates are necessary, for we never observe an event except at a certain time, and we are never cognizant of an instant of time without special reference to space. In discussing the laws of electromagnetic phenomena, it was shown by Minkowski that these laws could be represented by assuming a four-dimensional space-time and that the mathematical transformations of Lorentz and Einstein could be described by a rotation of a set of axes. In three-dimensional space of everyday happenings no one would be impertinent enough to say that there is any *absolute* direction of up and down, i.e., the vertical, at any one place on the earth's surface, although there are great numbers of people who believe that

New York City occupies a specially important place on the top of the world. The words "sooner or later" are likewise relative terms only and not absolute. We are all familiar with the mental process of assuming a set of three rectangular axes relating to the particular location of the observer on the earth's surface, a set, therefore, of relative coordinates, not absolute ones. If we make a section by a plane of three-dimensional space, we derive a two-dimensional space, a plane. Similarly, if a section is made through the four-dimensional world of events, a three-dimensional space will result. For an observer on the earth, there is one particular section only which will give the space of three dimensions with which he is familiar, so that the four-dimensional unity thus breaks up for him into space and ordinary time. A section through an observer on Mars would be unreal to us and unnatural in aspect. One of the postulates regarding four-dimensional space is that it must be entirely independent of the observer so that there cannot be any real difference between any two directions in an absolute sense. For any particular observer, as we have seen, the four-dimensional world of events may break up into ordinary space and time, but such a section would represent a particular case and not the most general solution.

We may draw a number of lines on a flat piece of paper or sheet of rubber. These lines intersect in many points. By taking the sheet of rubber into our hands and altering its shape, the lines on its surface become curves in three-dimensional space. A skillful mathematician could represent these curves by complicated equations of transformations involving x, y and z. By altering the shape of the piece of rubber we change the appearance of the lines on its surface, but we do not in the least change the number of the intersections made by the lines. The transformations expressing the intersection of two lines may vary greatly in their mathematical forms, but if they are true formulas, they cannot alter the actual intersections of the lines themselves. A simple case of change of axes of reference, a mathematical transformation, is familiar to everyone. While standing out in the rain, when no wind is blowing, one holds his umbrella vertically over his head if he wishes to keep dry. The rain-drops, obeying the law of gravitation, seem to fall vertically. If the observer starts to walk, the direction of the falling rain appears to have changed, the rain-drops come in a slanting direction. Of course, the direction has not actually changed, but the effect is just the same as if there had been a real change in the direction of gravity. Some obstinate person will say that he *knows* that gravity is acting in a vertical direction and that his motion has nothing at all to do with gravity. Under this assumption he persists in holding the handle of his umbrella vertical, for he has learned that when standing at rest he keeps dry by so doing. If he still insists in his foolishness when going at high speed in a swift-moving, open automobile, he will lose his umbrella and get dripping wet into the bargain.

In four-dimensional space the " world-lines " intersect in a series of events. For the description of the world-line we can choose any set of four-dimensional axes we wish. The event takes place, however, absolutely independently of any assumption of reference axes. If we change the axes, we at the same time change the coordinates with respect to these axes, but the interesting thing is the event and not the particular set of reference axes assumed. The world of nature is made known to us only through a series of observations or series of events. We learn, in fact, only through a series of coincidences. To represent the laws of nature our observations must be true no matter what selection we make of reference axes. In other words, the mathematical expression of the laws of nature must be such that their form does not change if we make a transformation of axes.

It is a curious fact that we know very little of the mechanism by which gravitation works. Is it propagated with the velocity of light, or does it act instantaneously at infinite speed? At a place on the earth's surface, gravity is the resultant of two forces, one the attraction of the earth, the other due to the centrifugal force of the earth's rotation. The former is spoken of generally as a " natural " force, the latter as an " artificial " one. Centrifugal force may be altered by taking different locations on the surface of the earth or may be made to vanish entirely by going to the north or south poles. For the purpose of simplicity in the mathematical treatment, centrifugal force is separated from gravity. We have no direct sensation for either force separately, and in fact there is no physical basis for the separation. So why not consider the " natural " and the " artificial " force on the same basis? The generalization of this notion led to Einstein's *principle of equivalence* that a gravitational field of force is precisely equivalent to an artificial field of force, so that it is impossible by any conceivable experiment to distinguish between them. Force, therefore, is relative and not absolute.

Guided by the two principles of relativity and equivalence, Einstein was led to assume that all gravitational fields of force must be illusions. He himself admits that he was brought to his new point of view by discussing the sensations of falling with a man who had just tumbled from a high building. No distinct consciousness of falling was actually experienced by the man. The simplest method after all for the consideration of gravitation may actually be the point of view of the falling man who was experiencing no peculiar sensations of his sudden flight. This was Einstein's method. The earth was surely rising to meet the man, and if he had not known of the sudden stoppage of his flight that was in store for him, he might have quietly theorized regarding the relativity of motion. Such an unconstrained body (as the falling man) if left to itself takes what we shall call a " straight " line, but which is actually a geodesic. The world-line is undeflected until the particle of matter comes into the vicinity of another particle. The world-line of each particle is bent in towards each other or deformed. A change of direction of the particle means that a field of force has been entered. Such a deformation can be brought into the equations by means of mathematical transformations.

Gravitation therefore needs no special treatment different from that of any other force. In developing the mathematical analysis it was not possible to follow the geometry of the flat space of Euclid where the straight line is the shortest distance between two points. The mathematician will assert that there is no reason to assume Euclidian geometry unless observation demands it. In the process of the difficult mathematical development, Einstein was at liberty to choose between several possibilities, and the decision reached seemed to give the simplest solution. In the final analysis his law of gravitation must produce the same effect as Newton's law for conditions of velocities small compared with that of light.

According to Einstein's views, gravitation is simply a geometrical deformation of unrestrained bodies, and hence we may regard the gravitational field as influencing or even determining the laws of space-time derived from measurements. The track of a ray of light therefore is deformed by the gravitational field. According to Eddington[1] "a ray of light passing near a heavy particle will be bent, firstly, owing to the non-Euclidian character of the combination of time with space. This bending is equivalent to that due to the Newtonian gravitation, and may be calculated in the ordinary way on the assumption that light has weight like a material body. Secondly, it will be bent owing to the non-Euclidian character of space alone, and this curvature is additional to that predicted by Newton's law. If then we can observe the amount of curvature of a ray of light, we can make a crucial test of whether Einstein's or Newton's theory is obeyed.

"This separation of the attraction into two parts is useful in a comparison of the new theory with the old, but from the point of view of relativity it is artificial. Our view is that light is bent in just the same manner as the track of a material particle moving with the same velocity would be bent. Both causes of bending may be ascribed either to weight or to non-Euclidian space-time, according to the nomenclature preferred. The only difference between the predictions of the old and new theories is that in one case the weight is calculated according to Newton's law of gravitation, in the other case according to Einstein's."

It has been repeatedly urged by many writers that New-

[1] *Space, Time and Gravitation,* p. 106.

ton's law is simpler than Einstein's. The former is indeed more familiar, but that does not necessarily mean that it is simpler. This depends on the point of view. The principle of relativity introduced into scientific thought, first destroyed the notion that space and time were absolute in character or objective in existence. This led to a consideration of the four-dimensional world of events and to the supposition that gravitational force is an illusion. To explain this " force " as an inherent curvature in space represents a further revolution in scientific thought. Gravitation is not the only " force " familiar to the physicist. Are these also illusions and must not the Einstein theory be generalized to include all of them? Such a generalization has been proposed by Weyl who of necessity was forced to introduce new curvatures in the four-dimensional space to represent the forces involved; in fact, the effects predicted by Weyl agree so perfectly with electromagnetic theory that no testing of the theory by experiment is possible. The only scientific test to which all theories and laws must submit is the fundamental one of whether they represent the observed facts. Einstein has not attempted an explanation of gravitation other than to assume that it is propagated with the velocity of light; he has only been occupied with the deduction of its laws.

In the foregoing chapter is given an inkling of the manner in which the philosophic trend of thought has been modified by the theory of relativity; in the following pages some few of the scientific consequences will be discussed.

CHAPTER XXIII

HAS THE EINSTEIN THEORY BEEN VERIFIED?

IF the path of a ray of light is identical with that of a material particle when passing close to a body of heavy mass like the sun, it is easy to see that the path will not be straight but will be a hyperbola. To calculate the amount of the deflection from the straight line course one might apply Newton's law; but according to Einstein's theory, this old and well established law is applicable only under the condition of small velocities. For an object traveling with the velocity of light, the Einstein formula gives a deflection twice that deduced from Newton's law.

Other important consequences follow as a direct result of the Einstein theory. According to the inverse-square law of gravitation, an isolated planet will describe about the sun an ellipse the direction of whose axis is fixed in space, but according to the Einstein law the path will be a spiral and not an ellipse. The direction of the major axis will therefore rotate in space instead of being stationary. The amount of rotation, calculated from the Einstein theory and derived without the introduction of any additional constants into the equations, is $43''$ per century for the planet Mercury.

A further consequence is found when comparing the time of vibration of an atom acting under the strong gravitational pull of the sun with that of a similar atom in a terrestrial laboratory. The atom in its vibrations always behaves very much like the pendulum of an ideal clock. Since the gravitational field of force is similar in its properties to that of a centrifugal field, it is comparatively easy to calculate the effect of the theory of relativity on the vibrations of light under the gravitational action of the sun. It has been found, in virtue of this theory, that the wave-lengths at the solar surface should be greater than those under terrestrial conditions in the ratio of 1.00000212:1. Each line in the solar spectrum should be shifted towards greater wave-lengths, or towards the red end of the spectrum, by an amount readily calculable. The shift at the F or Hβ line in the blue-green part of the solar spectrum should amount to 0.010 A, or one-hundreth of an angstrom.

There is a fourth consequence of the Einstein theory. If a massive particle is placed near the center of a circle and if the length of the circumference and the length of the diameter are accurately measured, it should be found that these two lengths will not be in the ratio of the famous number π. The experiment can hardly be regarded as a crucial one for testing the Einstein theory for the reason that a weight of five tons placed inside a circle of five yards radius would not change the value of π by more than one figure in the 25th place of decimals.

The astronomers at the eclipse of 1919 observed deflections in the rays of light from the stars in the neighborhood of the sun apparently in complete harmony with the prediction of Einstein. As already stated, the amount of bending predicted was $1''.75$ for a beam just grazing the edge of the sun. Was the observed deflection an actual verification of the Einstein theory or can it be accounted for by some other agency? After eliminating deflections from all known causes, the observations must necessarily conform to one of three possibilities. First, there may be no deflections whatever in the stellar rays, light traveling in exact straight lines under all possible conditions; second, the amount may be the so-called Newtonian deflection of $0''.87$; or third, it may be twice this size, as predicted by Einstein, *i.e.*, a curvature of space in the vicinity of the sun causing a bending of a light ray equal in amount to that demanded by the Newtonian theory. There are other deflections possible, but they need not be considered in the present status of the scientific theory.

Several books and articles have recently been written claiming that the gravitational pull of the sun has absolutely no effect on a beam of light. The authors confidently state that observation has abundantly verified the conclusion that light travels always in perfect straight lines and hence a deflection of light rays by the sun is impossible. The brave, or possibly foolhardy, writers apparently have failed to realize that with one fell swoop they utterly disregard not only the whole of the theory of relativity but at the same time they deny that light has weight. The consequences of relativity have been so thoroughly substantiated by observations that he who has a scientific reputation at stake must indeed be very rash to state that physicists must all be mistaken and that the whole theory of relativity is "tommy-rot."

That light has weight has been thoroughly and completely verified. Clerk Maxwell showed many years ago that light is electromagnetic in character. In fact, the waves producing X-rays, photographic light, visible light, heat and wireless waves, are exactly similar in character, the only real difference being the length of the waves. The wave-lengths of X-rays are ten thousand times smaller than those of light rays while wireless waves are just as much larger. Light unquestionably possesses mass or inertia like other forms of electromagnetic energy. The tiny waves of light exert a pressure on material objects in a manner quite similar to that manifested by waves in water that pound with such terrific force on the sea shore at times of storm. The pressure exerted by light is termed light-pressure or radiation-pressure. Its consequences are described in Chapter XXI, and its effects are manifested to the astronomer in the solar corona, in the tails of comets and in the conditions in the more diffuse giant stars. This pressure exerted by light has been found experimentally in the physical laboratory. It is a consequence of the orthodox electromagnetic theory. Recent discoveries, outlined in Chapter XIV, have shown that mass is electrical in character, and consequently energy is likewise electrical, and hence light which is also electrical must possess both mass and energy. If any writer denies these conclusions, he does not by so doing destroy the scientific facts, but succeeds only in displaying his own ignorance of these facts. Even Newton himself surmised that light has weight. The admirers of the great Englishman should therefore not dogmatically deny proven scientific knowledge and state that light moves under all circumstances in straight lines even when under the strong gravitational attraction of the sun. A beam of light should therefore be deflected from its straight line course like any material particle moving with the same velocity, and hence, in interpreting the results of the eclipse photographs we must rule out the possibility that no deflection can be caused by the sun. But which is the true deflection, the Newtonian or that predicted by Einstein?

The only possible method of testing the relative merits of the two theories observationally is by means of photographing the stars surrounding the sun. On account of the glare of the daylight sky, it has been found impossible to photograph the stars in the immediate vicinity of the sun, at any time except when the sun is totally eclipsed. The amount of the deflection depends on the angular distance of the star from the sun's edge, and at a distance of one solar radius the angular deflection is but half the value at the limb. At a distance of a solar diameter, the deflection is but one-third of the limb value. The most important stars for the purpose of testing the theory are therefore those which are closest to the sun. Naturally, the brighter the star the more readily can it be photographed. What is needed for a test of the theory is a group of bright stars distributed at various angles around the sun and close to its edge. Since the sun is always found in the plane of the ecliptic, it is easy to see at a glance from a celestial chart which part of the ecliptic contains the brightest groups of stars, and then to decide at what date of the year the sun is found at this favorable position. The date turns out to be May 29. The brightest assemblage of ecliptic stars is the group known as the Hyades in the constellation of Taurus.

The British observers in 1919 were especially favored, for the eclipse occurred on the twenty-ninth of May. Two expeditions were dispatched. Crommelin and Davidson representing the Greenwich Observatory went to Sobral in North

Brazil, while Eddington and Cottingham from Cambridge located at the Isle of Principe in the Gulf of Guinea, West Africa. The telescope used at Principe was the thirteen-inch "astrographic" telescope belonging to the Oxford Observatory. The telescope was used horizontally, the light from the sun being reflected down the tube by a coelostat mirror. The focal length of 11 feet 4 inches insures a scale of the photographs of one millimeter equal to one minute of arc. One second of arc is thus equal to one-sixtieth of a millimeter or one-fifteenth-hundreth of an inch. In order to secure increased sharpness of the star images, the aperture of the lens was stopped down to eight inches. On eclipse day clouds greatly interfered with the progress of the work. On account of the long duration of totality amounting to six minutes, sixteen photographs were obtained with exposures ranging from 2 to 20 seconds. Unfortunately on many of the plates, one or more of the essential stars were missing on account of clouds.

In Brazil the observers were favored with fine weather. They utilized two telescopes for securing their photographs, one similar in size to the one at Principe, the other with the greater focal length of 19 feet and aperture four inches. The photographs with the former instrument were a grave disappointment, for the star images were not in sharp focus, due unquestionably to the warping of the mirror by the heat of the sun. Seven of the plates taken with the four-inch lens were found satisfactory for measurement. Only two of the Principe plates were measured and on these the images were not particularly well-defined and moreover they were weak due to the interposing clouds. The sixteen plates secured at Sobral with the astrographic telescope were also measured in spite of their poor definition.

In order to secure from the eclipse plates results of the highest value, it is necessary to photograph as a check the same region in the sky but with the sun absent. To be of the greatest worth, the check plates thus obtained should reproduce as closely as possible the conditions under which the eclipse photographs are taken. The plates should, if possible be taken at the site of the eclipse, and with hour angle, temperature, etc., of the eclipse duplicated as nearly as possible in the check photographs. Such a procedure requires a stay of two or three months at the eclipse camp, frequently located thousands of miles from home. The Principe observers indeed secured check plates, but not at Principe, but with the telescope mounted in England. The conditions of latitude and temperatures were vastly different in the two series of plates, and furthermore the plates taken in England were secured without the interposition of the mirror used at eclipse time. In view of their faulty nature as check plates, an additional precaution was taken of securing an extra series of plates photographed at Principe and in England but portraying a different stellar region, since it was found impractical to remain at Principe. The Sobral observers did remain an additional two months after the eclipse to secure the desired check plates.

The values of the Einstein effect reduced to tangency to the sun's limb are:

From	7	Sobral	Plates	taken	with	4-inch	lens,	$1''.98 \pm 0''.12$
From	2	Principe	"	"	"	13 "	"	$1''.61 \pm 0''.30$
From	16	Sobral	"	"	"	13 "	"	$0''.86 \pm 0''.1$

On account of the poor definition of the sixteen plates taken with the Sobral 13-inch lens, the values of the Einstein deflection from these plates, ranging as they did from $0''.00$ to $1''.28$, were omitted from the final mean. Combining the two remaining series with weights depending on the probable errors, there results the deflection of $1''.93$. Giving equal weight to each of the series, the value becomes $1''.80$. The value predicted by Einstein is $1''.75$. The close agreement between observation and prediction was received with the greatest interest and remarkable enthusiasm throughout the whole of the thinking world.

The best series of the 1919 plates gives a result $1''.98 \pm 0''.12$. This certainly does not appear to confirm the half-deflection of $0''.87$, but on the other hand the poorest of the three series of plates does furnish a value closely agreeing with the Newtonian deflection. The remaining value, $1''.61 \pm 0''.30$, although it agrees closely with the full de-

LICK-CROCKER EXPEDITION TO AUSTRALIA, 1922
32 seconds exposure with the 40-foot camera.

Fine Structure in Corona, May 9, 1929
Photographed with the 63-foot camera of the Sproul Expedition.

flection, might, on account of the size of the probable error, possibly confirm the half amount. Although the plates taken at the 1918 eclipse by the Lick Observatory with the five-inch and forty-feet focus camera pointed directly at the sun gave no indication of any appreciable bending, the notion that light must necessarily move in straight lines was now forever gone from scientific thought.

In spite of the general good agreement of the photographic measures, no one, least of all the British observers themselves, will affirm that the Einstein deflection had been completely verified by the 1919 eclipse results. The work was pioneer in character; and improvements over the methods followed were readily apparent. First of all, the coelostat mirrors must be eliminated by placing the cameras on an equatorial mounting and pointing the lenses directly at the sun. Not only do the mirrors prevent a maintenance of constant focus, on account of the warping of their surfaces by the heat of the sun, but the mirrors may introduce distortions in the photographs. As Russell, Slocum and others have proved, evidences of such distortions are shown even in the published results of the British observers. It was immediately recognized that in the future much greater care must be exercised in securing photographs of the field of stars in the midst of which the sun would be found on the day of the eclipse. The conditions under which these plates are obtained must resemble as closely as possible those at the time of the eclipse, i.e., with same temperature, hour angle of telescope, etc. Scientific men are fully aware of the fact that errors may creep into measures no matter how carefully made and every effort must be made to minimize these errors or to eliminate them if possible. The chief place of importance at the 1922 eclipse was given to this problem.

It must not be thought that the amount of the deflection caused by the sun is an angle so minute in size that its measurement will tax the resources of modern astronomy. In determining the distances of the fixed stars, parallal observers are quite used to measuring displacements on their photographic plates less than one-hundredth of the Einstein deflection by the sun. There are half a dozen of the observatories in the United States alone which are devoting much of their energies to parallax determinations. Nor does the problem of measuring the eclipse plates contain anything startlingly new in the method. Astronomers for many years are quite familiar with all of the processes involved in the measurements and reductions, and moreover they are quite accustomed to securing an accuracy far greater than that demanded by the Einstein problem. In regular observatory work, such as required in the determination of the parallax of the stars, astronomers are confident that constancy of results can be expected from night to night in photographs taken with the telescope, and they have long series of observations to prove their conclusions; in fact, in parallax determinations an accuracy is secured which is measured by a probable error one two-hundredth part of the maximum deflection demanded by the Einstein theory.

If a total eclipse took place every day, the astronomer would have abundant opportunity to vary the observing conditions and thoroughly eliminate all possible sources of error, but with eclipses occurring so seldom he must do the best he can under unusual conditions. The check photographs are secured for this purpose. The scale of the photographs, the orientation and the various correcting factors that must be taken into account in the reductions are obtained from the check photographs. If these plates are photographed *through* the glass, which is possible by the use of plate glass, then the eclipse plates and the check plates may be placed film to film and the star images on the eclipse plates will nearly match in position those of the check plates. The measurement thus becomes a differential one of comparing the stars on the eclipse plate, image by image, with those of the check plates. The method of reduction makes due allowance for any changes in scale-value, orientation, refraction, etc., and we can confidently expect that the gravitational deflection will be readily separated from all other effects, provided care is taken in securing a satisfactory series of plates to act as checks. Before discussing the results of the 1922 eclipse we must survey the other consequences of the Einstein theory.

Of the eight major members of the solar system, Mercury is the exceptional planet. It is closest to the sun and has the greatest orbital motion, it has the smallest mass, its orbit has the greatest eccentricity and is inclined by the greatest angle to the ecliptic. Mainly on account of the great eccentricity and high inclination, the path of the planet has caused considerable worry to the mathematical astronomer. Mercury is never seen more than 28° from the sun and it is therefore only visible to the naked eye shortly after sunset or before sunrise projected on a strong twilight sky. With a telescope, however, the planet may be observed in the daylight sky. Tables of motion of Mercury were made by Kepler as early as 1627, by Halley about 1680 and by Lalande a hundred years later. The first really accurate tables were those of Leverrier based on meridian observations made at the Paris Observatory between the years 1801 and 1842, and supplemented by observations of transits of the planet across the face of the sun beginning with that of 1631. Leverrier discussed the observations in a thoroughly masterful manner, calculating the gravitational attractions of the other planets for Mercury whereby the orbit was perturbed. It was found by him that the observed motion of the axis of the planet's ellipse was moving 0″.38 per year, or 38″ per century, in excess of the amount calculated from the law of gravitation.

Leverrier discovered several possible reasons for the discrepancy, the most plausible being the presence of a planet, or group of planets, between the orbit of Mercury and the sun. On March 26, 1859, Lescarbault observed such a planet in transit across the face of the sun. It was found later that Lescarbault's planet did not follow the orbit calculated for it by its discoverer and that the transits predicted by him failed to materialize. Unquestionably, the original observation and those of many additional transits of other planets supposed to have been seen by others find a ready explanation in the appearance of spots on the face of the sun. The discovery of two intra-Mercurial planets at the eclipse of 1878 (see Chapter IX) has likewise never been verified.

Two decades after Leverrier had completed his work, the

same problem was attacked by Simon Newcomb; and naturally with much more material to his hands than was available earlier. No less than 5421 observations of Mercury entered into the discussion, and a total of over sixty thousand into the research on the motions of the four inner planets, Mercury, Venus, Earth and Mars.

The observed motion of the perihelion of Mercury deduced by Newcomb was $41''.6 \pm 1''.4$ in excess of that calculated from the law of gravitation. Such a large discrepancy appearing in Mercury, what of the other planets? The orbit of Venus is really too nearly circular to permit of the observation of the direction of the major axis of its orbit with accuracy and the same is true, but in lesser degree, for the earth. The values of the motions were calculated but the precision of the results for Venus and Earth were inferior to those of Mercury and Mars. This latter planet with its greater eccentricity of orbit shows an unexplained discrepancy of $8''.1$ in the motion of its perihelion.

According to Einstein's law of gravitation, the planetary orbits not being closed ellipses but spirals, the path is slightly advanced at each revolution in the direction of the motion of the planet. The prediction of Einstein gives a simple formula, namely, that the orbit of any planet will advance through a fraction of a revolution measured by $3v^2 / c^2$, where v is the velocity of the planet and c the velocity of light. The values of v for each of the planets is known with great exactness. For the four inner planets, the Einstein motions per century are: $42''.9$ for Mercury, $8''.6$ for Venus, $3''.8$ for the Earth and $1''.3$ for Mars; showing a remarkable agreement for Mercury but little improvement in the discrepancy of Mars. Further disagreements between observation and motion calculated on the Newtonian theory were found by Newcomb, the most important being $10''.2$ in the node of Venus and $0''.88$ in the eccentricity of Mercury. The Einstein theory does not apply to the two last discordances. Attention should thus be called to the fact that the motion of the perihelion of Mercury is not the only wide divergence from the law of gravitation found in the discussion of the motions of the planets. Its value is the largest, it is true, but the other disagreements are none the less important in spite of their smaller size.

Before we can accept the Einstein theory as the true explanation of the motion of the perihelion of Mercury, we must first exhaust all other explanations arising from the Newtonian law of gravitation, and all possible deflections in the photographs caused by refraction, etc. If the unexplained motion of the perihelion had been Leverrier's 38″ instead of Newcomb's 42″, the Einstein prediction of 43″ would hardly have been received with the universal acclaim accorded the almost perfect agreement. Newcomb, many years ago, showed that the motion of Mercury can be completely accounted for under several hypotheses, but unfortunately in explaining away the discordance in Mercury's motion, discordances were introduced into the other planets so that the last state appeared no better than the first. There are three possibilities:

(1) A non-spherical sun
(2) A ring of matter surrounding the sun
(3) The number 2 in Newton's law is only a first approximation.

According to Charles Lane Poor,[1] "the ordinary classical methods of physical and astronomical research can fully explain all of the observed phenomena; the motions of the planets can be fully accounted for by the presence of matter known to exist, and the light deflection, if real, can be explained as the result of refraction through the cosmic dust surrounding the sun."

It was shown by Poor that if the equatorial diameter of the sun exceeds the polar diameter by an angle of 0″.10, it will cause a motion of the perihelion of Mercury amounting to but 7″ per century. Although from theoretical grounds the rotation of the sun should cause a bulging of the sun at the equator, such an effect has never been detected with certainty (see Chapter VII). The excess of equatorial diameter of the sun can hardly amount to more than 0″.05, and this would affect the orbit of Mercury but little.

[1] *Gravitation versus Relativity*, 227, 1922.

It was next shown by Poor that the assumption of a lens-shaped material appendage surrounding the sun will be sufficient to explain not only the motion of Mercury, but practically all of the other divergences from the law of gravitation as well. The method adopted by him was one very familiar in practice, namely, that of introducing as many constants into the calculation as there are quantities to be allowed for. In the solar corona there is evidence of an appendage to the sun and, furthermore, the zodiacal light gives an indication of an envelope stretching out from the sun vastly farther than the corona. The corona is more or less symmetrical with respect to the sun's equator, while the plane of the zodiacal light is that of the earth's orbit. For the purpose of simplicity of computation, it was necessary to make several more or less arbitrary hypotheses.

According to Poor (*loc. cit.*, p. 237), " there is apparently no mechanical nor physical reason for the non-existence of a group, or groups, or bodies, sufficient to explain all the irregularities in the motions of the planets. Thus, all the discordances, including that of the perihelion of Mercury, can readily be accounted for by the action, under the Newtonian law, of matter known to be in the immediate vicinity of the sun and planets." Such a distribution of material would cause a true atmosphere surrounding the sun which would not be of the uniform density (assumed for ease in computation), but which would be very dense close to the sun's surface.

An atmosphere of this character would cause refraction on a beam of light when it passed through this atmosphere. In fact, Poor assumes that such a refraction is a sufficient cause to explain the whole of the bending of the rays of light at the 1919 eclipse. On the other hand, Eddington [1] shows that " at a height of 400,000 miles above the surface of the sun the refractive index required is 1.0000021. This corresponds to air at $\frac{1}{140}$ atmosphere, hydrogen at $\frac{1}{70}$ atmosphere, helium at $\frac{1}{20}$ atmospheric pressure. It seems obvious that there can be no material of this order of density at such a distance from the sun. If there were, the pressure on the

[1] *Space, Time and Gravitation,* p. 121.

sun's surface of the columns of material involved would be of the order of 10,000 atmospheres; and we know from spectroscopic evidence that there is no pressure of this order."

Dr. Harold Jeffreys[1] discussed the effect, on the perihelion of Mercury and on the shifting of star places, of matter known to be in the immediate vicinity of the sun and planets, namely, the solar corona and the zodiacal light, in the following manner: "You can determine the density perfectly from the luminosity. You know the intensity of light that falls on these masses; you know the distance; you have a fair idea of the physical constitution; and you need simply compare these with the light that you receive from the sky at midday to get a direct estimate of the amount of these two deflections that can be got from this mass that is known to exist. The results are rather startling. There is enough matter in the zodiacal light to account for one three-thousandth of the motion of the perihelion of Mercury. There is enough matter in the corona to account for about 10^{-8} of the motion of the perihelion of Mercury. The solar corona, by means of its refraction, is capable of producing one-millionth of the observed deflection of the star images. So I think it is clear that the only reasonable interpretation one can make of these two shifts is that they are due to some property of the law of gravitation."

Another method of explaining the motion of Mercury under Newton's law was that suggested many years ago by Hall. If the exponent 2 in the inverse square law is increased until it becomes 2.00000016, then the effect on the perihelia of the inner planets as computed by Newcomb becomes: Mercury 43″.37, Venus 16″.98, Earth 10″.45 and Mars 5″.55. While the values of Mercury and Mars agree well with the observed displacements, those of Venus and the Earth are very wide of the true values.

Of the utmost importance to the Einstein theory is the accurate determination of the shift of the spectral lines in the sun towards the red when the wave-lengths are compared with those derived under terrestrial conditions. This spectroscopic test is not based on mathematical reasoning of such

[1] *Monthly Notices, R. A. S., 80,* 116, 138, 1919.

difficulty that few can follow the methods, but simply and solely on the theory of relativity that motions are relative. According to the principle of equivalence, Einstein asserts that a ray of light falling from the ceiling to the floor of a room moves exactly the same, no matter whether it is assumed that the light moves in the earth's gravitational field or with respect to a system of axes of reference moving at the rate of 32 feet per second upwards. The result of light moving in a gravitational field cannot be distinguished from a Doppler effect of motion in the line of sight. Consequently, it is easy to calculate the amount of the shift caused by the gravitational field in the sun, and this is equivalent to a Doppler effect of 0.634 km per second. If the relativity shift is verified in the sun, this same principle becomes of the greatest importance in the investigations of other suns, the more distant stars. If we know the mass and density of a star, from which can be found the gravitational attraction, then the star may be submitted to the Einstein test. As will be shown later, this test has been carried out with great success on the feeble companion of the brilliant Sirius.

At first sight the problem seems to admit of a ready solution. All that appears necessary is that the wave-lengths of the arc spectrum be determined in the laboratory with the highest degree of accuracy attainable and that these wave-lengths be compared directly with solar values. Let us look back and see what the earlier chapters have shown concerning the history of spectroscopy.

As early as 1890, in the process of deriving his famous table of solar wave-lengths, Rowland had noticed that comparisons of the solar spectrum with that of the electric arc showed that the solar lines were nearly always shifted to the red. It was thought that at the time this shift was probably accidental, due unquestionably to some systematic errors called in by the apparatus. However, a further study of the problem by Jewell convinced him that the cause of the shift to the red could not be instrumental for the reason that the shift was far from being a constant, since it varied in amount with different chemical elements and even different amounts of shift appeared for different lines of the same element.

Not being instrumental, the cause of the displacements seemed to be due to the fact that the temperatures and pressures of the vapors in the solar atmosphere were vastly different from those found in the vapors rendered luminous by the electric arc. It seemed impossible for temperature to have any influence whatever on the wave-lengths either in th arc or in the sun, and hence the cause unquestionably was that of pressure. Accordingly, extensive series of investigations were carried out by Jewell, Humphreys and Mohler [1] who came to the conclusion that the pressure in the sun's reversing layer varied somewhere between two and seven terrestrial atmospheres with the mean value of five to six atmospheres. The surprising feature of the investigation was that the pressures derived were so small, being in fact much less than one might have supposed to exist in an enormous body like the sun.

In 1909, Fabry and Buisson [2] applied the interferometer to the problem and succeeded in deriving wave-lengths with a greater accuracy than had been possible with a concave grating alone. In investigating the spectrum of iron in the electric arc, it was found that as a whole there was a shift to the red in the sun compared with wave-lengths of iron but that the amounts of the shifts varied greatly, some of the lines even showing a shift to the violet. In comparing directly the wave-lengths of the solar rays and the corresponding rays of the iron arc in a vacuum, Fabry and Buisson found that the solar rays always showed this shift towards the red, which must be the case if the displacement is the effect of pressure in the sun's reversing layer, provided the pressure in the sun is greater than that found at the earth's surface at sea-level. The explanation seemed satisfactory and sufficient, and as a result of a long series of observations very carefully executed the conclusion reached was that the solar pressures were approximately five atmospheres, thus being in good agreement with the values of Jewell, Humphreys and Mohler. The shift of the solar lines to the red amounted to 0.008 A at wave-length 4500 A.

[1] *Astrophysical Journal, 3,* 114, 1896.
[2] *Journal de Physique,* (4), 9, 298, 1910.

In 1911, Einstein[1] called attention to the amount of this shift and showed that if it could be assumed that the pressure of the reversing layer was less than one-tenth of a terrestrial atmosphere, then the displacement of 0.008 A was readily explainable on the theory of relativity just then being developed by him. Unfortunately, all of the observations pointed to much higher pressures at the photosphere and to add to the complications Halm[2] had found (1907) that the wave-lengths from the edge of the sun were displaced towards the red with respect to those from the center of the sun. Neither pressure nor Einstein shift could explain these difficulties nor was the theory of anomalous dispersion more successful when brought forward by Julius.[3]

There seemed nothing to do but to secure more observations and with greater precision. These observations came principally from three observatories, from Kodaikanal, from Allegheny and from the Mt. Wilson Observatory.

All observers were now agreed in finding an unmistakable shift to the red of wave-lengths in the sun when compared with terrestrial standards. By the year 1912, four different explanations were offered: (1) pressure, (2) anomalous dispersion, (3) relativity shift and (4) a Doppler effect of motion of the solar vapors, advanced by Evershed. Which of the four would survive the test of precise observations?

At Mt. Wilson an extended series of researches were undertaken with the electric furnace wherein both the temperature and pressure could be controlled at will. These researches, continued up to the present, have added enormously to our knowledge of spectra. In the investigation of the spectrum of iron, Gale and Adams[4] found it possible to divide the iron-lines into four groups, designated by the symbols *a, b, c,* and *d.* In these four groups the amounts of the displacements, for the same value of the wave-length and same increase in pressure, are proportional to the numbers 1 : 1.5 : 3.4 : 6.6. In passing from the arc in a vacuum to the arc at ordinary atmospheric pressure these investiga-

1 *Annalen de Physik,* (4), *35,* 905, 1911.
2 *Astronomische Nachrichten, 173,* 273, 1907.
3 *Rev. Gen. des Sciences, 15,* 480, 1904.
4 *Astrophysical Journal, 35,* 10, 1912.

tors confirmed the conclusions already advanced by Fabry and Buisson regarding the pressures in the photosphere.

In 1913, still another difficulty appeared. Wave-lengths secured with different strengths of current and with the poles of the electric arc placed at different distances apart showed wide variations. Hence there arose what is now known to spectroscopists the world over as the "pole-effect," to correct which was devised the "Pfund arc." The pole-effect at times may be quite large. For the sodium pair at 6154 A and 6160 A, for instance, the displacement may amount to as much as 0.52 A. The Stark-effect and the Zeeman-effect added still further to the confusion. The problem had thus become a frightfully complicated one.

Still another cause of discord appeared in the year 1915 when it was recognized by all observers that the amounts of shifts varied with the intensities of the solar lines. Not only were the wave-lengths at the *edge* of the sun shifted towards the red with varying amounts depending on intensity when compared with lines from the *center* of the sun, but also the lines from the center of the sun were likewise shifted towards the red compared with the corresponding lines of the electric arc, but these displacements varied in an entirely different manner with change in intensity.

Theoretically, the amounts of the pressure shifts should vary directly as the cube of the wave-lengths. The observed displacements certainly did not exhibit any verification of this law, and hence of necessity it was recognized either that the pressures at the solar surface were very small or else their effects were masked by other causes. But what were the other causes? A close examination of the observed quantities brought to light the fact that the displacements from the same line of the spectrum gave very different results on different days of observation, and moreover even the lines of the same group with same intensities and same wave-lengths gave discordant values. It was evidently not sufficient for the purpose to be satisfied with having a stable form of apparatus. The temperature must be kept rigorously constant during the observations, the grating must be covered in precisely the same manner by the rays of light in each exposure, but most important of all, it was no longer sufficient to expose first to the sun and then to the electric arc, but the exposures had to go on strictly simultaneously. These refinements when carried out imposed on the observations a much more satisfactory internal agreement than had hitherto been found.

But what was the cause of the systematic deflections towards the red? Apparently, pressure was not the chief cause. The shifts towards the red throughout all wavelengths seemed to behave exactly like a Doppler effect of motion in the line of sight. It therefore seemed possible that these motions were actually caused by the vapors descending towards the surface of the sun or else withdrawing away from the earth. When comparing the center of the sun and the lines of the electric arc, the dependence of the amounts of the shifts on the intensities of the lines could be readily explained by making plausible assumptions regarding the heights of the various layers in which the solar lines originate, but it was difficult to apply this same explanation to the shifts of the lines at the edge of the sun when compared with those in the center. Evershed and Royds [1] were therefore forced to the bizarre conclusion that the *earth* was the cause of the displacements and that this small planet of ours exerted a repulsion on vapors in the sun's atmosphere. The authors themselves recognized the absurdity of the earth-repulsion theory for there seemed no reason why the other planets should not give evidence of similar repulsions. Evershed [2] therefore put this theory to a crucial test to discover whether there was a similar effect from the earth's twin-sister, Venus. For this purpose, a comparison was made between the spectrum of Venus and the spectrum of daylight, that is to say, the whole of the sun's hemisphere turned towards the earth. The observations were carried out in 1918 while the angle from Venus to sun to earth changed from 45° to 135°. These observations seemed to give satisfactory proof to Evershed of a Venus-effect, but these conclusions were not confirmed by the Mt. Wilson observers [3]

[1] *Kodaikanal Bulletin, 49,* 1916.
[2] *Observatory, 42,* 51, 1919.
[3] *Astrophysical Journal, 53,* 380, 1921.

LICK OBSERVATORY CAMERA FOR TESTING THE EINSTEIN EFFECT, AUSTRALIA, 1922

SAILORS OF THE AUSTRALIAN NAVY ASSISTING THEIR COUSINS FROM CANADA IN ERECTING THE EINSTEIN CAMERA

who in 1919 and 1920 made special observations towards the same end.

Meanwhile the whole world had been electrified by the results of the 1919 eclipse which showed stellar deflections confirming in such splendid manner the Einstein theory. It was but natural that many scientists would claim that they had discovered the Einstein shift in the sun — but many of these contributions exhibited little knowledge of the difficulties underlying the problem. For lack of space it will be impossible here to go into details.

When one refers to the widely different values of the solar rotation obtained as the result of an extended series of most careful and refined measures carried out by the most skillful of the world's solar spectroscopists, as given in Chapter XVIII, one glimpses some of the difficulties of the problem. The verification of the Einstein prediction by sun-arc displacements is even a more difficult task than that of measuring the solar rotation by spectroscopic methods.

The problem, however, became much simplified in 1923 and 1924 by the recognition by a number of competent authorities, almost simultaneously and from a number of different lines of attack, that the pressures of the photosphere were much less than previously had been supposed. It had already become evident that the relativity shift could not be determined by measuring a few lines and making plausible assumptions regarding pressures and various other effects. To reach a definite conclusion it was manifestly necessary to carry out a much more extensive program of work, both on the sun and in the laboratory. The goal of such work is indeed the determination of a satisfactory system of arc and solar wave-lengths each reduced to the *international standard,* and with the wave-lengths in sun and arc freed as much as possible from possible sources of errors. Such investigations, demanding the most refined instrumental equipment and specialized training of the highest order that can come only from long years of practical research, seem to have been realized.

Wave-lengths of the highest degree of reliability have been determined at the Allegheny Observatory in coöperation with the Bureau of Standards. The method of securing the photographs was by means of a powerful grating spectrograph and interferometer, the wave-lengths being accurate within one part in five million to one part in eight million. These measures can readily detect a shift one-tenth of the Einstein prediction. The results for neutral iron are more exact than for any of the other elements. Wave-lengths in sun and vacuum arc must be carefully reduced to one system. From the differences sun *minus* vacuum arc are then subtracted the predicted relativity shift $2.13 \times 10^{-6} \times \lambda$.

SYSTEMATIC DIFFERENCES IN HEIGHTS AND WAVE-LENGTHS*
WITHIN MULTIPLET GROUPS

E.P.	Intensity	Main Diagonal			Side Diagonals			Whole Multiplet		
		No.	Δλ	Hgt.	No.	Δλ	Hgt.	No.	Δλ	Hgt.
0.07......	6–40	9	+3.0	1722	10	+1.4	1300	19	+2.2	1500
0.93......	4–40	15	+2.0	1157	20	+0.5	792	35	+1.2	948
1.54......	2–30	12	+2.0	1400	13	−0.1	823	25	+0.9	1100
0.93......	4– 8	5	+1.2	540	6	−0.6	500	11	+0.2	518
2.2–2.8...	3–10	10	+0.9	625	17	−1.0	500	27	−0.3	548
2.4–3.2...	1– 8	17	−0.9	497	25	−1.2	409	42	−1.1	445
2.2–2.8...	1– 6	27	−2.6	525	34	−2.5	518	61	−2.6	522
3.2–4.8...	0– 6	40	−2.9	406	49	−2.4	361	89	−2.6	381

* The unit for Δλ is 0.001 A.

For all of the individual multiplets of *Fe,* where there are enough spectral lines to make it worth while, the lines can be divided into two groups, depending on whether the lines are on the main or on the side diagonals of the multiplet. Altogether a total of 48 multiplets involving 309 *Fe* lines were discussed, the results in the table being separated into three series, representing strong, medium and weak multiplets. In each case are given: the number of lines, the height in kilometers from the flash spectrum, and Δλ (in units of 0.001 A), which is the equivalent to the difference sun *minus* vacuum arc from which has been subtracted the relativity shift.

It is evident from the abundant material that went into the table that the value of Δλ on the main diagonals of the multiplets averages greater in size than the value on the side diagonals. For obvious reasons, the relativity shift in each multiplet must be practically identical for main and side diagonals. Although the values of $\Delta\lambda$ for all of the lines in the table will average nearly zero, yet it is clear that there are systematic differences, the strongest lines, those which reach the greatest heights, have the largest plus residuals, while on the other hand the weakest lines or those which reach the least elevations, have the largest minus residuals.

Apparently therefore the wave-lengths in the sun are greater than those in the laboratory, the average difference being approximately that of the Einstein relativity shift.

Several attempts have been made to explain the residual differences. It has been assumed by St. John that the positive and negative residuals represent a Doppler shift, and in consequence the atoms forming the strong lines at great heights above the photosphere are falling toward the photosphere, while on the contrary the atoms forming the weak lines of less heights are ascending from the solar surface. To this very interesting hypothesis, Milne takes exception and points out the various difficulties. He himself advances the hypothesis of an asymmetrical distribution of velocity among the velocities of agitation of the individual atoms. Burns [1] favors Milne's point of view. The present writer believes that the simplest explanation is to be found in the St. John-Babcock interpretation, which is nothing more or less than a circulation of *Fe* vapor in the chromosphere. On account of the higher temperatures and higher pressures near the photosphere, the solar activity causes the *Fe* atoms to ascend through the medium of thousands of weak lines of high excitation potential, the maximum velocity found from flash spectrum for *Fe* being 0.2 km per second. The heights to which the *Fe* atoms are ejected are much greater than can be measured by the lines in the flash spectrum. In the upper reaches of the chromosphere, especially where the pressures are minute, some of the atoms lose an external electron and become ionized. From a law of nature that "whatever goes up must come down" the *Fe* atoms descend from the maxi-

[1] *Journal Optical Society of America, 20,* 212, 1930.

mum heights. In their descent some of the ionized atoms gain an external electron and again become neutral. At less heights the atoms descend through the medium of comparatively few lines of great strength but of low value of the excitation potential. For the *Fe* lines of highest level in the flash spectrum the velocity is 0.4 km per second, while for the H and K lines which reach the greatest heights in the chromosphere, Evershed[1] finds a residual value of $\Delta\lambda$ equal to 0.0070 A, equivalent to a velocity of + 0.5 km per second.

From information given in earlier chapters, it is possible to make a calculation of the number of atoms crossing a given level per second of time in their ascents and descents. The numbers appear to be roughly equal.

Consequently, the greater wave-lengths at the center of the sun when compared with the arc appear to be satisfactorily explained. But with the observed shifts to the red at the *sun's edge* " it is another story." Motions of circulation assumed in the sun must be directed toward or away from the center of the sun. At the sun's limb these motions are *at right angles to the line of sight* to the terrestrial observer, and hence they cannot cause a Doppler effect of motion in the line of sight. As a consequence, the spectrum of the limb of the sun compared with the iron arc in vacuum *should* exhibit a constant deflection, independent of the intensities of the solar lines under consideration, and this constant value should be that demanded by the relativity prediction; which, however, observations do not confirm. Hence St. John is forced to assume a " limb effect " caused by molecular scattering as a result of the great differences in the lengths of paths traversed by rays coming tangentially from the edge of the sun compared with those from the sun's center.

This explanation of the limb effect seems to be a makeshift and no great confidence can be felt in it until verified by more complete observations. Unfortunately, the necessity of assuming such an effect enormously weakens all of the arguments tending to prove that the Einstein prediction has been verified to exist in the solar spectrum. Unques-

[1] *Monthly Notices, R. A. S., 90,* 186, 1929.

tionably the wave-lengths in the sun are greater than those in the terrestrial laboratory, unquestionably the chief cause of the greater wave-lengths is the slowing up of the atomic clock in the sun — but there are minor differences in wave-length for which there is now no adequate explanation. Fortunately, Adams [1] at Mt. Wilson, from measures of the spectra of the companion of Sirius, in a spectacular manner has not only confirmed the validity of the theory of generalized relativity, but at the same time has shown that densities 50,000 times the density of water can exist in the atmospheres of dwarf stars of early-type spectra.

Since the first burst of enthusiasm in 1919, following the measures made by the British astronomers at the total eclipse of that year, scientists have had a chance to examine more fully into the theory of the bending of the light of a star when the stellar beam passes close to the sun. The problem is after all not a novel one in scientific thought, but dates back more than one hundred years to the celebrated English physicist Cavendish who calculated, on the basis of the corpuscular theory, the amount that the corpuscles of light are bent towards the sun in passing it. As long ago as 1801, Dr. J. von Soldner derived formulas, which were published in Bode's *Astron. Jahrbuch* for 1804, from which the deflection of light at the edge of the sun on the corpuscular theory was computed to be 0.″84. (Using modern values of the astronomical constants, this value is changed to 0.″87.) This deflection, based on an erroneous theory of light, is the so-called Newtonian deflection. The mathematical discussion should not under any circumstances be confused with the prediction by Einstein resulting from his splendid development of the general theory of relativity.

It should be clearly borne in mind, however, that in the progress of the mathematical treatment by Einstein the equations became very complicated; and for the sake of simplicity, of securing a set of equations that could be manipulated, it was necessary to make certain assumptions. The theory coming from his hands is therefore not in its most generalized form, and it is true only in so far as the restric-

[1] *Proc. Nat. Acad. Sciences*, *11*, 382; *Observatory*, *48*, 337, 1925.

tions imposed by the selection of the set of hypotheses can represent the observed facts. These limitations curtail the applicability of the theory — but the method, rendered necessary for the purpose of securing results, is almost as old as science itself. Since knowledge can be obtained only as the result of observations, it is evident that the velocity of light is the maximum velocity that can enter into any observation.

Two great difficulties present themselves to the average mind when attempting to grasp the meaning of the Einstein theory. The first of these two is that it is almost impossible to conceive that the coordinate time differs in no essential particulars from the three dimensions of length, breadth and thickness. The upholders of the relativity theory assert that the world of events is rendered the simpler by this consideration of a space-time concept of four dimensions, even though it is necessary, as a consequence, to regard the whole volume of space as finite, but nevertheless without a boundary. As a matter of fact, no one can possibly picture the curvature of a four-dimensional continuum for the reason that there is nothing at all in our experience comparable with it.

There are some individuals with mathematically trained minds who have persuaded themselves that they can really grasp the conception of the four-dimensional world of space-time; but most of these rebel and refuse to be convinced when, as a second consequence of the Einstein theory, it is necessary to assume that gravitation should receive no treatment different from that accorded the other "forces" of nature. Gravitation is therefore *explained* by Einstein as being due to the curvature of space and time. Certain master minds, like Eddington,[1] state that a great simplification has taken place; and that "we have realized for the first time that a world without gravitation (without curvature) would be more specialized and stand more in need of explanation than the actual world disturbed by gravitation."

Eddington's colleague at Cambridge, Sir Joseph Larmor,[2] finds difficulty in believing that "the world of so-called classical physical science is an illusion, an imperfect picture of a

[1] *Scientia,* 33, 313, 1923. [2] *Phil. Mag.* 45, 243, 1923.

reality depending symbolically on a field of fourfold quasi-geometrical algebra, of a presumed cosmos deep below the sensual sources of our concrete physical knowledge, of which we may obtain shadowy glimpses, but which is as yet quite beyond any direct grasp or intuition." Furthermore, in this important paper (*loc. cit.*) Larmor states the belief that the *principle of equivalence* between a field of gravitation and a field of acceleration seems definitely to have failed. Larmor makes a careful study of the whole problem and comes to the definite conclusion (derived before the date of the 1922 eclipse) that the maximum deflection experienced by the stars surrounding the sun at an eclipse should amount to one-half that predicted by the Einstein theory, or in other words, $0''.87$ at the limb of the sun.

The eclipse of 1919 certainly did not furnish a decisive answer regarding the verification of the Einstein problem. The work at best was pioneer in its nature, but it did give invaluable information regarding the best method of procedure for researches of the future. At the eclipse of 1922, four improvements over the methods adopted in 1919 were incorporated. First, the coelostat mirrors were eliminated, the cameras being pointed directly at the sun. Second, the lenses employed were quadruple, and not the ordinary achromatic combination of flint and crown glass lenses; thus ensuring a flat field over a very large area. Third, instead of trusting to luck that the driving mechanism would perform its duties perfectly, auxiliary guiding on a star during the eclipse exposures was attempted. Fourth, the securing of check plates was thoroughly accomplished.

The observers devoting themselves mainly to the Einstein problem were the British party of Jones and Melotte who located on Christmas Island in the Indian Ocean, and a combined German-Dutch expedition with headquarters also on Christmas Island. In Northwest Australia was the Lick-Crocker expedition under the direction of Dr. W. W. Campbell, a Canadian party from the University of Toronto consisting of Dr. C. A. Chant and Dr. R. K. Young, and Mr. J. Evershed from the Kodiakanal Observatory of India.

The British observers at Christmas Island spent six months

at the eclipse site in order to secure the check plates in advance of the eclipse and also that photographs might be secured of the southern skies to be used for measuring the magnitudes of the southern stars. The observers unfortunately found conditions nearly identical with those experienced in Sumatra at the eclipse of 1901. The skies were so covered with thin haze or heavier clouds almost every night that neither the check plates nor the other stellar photographs so much desired could be secured. To add to the continued disappointment of the observers, the sky was clouded during the critical minutes on September 21 and no eclipse photographs were secured either by the English or by the German-Dutch parties.

The Einstein equipment of the Lick Observatory party consisted of one pair of lenses each of 5 inches aperture and 15 feet in focal length, and another pair of 4 inches aperture and 5 feet in focal length. Each lens was a quadruplet and the specifications required a flat field giving good star images over plates 17 x 17 inches. The photographic films were coated on plate glass a quarter of an inch in thickness.

Wallal, the headquarters of a sheep ranch on the bleak shores of Northwest Australia, was chosen as the scene of operations by the Lick and Canadian observers. It was found not to be practicable to send an observer with the Einstein cameras to Wallal three months or more before the eclipse in order to mount and adjust the cameras and secure the necessary check plates in the night skies, and consequently it was decided that Trumpler should secure these plates in Tahiti in the Southern Pacific where the living conditions were more kindly. The latitude of Tahiti (17° 32′ S) differed little from that of Wallal (19° 45′ S) and it was thought that the temperatures at night at Tahiti would differ little from those by day at Wallal. Dr. Trumpler, in spite of much interference from clouds, secured the necessary plates, not only with each of the four Lick objectives but with the two Canadian cameras as well.

As an extra precaution, it was decided to photograph on each plate taken at Tahiti not only the critical star group where the sun would be projected at the time of the eclipse, but also an auxiliary group having the same declination but with a right ascension six hours farther east. The night before the eclipse, the photographic plates to be used for the first half of the eclipse set were exposed to the auxiliary group, while the second half of the eclipse plates were exposed to the star group at night following the eclipse. The use of the second star group thus furnished a double check on the accuracy of the eclipse photographs and by their means it was hoped to eliminate completely any systematic errors that might possibly creep into the Tahiti plates because of their not having been actually taken at the eclipse site.

The Australian government set a high standard of scientific cooperation. All eclipse observers were carried by railroad free of charge from Sydney to Freemantle and return. They were transported by ship from Freemantle to Broome and then to Wallal and return. The heavy equipment, weighing about thirty-five tons, was landed on the beach and transferred a mile and a half inland to the observing camp. The labor needed for this and for most of the operations required by nearly a month's stay at Wallal, including complete sleeping and dining service, was rendered entirely gratis. In the estimation of Director Campbell[1] "the hospitality extended, the interest shown, and the assistance offered, were of a standard higher than I have ever observed in any other part of the world on any occasion."

The population of whites in the Wallal region numbered only six persons. Many serious obstacles had to be overcome in the short time available before the instruments were erected and accurately adjusted. Perhaps the greatest trials were the dust and the flies. The eclipse site was in the bed of an extinct lake, and on account of the small annual rainfall the ground was very dry and a cloud of fine dust was raised at each step by a person walking, and a gentle wind carried it to all portions of the plate holders and optical parts of the delicate apparatus. In order to attempt to keep down the dust, five blacks were continuously employed throughout the whole stay, carrying coarse sand and distributing it on the

[1] *Publ. A. S. P.*, 35, 11, 1923.

ground surrounding the instruments. On the morning of eclipse day, September 21, green boughs were cut from trees and placed on the ground and water plentifully sprinkled around in order to attempt to diminish the radiation from the heated soil. Fortunately for the observers, the work of adjusting and erecting the apparatus was not interfered with by clouds. Only occasionally during the month's stay at Wallal were clouds seen and these never appeared far above the horizon.

The account (*loc. cit.*) by the director of the Lick Observatory of the preparation for the eclipse gives one an inkling of the reasons why the Lick-Crocker expeditions have established such a splendid record for scientific work accomplished at eclipses. The Lick program is always carefully planned, the equipment is the very best that modern engineering can supply, the apparatus is thoroughly tested at home so that when it arrives at the eclipse site it can be readily erected and adjusted. Added to the above is the vast experience gained by the successful observations of eclipses in all quarters of the globe during the past thirty years. The author wishes to pay high tribute to his friend, the director emeritus of the Lick Observatory and also to the friends of the Observatory whose scientific interest and financial backing have made the Lick-Crocker expeditions possible.

On eclipse day, September 21, in Wallal not a single cloud was visible from morning to night — the astronomers could not therefore lay the blame on the weather if their photographs turned out unsuccessfully.

The most important problem for this eclipse was unquestionably the photographs for testing the Einstein problem. The measurements of the Lick plates by Campbell and Trumpler (*Lick Observatory Bulletin,* No. 346) are of the high precision one naturally expects in work emanating from the Lick Observatory. The methods adopted were the same as those of the British astronomers in 1919 and the Canadians in 1922 and consisted essentially in comparing each of the eclipse plates with a check plate of the same stellar region but with the sun absent. The most important sources of error that could influence the results were:

1. Distortions of the star places by the optical parts of the apparatus.

2. Disturbing effects in measuring the star images of the eclipse plates on account of being superimposed on the background of the corona.

3. Systematic distortion in the film of the eclipse photographs. Ross [1] has shown that the blackened part of the corona should dry more quickly than the balance of the plate, possibly resulting in a contraction of the film towards the inner corona.

4. Abnormal refraction [2] in the earth's atmosphere on account of the cooler temperature inside the shadow-cone of the moon.

5. Refraction [3] in an extended solar atmosphere.

6. A " yearly refraction " suggested by Courvoisier.[4]

Unfortunately, the stars surrounding the sun were very weak at the 1922 eclipse, there being only two stars brighter than the ninth magnitude within 1° of the sun's center, while in 1919 there were seven stars brighter than the sixth magnitude in a similar area. It was necessary to prolong the exposures to secure as faint stars as possible. Exposures of 2 mins. and 2 mins. 5 secs. respectively, were given for each of the Einstein cameras. During totality two plates were secured with each camera, there being two cameras of 15-foot and two of 5-foot focus, the exposures starting 10 secs. after the beginning and stopping 10 secs. before the end of totality. The check plates secured in the night skies in Tahiti were given exposures of three minutes. The results given below are those from the measures of the four plates with the 15-foot camera (ratio of aperture to focal length 1:36). Stars of photographic magnitude 10.5 showed measurable images, the first time in the history of eclipse work that such faint stars have been photographed. In the process of the work the positions of 118 separate stars altogether were measured, while about 50,000 settings were necessary to accomplish this.

[1] *Astrophysical Journal, 52,* 98, 1920.
[2] A. Anderson, *Nature, 104,* 1919–20; Poor, *Science, 57,* 613, 1923.
[3] H. F. Newall, *Monthly Notices, R. A. S., 80,* 22, 1919.
[4] *Beob. Ergebn. Sternw. Berlin, 15,* 1913; *A. N., 211,* 205, 1920.

The star images on the eclipse plates were round and symmetrical. The present writer has found in his experience in measuring stellar photographs that a fogging of the plates by daylight, moonlight, or other causes, renders the edges of the star images ill-defined and fuzzy in appearance; consequently, one is not surprised to learn from the Lick report that the stars on the eclipse plates were not as sharp in appearance as those on the check photographs. The focus, however, was of the very best.

The details of the measurements and reductions cannot be given here. It was necessary to correct the measures for the effects of proper motion and annual parallax of the stars, for differential refraction and aberration, and for the inclination of the plates to the optical axis. Ten stars of the eclipse field and three of the auxiliary check field had proper motions of known amount and of sensible size. One star (Beta Virginis) has a measured parallax of 0″.10, and three other stars had hypothetical parallaxes which were derived from their known spectral classes and proper motions.

Arranging the stars according to their distances from the sun's center, the group means were formed and are given in the following table:

GROUP MEANS OF OBSERVED RADIAL DISPLACEMENTS

Group	*Stars*	*Mean Dist. from Sun* °	*Observ'd rad. displ.* ″	*Correct'd rad. displ.* ″	*Einstein Theory* ″
1	8	0.64	+ .64	+ .69	+ .70
2	11	1.06	+ .35	+ .46	+ .37
3	10	1.40	+ .30	+ .39	+ .24
4	8	1.66	+ .16	+ .22	+ .17
5	9	1.90	+ .17	+ .21	+ .13
6	8	2.00	+ .15	+ .17	+ .11
7	11	2.22	+ .08	+ .08	+ .08
8	13	2.55	− .09	− .14	+ .02
9	14	2.97	− .04	− .08	− .03

It is at once noticeable that the measures are in remarkable agreement with the values predicted from the Einstein theory; even the stars, far removed from the sun's center and well outside the region of the denser corona, showing the light-deflections well marked.

The Einstein deflections are inversely proportional to the angular distance from the sun's center, and reducing the values for the separate stars to the amount at the sun's limb, then there results for the four pairs of plates the values given below:

LIGHT DEFLECTION AT THE SUN'S LIMB

Pair	*Campbell*	*No. of Stars*	*Trumpler*	*No. of Stars*	*Plate Mean*
1	1″.72±″.32	62	1″.88±″.27	69	1″.80
2	1.35 .22	77	1.62 .22	81	1.48
3	1.78 .22	80	1.91 .19	84	1.89
4			1.76 .22	85	1.76
Mean for each observer	1″.60±″.14		1″.78±″.11		
Mean from four plates					**1″.72**.±″11
Einstein value					1″.745

In addition, the star field was divided into four quadrants so that the sun's equator passed through the middle of two of the quadrants and the sun's axis of rotation through the other two. The stars in each quadrant were solved separately by least squares with the results below:

Quadrant	*No. of Stars*	*Light deflection at Sun's limb*
Preceding sun's equator	24	1″.61±″.28
Following sun's equator	29	1.68 .25
At sun's north pole	18	1.76 .26
At sun's south pole	21	1.73 .24
Mean of equatorial quadrants ...	53	1.63 .15
Mean of polar quadrants	39	1.76 .18
Mean of plates.................	92	1.72 .11

These magnificent results obtained with the two cameras of 15-foot focus furnish unmistakable evidence that there is a bending of the light rays from the stars as these rays pass close to the edge of the sun and that the amount of the deflection agrees remarkably closely with that predicted by the Einstein theory.

This theory, however, not only predicts a total deflection of 1″.75 for a ray of light just grazing the sun's surface, but it also requires that the deflections from the other rays shall vary inversely proportional to the angular distances from the center of the sun. It was for the purpose of testing this distance law that Campbell obtained photographs with the two cameras of 5-foot focus. These lenses were built on lines similar to those of 15-foot focus. The plates of 17 x 17 inches covered a very large field of fifteen degrees square. Six plates were secured during totality. The eclipse photographs show between 400 and 500 stars, of which only 135 to 140 of the best defined images suitably distributed over the plates were selected for measurement. Although the scale is only one-third of that secured with the plates of 15-foot focus the observed probable error is only 1.6 times greater.

The difficulties encountered in trying to establish the law of distances is two-fold. In the first place, the corona surrounding the sun is so bright that it obscures and hence prevents the obtaining of observations close to the edge of the sun where the deflections are large. However, the chief difficulty is that the exact scale of the eclipse plates is unknown and can be obtained only from the observations themselves. Since the scale of the plate is directly proportional to the distances measured, the distance law is superposed on that of scale, and the two cannot be separated unless there is a large range of distances to be considered.

A masterly discussion [1] by Campbell and Trumpler of the plates taken by the 5-foot cameras gives the deflection at the sun's limb of $1''.82 \pm ''.15$. Combined with the results from the 15-foot plates, noted above, the measured deflection of the Lick 1922 photographs turns out to be $1''.75 \pm ''.09$; in exact agreement with the value predicted by the generalized theory of relativity.

In consequence of the difficulties of separating the Einstein deflection from the scale of the photographs, noted above, confirmation at other eclipses is desirable. At the eclipse of 1929, the expedition from Potsdam [2] employed two cameras for measuring the relativity shift. The results from four plates from the 28-foot horizontal camera used with coelo-

[1] *Lick Observatory Bulletin, 13,* 130, 1928.
[2] *Zeit. für Astrophysik, 3,* 171, 1931.

stat have been published. Freundlich, von Klüber and von Brunn find from their plates a relativity displacement at the limb of the sun amounting to 2″.24, a value which not only differs greatly from the 1922 Lick results but also contradicts the value predicted by Einstein. Moreover, in rediscussing the measures of the 15-foot Lick photographs, the Potsdam astronomers find a value of 2″.05, greatly in excess of the deflection 1″.72 derived by the Lick observers from their own plates. The larger value obtained from the 1922 plates seems possibly to confirm the Potsdam results from the 1929 eclipse.

It is very unfortunate that the star field at the 1929 eclipse was ill suited to attain the highest accuracy, for the reason that of the total of 18 measurable star images on the Potsdam plates, 17 of the stars were on one side of the sun and only one star on the other side. This poor distribution brought serious complications into the determination of the exact scale value of the plates. A rediscussion of the 1929 results by Jackson [1] gave 1″.98 as the value of the Einstein term. A more thorough discussion by Trumpler derives the deflection from the Potsdam plates of 1″.75 instead of the much larger value of 2″.24. This large difference does not come from a radical change in methods of reduction, which are practically identical, but rather from the different method of deriving the scale value of the plates.

After reviewing all of the computations for the eclipses of 1922 and 1929, Trumpler [2] comes to the conclusion that the Lick discussion of the 1922 plates is the more reliable, and also that the value of 1″.75 for the 1929 eclipse is entitled to greater confidence than the larger Potsdam value for the reason that the systematic trends in the Potsdam residuals are thereby eliminated.

Adopting Trumpler's reductions, the following table gives the results for the 1919, 1922 and 1929 eclipses for the observed deflections reduced to the sun's limb. The extreme distances from the center of the sun at which stars were measured are also given. The values are in excellent agreement with the prediction of Einstein, for not a single ob-

[1] *Observatory*, *54*, 292, 1931.
[2] *Zeit. für Astrophysik*, *4*, 208; *Publ. A. S. P.*, *44*, 167, 1932.

servation, except the plates of poor definition obtained with the astrographic lens at Sobral (p. 448), differs from the predicted value by more than twice the probable error.

STELLAR DEFLECTIONS OBSERVED AT THE 1919 AND 1922 ECLIPSES

Eclipse	Expedition	Focal Length	No. of Plates	No. of Stars	Ang. Dist. from Center	Deflection at Limb	Authors
1919 May 29	Sobral	19 ft.	7	7	1°.5	1″.98±″.12	Dyson
	Sobral	11	16	6–12	1.5	0.86± .1	Eddington
	Principe	11	2	5	1.5	1.61± .3	Davidson
1922 Sept. 21	Wallal	10	2	18	2.6	1.75± .3	Young, Chant
	Cordillo Downs	5	2	14	2.9	1.77± .3	Dodwell, Davidson
	Wallal	15	4	62–85	3.4	1.72± .11	Campbell
	Wallal	5	6	135–140	10.4	1.82± .15	Trumpler
1929 May 9	Takengon	28	4	17–18	2.0	1.75± .13	Freundlich, von Klüber, von Brunn

Neglecting the one value at Sobral (as was done by the British astronomers), the weighted mean value for the relativity deflection reduced to the sun's edge is:

1″.79±″.06

No other effect or combination of effects can confirm in such an adequate manner the observed deflections of eclipse plates as can the Einstein prediction. Hence it must be concluded that eclipse plates have abundantly *verified* the theory of relativity. It must not be assumed, however, that thereby the theory has been *proved* to be true. No amount of observational confirmation can furnish indisputable *proof* of a law of Nature. Even one exception to the great law of gravitation would make it necessary to discard the law or to revise it.

The theory of relativity is based on the notion that it is impossible to observe *absolute* motions and that all of our knowledge is derived from motions which are *relative.* The theory has been very successful in interpreting observations in many different branches of physical science. Einstein

1932 CORONA
Photographed by Wright of the Lick Expedition.

Einstein Camera of the Kyoto Expedition at the 1934 Eclipse

extended the theory to the more precise observations furnished by astronomy and derived a new law of gravitation. The motions predicted by this new law differ from those derived from the Newtonian law by quantities that are so very minute that they can be tested only as the result of very refined observations. The Einstein law has been superior to the Newtonian law (1) in explaining the motion of the perihelion of Mercury, (2) in interpreting the stellar deflections observed on eclipse plates, and (3) in explaining the observed shifts to the red of lines in the solar spectrum compared with terrestrial standards. The spectroscopic problem is a very complicated one with many different factors to be considered. It will be necessary to obtain additional observations before the spectroscopic test is regarded as entirely conclusive.

The question now (1935) before astronomers is whether we shall accept the observed value of the deflection at the sun's limb of 1″.79± ″.06, found on page 476, as a satisfactory confirmation of the value 1″.75 given by the relativity theory. In tabulating these values the author has considered them worthy of greater confidence than the larger displacements, between 2″.0 and 2″.2 at the sun's limb, derived (*loc. cit.*) by Freundlich, von Klüber and von Brunn.

The differences between the two sets of reductions show the underlying difficulty of the whole problem, namely, the practical impossibility of adequately separating the scale of the plates from the star deflections. According to the opinion expressed by Freundlich and von Brunn,[1] the problem demands for the scale of the plates the excessively high accuracy of no less than one part in one hundred thousand! It need hardly be added that nowhere has this accuracy been attained in the past. Therefore it would seem to the man in the street an evident consequence that the work must be repeated at future eclipses and with far greater accuracy than in the past. Such a recommendation appears obvious — but unfortunately Wright [2] has found that no one of the star fields of any total eclipse of the future including that of February 25, 1952,

[1] *Zeit für Astrophysik*, *6*, 218, 1933.
[2] *Publ. A. S. P.*, *45*, 25, 1933.

is adequate to give reliable information concerning the relativity displacement.

Measures by Wright [1] of photographs of the eclipse of April 28, 1930, by the 36-inch Lick refractor offer " no support for the suggestion that has been made that the Einstein effect may be a consequence of abnormal refraction within the cold shadow of the moon at the time of a total eclipse of the sun."

There are still many doubting Thomases for whose benefit it will be necessary to accumulate additional observations. The author wishes to state that he is one of those who believe that the Einstein theory of relativity has been abundantly verified by astronomical observations.

Accepting the truth of the relativity theory, de Sitter in 1917 brought forward his famous prediction that the velocities of very distant objects must show motions of recession. At the time of his prediction, a few motions of distant spirals were known as the result of observations made at the Lowell Observatory. Since then many velocities have been accumulated at Mt. Wilson, and they show almost without exception that the spiral nebulae are receding from us with velocities that apparently depend on the distances, the greater the distance, the greater the velocity. The largest velocity so far found among the extra-galactic nebulae is + 24,500 miles per second.

The important point of de Sitter's theory was that it predicted the tendency of matter to scatter. This theory, however, was based on the assumption that space is utterly devoid of matter — which is hardly in harmony with our ideas of the physical universe. Opposed to the " de Sitter universe " is the " Einstein universe." The theory of the latter assumes a uniform distribution of matter from which was concluded a curvature of space so that the universe appeared to be a sphere of finite radius. In such a universe the gravitational attraction would exactly balance the tendency of the spiral nebulae to scatter, producing static conditions. The fact that the nebulae are receding with such high velocities is directly contrary to the static universe of Einstein.

[1] *Publ. Amer. Astron. Socy., 8,* 79, 1934.

Apparently, therefore, the nature of our universe may be intermediate between Einstein's "spherical universe," containing matter, and the "expanding world" of de Sitter, devoid of matter. According to the theories of Lemaître, as developed by Eddington,[1] the universe must be expanding at a rate which is constantly increasing and ultimately the nebulae must be separating at a speed greater than that of light itself. Under these circumstances, they will be invisible to us and will be "below our horizon." Possibly a great part of the universe is already below our horizon.

The great difficulty is that the enormous velocities of the nebulae now observed permit a time scale of a mere 10^{10} years. At the expiration of this time, matter will cease to exist. We might overcome the difficulty involved in the theory of the expanding universe by assuming (since all motions are relative) that we ourselves are becoming smaller and smaller. On this assumption, our terrestrial yardsticks, the earth and the moon and the sun, and even the stars and nebulae of our galactic system, are dwindling in size in the same progression and will ultimately cease to exist.

Unfortunately, the time scale of years, colossal though it be when compared with the span of human life, is far too small to explain the life of the stars, or even of that star of great importance to us, the sun.

In an important paper [2] written jointly by Einstein and de Sitter, the conclusion is reached that an increase in the precision of the observations may enable us in the future to determine the density of matter in the universe, and to fix the sign and determine the value of the radius of curvature. Unfortunately, any further discussion of these problems now of such vital interest to astronomers will carry us too far afield.

[1] *Monthly Notices, R. A. S., 91,* 413, 1931.
[2] *Proc. Nat. Acad. Sciences, 18,* 213, 1932.

CHAPTER XXIV

SHADOW BANDS

SHADOW bands have no astrophysical importance and for that reason they will be treated here very briefly. They are seen at the beginning and at the end of the total phase of a solar eclipse, and they should, perhaps, be mentioned in a treatise on eclipses even though they involve meteorological rather than astrophysical problems.

The most complete observations of shadow bands at an eclipse are those made on January 24, 1925. The reason is that snow covered the ground throughout the whole of the belt of totality with the result that each and every observer was furnished with a white sheet on which the bands were portrayed. Excellent descriptions of the observations are given by Bassett.[1]

The 1925 bands were probably intensified due to the fact that the time of the eclipse was early in the morning when convection was active but probably confined to the lower layers of the atmosphere, within 300 yards of the earth.

The general features of the observations agree with those of earlier eclipses. The shadow bands were seen for two or more minutes before totality and the same length of time after totality. Some observers reported seeing bands twenty minutes before totality, but such bands were probably caused by heated air rising from furnace chimneys, and in several cases this was proven at the time to be the cause. No observers regarded to be reliable reported the shadow bands during totality.

The bright bands were generally observed to be broken and confused arcs or short blotches of light, a third, a half, or even a full meter in length. The dark band was usually reported as having a width one-fifth that of the bright band, the dark

[1] *Popular Astronomy*, 33, 232, 1925.

band being observed by different persons from ½ inch to 2 inches in width, while the bright areas had a width varying from 6 inches to 1 yard.

Naturally, in the estimation of the speed of travel of the shadow bands, there was a wide range from stationary with a shimmering light, to very fast motions of 30 miles per hour.

The general consensus of opinion of the very large number of observers taking part was that the shadow bands were narrowest, most distinct, and most sharply defined immediately before and after totality, and that they became broader, farther apart, and more indistinct until they disappeared, with increase of width of the solar crescent.

At the October, 1930, eclipse, shadow bands [1] were observed. They were first seen about one minute before totality and were visible for about twenty seconds. The dark bands were about 1½ inches in width and were separated 6 inches. Their motion was at the rate of 10 miles per hour in a west by north direction which had no connection with the direction of the wind at the time. They were seen also after totality but not during totality.

Hastings pointed out many years ago that the motions of the shadow bands must be perpendicular to their lengths, otherwise the motions could not be observed, and that they must lie approximately parallel to the crescent of the sun. The shadow bands therefore can have no connection with the direction of the wind, a fact which all observations seem to confirm.

Most authorities agree in believing that the shadow bands are caused by irregularities in the optical density of the atmosphere, the crescent character of the sun magnifying the effects of contrast in the bands.

In *Handbuch der Astrophysik, 4,* 353, 1929, a detailed discussion is given. As shown there, Humphreys considered four theories for explaining the bands, only one of which will be mentioned here.

The heated air commonly rises from the earth in numerous isolated masses, and experience teaches us that when this lighter air ascends in a connected column a whirlwind is es-

[1] *Popular Astronomy, 39,* 241, 1931.

tablished. The passage of light from the denser to the rarer air may cause total reflection over a small area of the interface. In reality there is a deflectional bending of the rays as they pass through the shell exactly in the manner of the inferior mirage, so commonly seen in summer over heated streets and smooth roads. All the light from every object included within the angle of total reflection would be wholly excluded from that place on the surface of the earth which it otherwise would have reached, and the light thus excluded would be added to that in the adjoining place.

In this way, the crescent of the sun a few minutes before and after totality must produce narrow dark bands which become more distinct and more numerous, since smaller inhomogeneities would then suffice, as totality is approached. Each dark band should be bordered on one side by a narrow strip of extra brightness. The dark band with bright border would be much narrower than that of the intermediary lighted strip between successive bands. Moreover, there would be no uniformity in either the widths of the bands or the distances between them. The bright edge of the dark band would probably be difficult to observe on account of the shimmering which is constantly in play.

Apparently, the shadow bands are caused by pseudo-total reflections, or mirage effects, produced by the transition shells between warmer and cooler adjacent masses of air in a state of thermal convection.

CHAPTER XXV

THE ECLIPSES OF 1932 AND 1934

THE great majority of eclipse astronomers who attempted observations on August 31, 1932, succeeded only in witnessing sky conditions as if sunset, midnight and dawn had all come within the brief compass of two minutes of time.

The eclipse took place in the midst of the holiday season of the year along an easily accessible track passing through a beautiful section of the country. Moreover, the meetings of the International Astronomical Union had been timed to begin, in Cambridge, Massachusetts, two days after the eclipse had occurred. The result was that more astronomers attempted to see and to observe this phenomenon than ever before in eclipse history. Every few miles along the moon's shadow path, from Parent away north of the St. Lawrence River in Quebec and eastward to the tip of Cape Cod, there were expeditions of all sizes and descriptions, from amateur equipment of small dimensions to the largest and most refined of professional instruments, cameras, spectrographs, photometers, polarimeters, etc.

As the eclipse track lay in the St. Lawrence River Valley, the expectations for clear skies were none too high. The areas of low barometric pressures which bring clouds and rain in their passage eastward over the great American continent seem to acquire the habit of passing out through the St. Lawrence Valley. The weather records for five years carefully assembled for the purpose seemed to show that the mountains in Vermont and New Hampshire possibly might attract clouds and the Atlantic Ocean bring fog. In planning the details of an eclipse expedition, an observer of eclipses must be a confirmed optimist in hoping always for clear skies. The three astronomers, each with several eclipses to his credit, who appear to have had the best average of luck with clear skies at eclipses in the past were my friends Campbell, Miller and myself. As Campbell was not to observe the 1932 eclipse, the interesting question was, would it be a wise plan to locate near Miller or myself? Or would the law of averages work out with the result that we both should be avoided? Of course, the careful scientist never believes in luck!

Unfortunately, the weather experienced at eclipse time, particularly near totality, did not live up to the fifty-fifty expectations, clouds being the rule rather than the exception. In fact, the skies were pretty generally overcast throughout the whole of the eclipse path except for a hole here and there through the clouds near where the sun was found at totality. A partly clear patch greeted the expedition from Greenwich Observatory at Parent. Similar conditions, a little better or a little worse were experienced as far east as the St. Lawrence River. It was cloudy in Montreal and cloudy generally eastward where there were large well-equipped expeditions throughout Quebec, Vermont and New Hampshire. At the boundary between New Hampshire and Maine a few clear patches of sky brought partial success to parties from Lick, Michigan, Van Vleck and Georgetown Observatories. Nearer the Atlantic, where conditions promised less, were found the clearest skies of all.

As a whole, the well-prepared plans of the various expeditions accomplished comparatively little, though success through propitious skies was met here and there. Perhaps with the irony of fate, many of the astronomers assembling for the meetings of the International Astronomical Union who had little or no eclipse experience and who wished merely to see the gorgeous spectacle were given clear skies, though at times rapid trips in automobiles had to be made in directions towards which the clouds appeared to be breaking, while veteran observers with refined equipment and big programs were doomed to disappointment. The largest group of instruments at any one location was at Magog, 100 miles east of Montreal. Here were the British party under the direction of Professor Stratton of Cambridge, a Canadian party from McGill University, Dr. Minnaert from Holland and an Ameri-

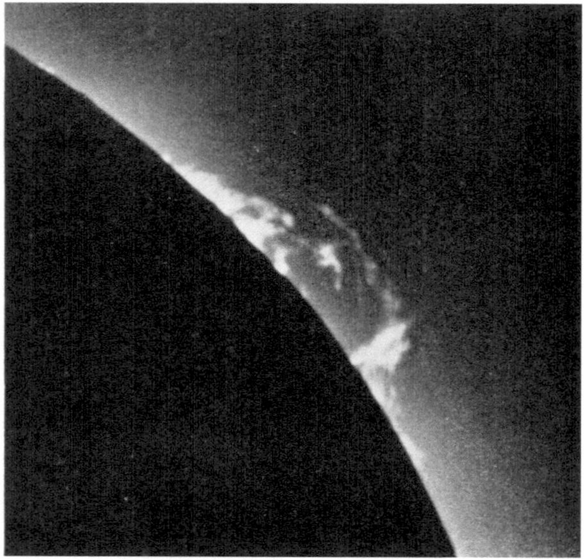

Clouds and Prominences in 1932
Photographed by the Lick Expedition at Fryeburg.

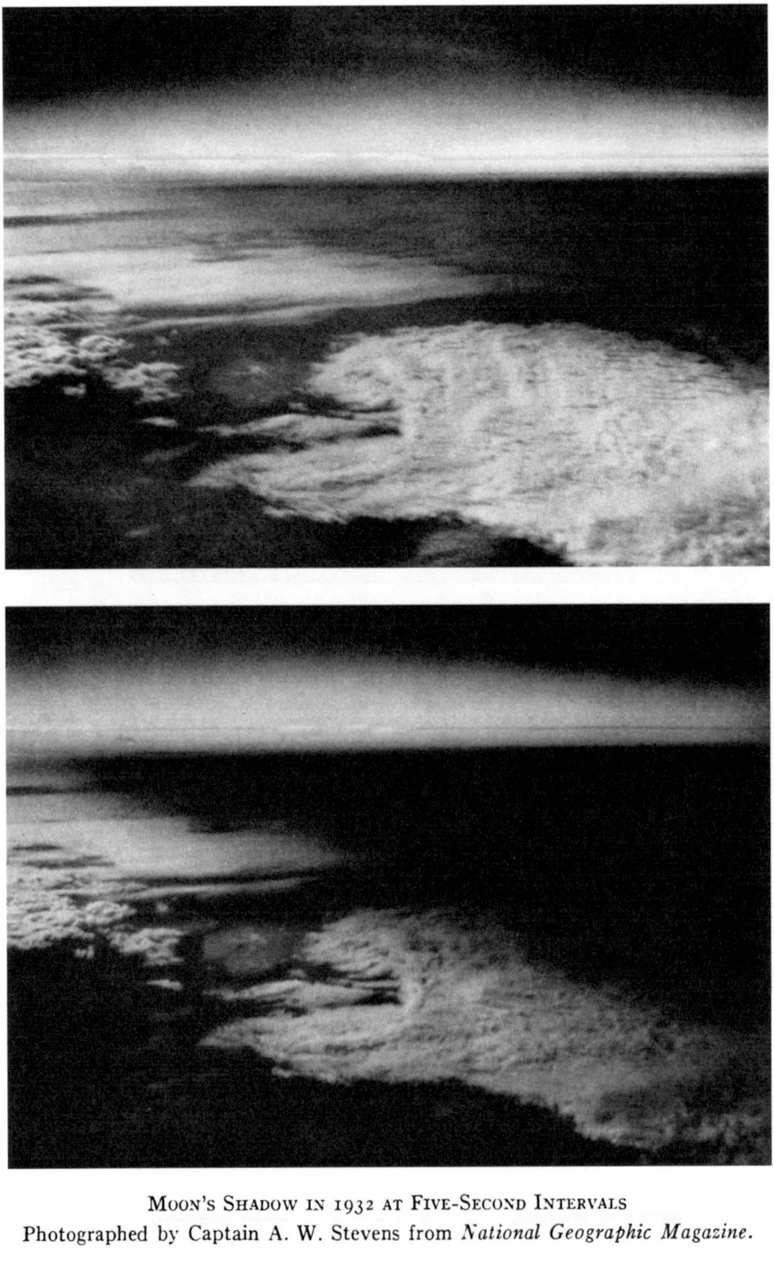

MOON'S SHADOW IN 1932 AT FIVE-SECOND INTERVALS
Photographed by Captain A. W. Stevens from *National Geographic Magazine*.

can expedition from the Leander McCormick Observatory. Four nations were represented, a total of thirty instruments and sixty astronomers on eclipse day. Many of us had been there a month living at the Hermitage Country Club. On the rough ground between two fairways of the golf course, we had wielded pick and shovel, and hammer and saw. We had been civil engineer, electrical engineer, photographer, physicist and astronomer all rolled into one.

On eclipse day clear skies remained with us all morning. At noon clouds began to gather. Would they clear away in the hours before totality? Unfortunately, as time progressed, conditions went from bad to worse. First contact was observed through heavy clouds. At times a little thinning of the clouds appeared now and then. Two minutes before totality, the sun's crescent appeared which permitted the solar beam to be adjusted into position in the powerful concave grating spectrographs. Would this semi-clear spot expand and bring success or would it close up and bring failure? No matter what the conditions, we decided to carry through our programs. I was to give the signal " Go! " for the combined eclipse camps. At the calculated time of totality, the sun's crescent was still visible. Three seconds later this had disappeared and the signal " Go! " rang out. Totality had come, and the steady count was heard: " one," " two," " three." . . . There was a clear patch of sky directly overhead but there was nothing but heavy clouds in the direction towards the eclipsed sun. I in Quebec, and, a little farther along the eclipse track across the international boundary in northern Vermont, my good friend Miller were both experiencing the thickest clouds either had ever met in the combined seventeen eclipses witnessed. It has so happened that we have never observed the same eclipse at the same spot and in 1932 we were located closer together than ever before.

It would serve no useful purpose to attempt to give here a detailed account of the many expeditions in the field or the programs of observations attempted. Those interested may readily find this information from the contemporary astronomical journals like *Popular Astronomy, Journal of the Royal Astronomical Society of Canada, Publication of the*

Astronomical Society of the Pacific, Monthly Notices, etc. An attempt will be made merely to give the outstanding accomplishments, arranging these from northwest to southeast as the eclipse passed over the country.

Farthest north was the party from Greenwich Observatory at Parent. On eclipse day it was cloudy practically continuously. Twice during the morning the sun was glimpsed and an opportunity was given for checking the position of the 45-foot camera. After noon there were one or two breaks in the clouds, one of them about five minutes before totality. "The corona [1] was seen before second contact, but the clouds were thickening, and after mid-totality nothing was visible until near the end when prominences and corona were again faintly visible. . . . When the eclipse was over and before the photographs were developed, it was expected that no results of value had been obtained." Fortunately, the photographs with the 45-foot camera and 21-foot prismatic camera far exceeded expectations.

A reproduction of the flash spectrum at second contact given facing page 492 shows a sudden termination of the crescent lines of the spectrum particularly at the upper ends of the arcs. The direct photographs show a distinct flattening of the edge of the moon. To prove that this effect was real and not apparent caused by irradiation, other photographs were measured. It was concluded that "there was a depression amounting to 2″ extending about 8° on each side of position angle 132° — the point of second contact." On account of the increasing density of the clouds, the photographs taken toward the end of totality showed nothing.

Jackson observed second contact visually three or four seconds later than the corrected time furnished by the director of the American Ephemeris shortly before the eclipse, while the observed duration of totality was 99 secs. against 102 secs. calculated. The retardation of second contact and the shorter duration of totality are both readily explained by the flattening of the moon's eastern limb.

Concerning the photograph of the flash spectrum, it will be best to let the Greenwich astronomers, Davidson and Jack-

[1] *Monthly Notices, 93,* 3, 1932.

son, speak for themselves (*loc. cit.*). " The photograph taken at the first flash is very good. The lines are numerous and in good focus from 3590 A to the limit of sensitivity of the panchromatic plate, the helium line at 7065 A being shown. It is not considered, however, that it will show any lines which have not been photographed by one or another of the past observers of flash spectra. Although heights might be deduced from the lengths of the arcs, these would be very uncertain for the lower lines, on account of the irregularity of the moon near the point of second contact. Many high level lines which are visible 30° from the geometrical point of contact are invisible 20° from it.

" The objective prism photographs were taken to show the distribution of the various elements in the chromosphere and prominences, and for comparison with the direct photographs of the corona secured with the large coronagraph. As regards wave-length determinations and the identifications of the elements in the chromosphere, the best spectra already secured with objective prisms and gratings or with slit spectrographs cannot greatly be improved on in the region covered by the panchromatic plate except with more powerful instruments under favorable circumstances."

The spectrum photograph immediately following the first flash showed the high level chromospheric lines and also the coronal lines at 5303 A in the green and 6374 A in the red. The latter line is shown as a nearly complete ring of practically uniform intensity while the green line has its light concentrated near prominences. The 1932 eclipse thus confirms the conclusions found on page 380 that these two strong coronal lines do not have a common origin.

In addition to the excellent spectra, good photographs were obtained with the 45-foot camera used with coelostat showing the prominences and the brighter parts of the inner corona.

The holes in the clouds were more favorably placed near the St. Lawrence River especially at Louiseville and Actonvale. The most powerful equipment in this region was that of the French party under the direction of the veteran eclipse observer, de la Baume Pluvinel. The clouds broke immediately before totality, permitting valuable observations on polarization by Dufay and Grouiller, to which reference is made on page 509. Not far away, parties consisting largely of amateurs organized by the Royal Astronomical Society of Canada made direct photographs of the corona and also observations to locate the edge of the moon's shadow path.

At Montreal, and south and east of the St. Lawrence through Quebec, Vermont and most of New Hampshire where expeditions were located, the clouds permitted at best only fleeting glimpses of the corona. East of the White Mountains at North Conway, the clouds were just about as thick as those experienced south from Montreal. A few miles farther southeast at Center Conway and Fryeburg, totality began through clouds but fortunately conditions greatly improved, though thin clouds still persisted to the end of totality. Under such conditions, happening during the excitement of a most unusual event, it is impossible for the eclipse observer to know whether the clouds have actually thinned or whether the apparent improvement results only from the fact that there is no strong sunlight shining on the clouds to make them visible. Few astronomers have had experience sufficient to serve as a guide — but the developed photographs tell the tale. Fortunately, the corona was photographed in excellent detail by the Lick expedition (reproduced facing page 476) and by the Van Vleck, Georgetown and University of Michigan expeditions. Closer to the Atlantic Ocean, in the state of Maine, the weather defied all rules with the result that conditions were much better than expected. At Limerick, the U. S. Naval Observatory party secured successful photographs of the corona through thin cirrus clouds with 65-foot, 15-foot, and with 38- and 33-inch cameras. At Alfred, conditions were even better, permitting the expedition from Japan to obtain photographs of the corona and flash spectrum. The Harvard party at Gray had the best luck of all the expeditions in the field. Thin cirrus clouds were present until noon, but a perfectly clear sky prevailed throughout the whole of the eclipse. The program was chiefly photometric and was concerned with measuring the relative proportions of scattering and polarization in the corona. The work was done both visually and photographically, plates being secured in six different spectral regions by using appropriate filters. Three out of four pinhole photometers at different localities were operated under clear skies.

The Lick-Crocker expedition [1] at Fryeburg, with interesting spectroscopic programs planned with powerful and varied equipment, fortunately had good luck with the weather. A total of ten spectrographs were employed, four for the chromosphere and six for the corona. Although the sky was never clear, conditions at the end of totality were much better than at the beginning, with the result that the Lick spectra, particularly those of the chromosphere, supplement those of the Greenwich party far north at Parent where the clearest sky was experienced at the beginning of totality.

For photographing the flash spectrum without slit, two methods are described in Chapter XV, with fixed and with moving plate. Each method has its advantages and its disadvantages. At Fryeburg it was wisely decided to try both methods. In order to photograph as great a range of wavelengths as possible, two spectrographs were used with both moving and fixed plates. Experience at former eclipses has shown that with an objective prism, well-timed chromospheric spectra may be secured with exposures of one second duration. In place of the usual type of objective spectrograph with fixed plate heretofore used at eclipses, a new instrument (*loc. cit.*) was designed for use at Fryeburg called a " jumping film spectrograph." Starting one minute before totality, exposures each of one second were made every alternate second. With a break for one long exposure lasting 30 seconds at mid-totality, the film was continued in its steady and jump motion until one minute after totality.

It is evident that with a long series of successful photographs taken with a jumping film, heights in the chromosphere can be obtained by two methods: by measuring the angular lengths of the chromospheric arcs or by noting the appearance or disappearance of spectral lines on successive photographs. This mechanical method of exposure has distinct advantages over that always followed by the writer of observing with a direct-vision spectroscope and timing the ex-

[1] *Publ. A. S. P.*, *44*, 341, 1932.

posures for the flash spectrum accordingly. At Fryeburg, the mechanism and the observer did their work as planned, with beautiful photographs as the result, though thin clouds of variable thickness unfortunately covered the eclipsed sun.

Measures of the flash spectrum at previous eclipses have shown that one-quarter of all chromospheric crescents visible on the best photographs extend to the moderate heights of 300 kms or less above the photosphere and hence are covered in one second of time by the advancing moon; two-thirds of all lines of the flash spectrum are covered in the brief interval of one second and a half. Evidently, if the one second visibility for the low-lying lines happened to exactly coincide with the exposure time of one second given by the jumping film spectrograph, the observer should be congratulated on his rare good luck. If, on the other hand, the film was being changed during this one particular second, the developed photographs would show none of the lines of the lowest level. As either of these two methods of exposure are equally probable, it is obvious that this hit-and-miss method of making exposures, on the average can not photograph the low-level lines; also, notwithstanding the quick succession of photographs, the observer will not know, and the photographs can not show, just when the effective existence of the several chromospheric strata began and ended.

The spectra taken by the jumping film at Fryeburg are superior in quality to those taken with the moving plate. Facing page 493 is a reproduction of the flash spectrum taken at the end of totality by Menzel. The very rough character of the moon's edge shown by this photograph will make the determination of heights, particularly for lines of low level, a difficult problem. It is unfortunate that at the 1932 eclipse, the limbs of the moon both at second and third contacts were so jagged. The long exposure spectrum taken at mid-totality shows the green and red coronal lines with distributions of radiation exactly similar in character to that found on the Greenwich plates.

For the corona, six spectrographs were used by the Lick party. Three instruments with slit placed east and west of one, two and three prisms, respectively, permitted spectrograms of excellent definition which record the spectrum of the corona, especially that of its brighter parts, in great strength. A marked feature of the three is the great intensity of the continuous spectrum of the inner corona and the number and strength of the prominence lines. Only the stronger emission lines are visible. For photographing in the red, two grating spectrographs were used, one of concave, the other of plane grating. With these no results were obtained. The sixth was an interferometer spectrograph for the wave-length of the green line with which successful photographs were obtained. The results obtained from measures of these spectra are discussed on page 502.

Also at Fryeburg, the University of Michigan expedition secured excellent photographs of the corona with a 40-foot focus camera pointed directly at the sun and also of the flash spectrum with two-prism dispersion. In addition, large-scale movie photographs were secured by the McNath-Hulbert branch of the University Observatory with cameras of 74, 40 and 14 inches focus, respectively. Nearby, the party from the Van Vleck Observatory successfully photographed the corona with 25-foot focus and with smaller cameras. Photographs obtained by these two expeditions are given in *Popular Astronomy* for October, 1932. A beautiful photograph obtained by the Georgetown Observatory expedition is reproduced in the *National Geographic Magazine*, *57*, 600, 1932.

As was predicted on page 404, the corona was of the minimum type with long equatorial streamers and pronounced polar brushes.

The distinguished air-pilot, Captain Albert W. Stevens, writes a thrilling account[1] of photographing the moon's shadow at a height of five miles above the earth. Facing page 485 are two of his remarkable photographs showing the motion of the edge of the moon's shadow as projected on the clouds at 5-second intervals. The end of totality is described in the following words: "At this time the most remarkable sight, except the corona itself, was the horizon, beyond the shadow, to the northwest. The far edge of the shadow was defined by a narrow band of a saffron color and was distinctly

[1] *National Geographic Magazine, 57,* 581, 1932.

seen to be curved. The air between seemed to be of an indigo color. Under the airplane, through openings in the clouds, the earth's surface, five miles below, could be dimly seen."

In addition to the observations made by the U. S. Army pilot, the U. S. Navy also photographed [1] the moon's shadow above the clouds. Among the other observers from airplanes, mention will be made only of Clyde Fisher who states,[2] " The shadow of the moon was first seen, just a little before totality, as a narrow band on this level sea of clouds, at a distance of probably more than one hundred miles. As it approached it became less distinct, less definite. . . . Upon development, the picture seems to be well timed and it shows just enough diminution of light to indicate the passing of the moon's shadow. However, no edge is to be seen in the pictures, and none was seen in our observation."

Thermometers were read at more than 200 stations in order to record the drop in temperature during the progress of the eclipse, the work being organized by the Blue Hill Meteorological Observatory.[3] The widespread clouds which existed at five different levels and played such havoc with the astronomical observations were a blessing in disguise for the temperature records. It was found that temperatures fell from 2 to 11 degrees F., the smaller values occurring where clouds were dense, the larger where the weather was clear and the surface of the ground sandy. With the assistance of three U. S. Navy airplanes, temperature records were made at Fryeburg [4] at eight different levels above the ground at altitudes of 5, 25, 50, 75, 100, 866, 3261 and 10122 feet, respectively. The observations show (1) that the greatest range in temperature occurred at the earth's surface, (2) that the ranges decreased rapidly with altitude and (3) that the effect of atmospheric refraction upon the deflection of starlight by the sun during an eclipse is entirely negligible.

The total eclipse of February 14, 1934 was successfully observed [5] from two small coral reefs, Losap and Laol, in the

1 *Popular Astronomy, 41,* 202, 1933.
2 *Ibid., 40,* 467, 1932.
3 *Bull. Amer. Meteorol. Socy., 13,* 159, 1932.
4 *Annals of the Dearborn Observatory, 4,* 27, 1933.
5 *Popular Astronomy, 42,* 307, 1934.

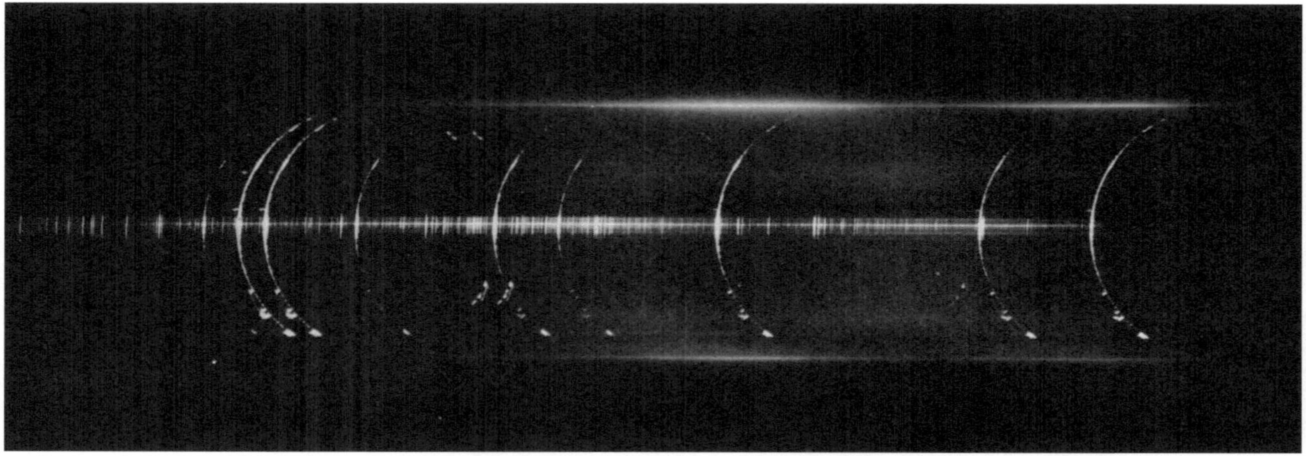

The Flash Spectrum in 1932 with Objective Prism

Photographed by the Greenwich expedition at the beginning of totality.

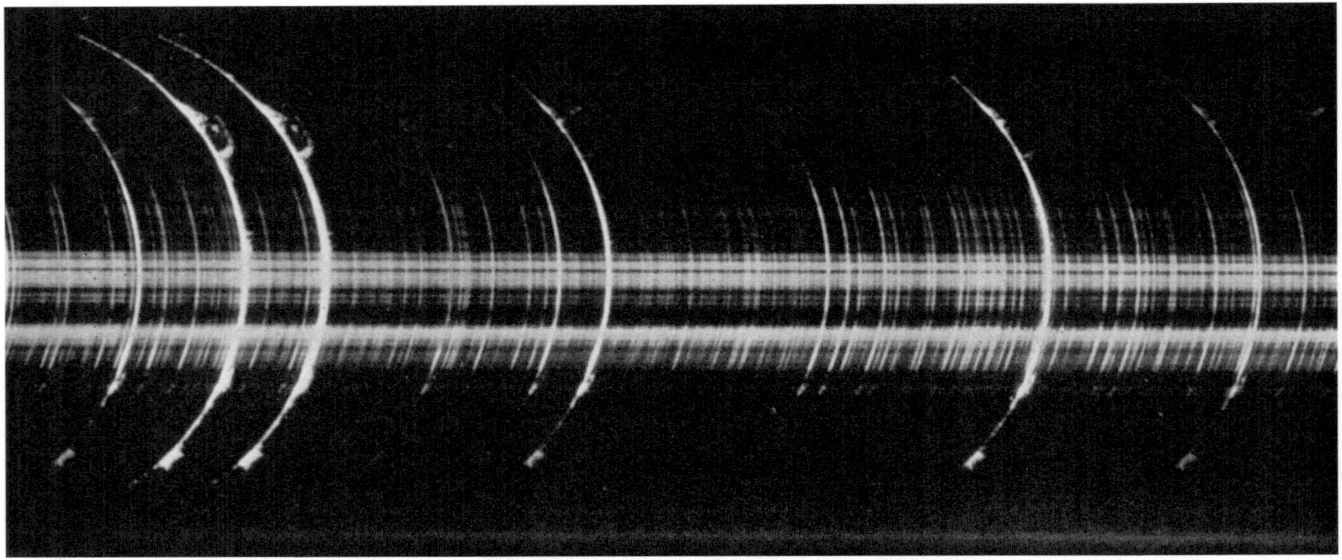

Spectrum of the Chromosphere in 1932 at the End of Totality
Photographed by Menzel of the Lick Expedition.

far-off Caroline Islands. The National Research Council had been authorized by the Imperial Japanese government to invite astronomers the world over to be the guests of Japan for the observation of the eclipse. Only two foreigners, however, availed themselves of this generous offer, Dr. Willi M. Cohn and J. J. Johnson, both of California. The Cruiser " Kasuga " of the Imperial Japanese Navy conveyed the eclipse party on its eight-day voyage from Yokohama. Three busy weeks were spent on the islands in preparation for the great event. The largest part of the expedition was from the Tokyo Imperial University under the direction of Dr. Satome. A well-planned series of observations was carried out covering the most important problems dealing with eclipses. Special stress was placed on securing photographs to test the relativity displacement of star images, the 1934 eclipse being specially favorable [1] because the sun was projected against an unusually rich field of stars. In fact, for the next two decades there will not be as favorable a star field for measuring the so-called Einstein displacement. The party from the Kyoto Imperial University duplicated most of the work on the Tokyo program, Dr. Ueta secured the Einstein plates and large-scale photographs of the corona.

Dr. Cohn used several cameras on a polar axis for measuring the polarization of the sky near the eclipsed sun. Visual observations were also made. Johnson's observations were visual with a modified Lyot-type polarimeter.

As is usually the case with eclipse expeditions in the tropics, there was a superabundance of clouds and rain during the weeks of preparation. The morning of eclipse day brought a great surprise. In describing the eclipse, Johnson writes, " It was our good fortune to observe one of nature's rarest treats, a total eclipse of the sun under absolutely perfect conditions: clear blue sky for miles around, not the slightest suggestion of haze or mist, a steady atmosphere — in short the kind of eclipse which the astronomer dreams about but seldom sees. The scientific results obtained from this latest total eclipse are discussed in the following pages.

American astronomers can look back with great pride to the

[1] *Ibid.*, *40*, 623, 1932.

enviable record of achievements made in eclipse history as recorded in the foregoing pages. Science, however, knows no international boundaries, with the result that all eclipse astronomers no matter what their nationality look confidently to the future. Who are to be the fortunate individuals who will observe the 1936 total eclipse? Will it be possible to observe, under conditions that promise success, the eclipse of 1937 with its seven-minute totality? Will South Africa live up to its good reputation for clear skies in 1940? Discoveries of great importance to solar physics will come only as the reward of much careful planning and concentrated effort. In the following pages a brief summary is given of the results obtained from eclipse photographs in the past decade, or since the appearance of the first edition of this book, and at the same time there are outlined the main problems of solar physics now awaiting solution.

Since 1923 with its disappointing eclipse, observations have been made in 1925, 1926, 1927, 1929, 1930 (April 28 and October 21), 1932 and 1934. In the dozen years, there have been nine total eclipses, one more than the average (see page 60). Four of the nine have been visible in the United States. The weather conditions experienced for the nine eclipses have probably been about average. The four eclipses observed in the tropics, 1926, 1929, 1930 (Tin-Can Island) and 1934, contrary to expectations, experienced better average conditions of sky than the balance of five observed in temperate zones.

The spectacular beauty of a total eclipse, observable at times only by traveling, at great expense, great distances from home, brings in its train a large body of observers, professional and amateur when, as in 1932, the eclipse track is near at hand. One of the simplest problems to be attacked is the direct photography of the corona. The camera employed may be any size from the short focal length of a kodak to the huge proportions used by the expert eclipse astronomer, the greatest length heretofore used being 135 feet at the 1900 eclipse. For such large dimensions a coelostat is imperative. At the 1919 eclipse, the relativity photographs showed quite conclusively that coelostat mirrors may bring distortions.

Hence the tendency in recent years has been to dispense with the mirror. For focal lengths greater than fifteen feet, the mounting has been the tower telescope. On Sproul Observatory expeditions Miller and Marriott have used in this manner focal lengths of 63 and 65 feet. In my opinion the definition secured at the Tin-Can Island eclipse of 1930 had never been surpassed. As is well known to expert photographers, the detail appearing on plates of the corona depends on securing well-timed exposures but also correct photographic development of the photographic plates.

In preparing plans for the 1932 eclipse, Wright [1] decided to abandon the long-focus tower telescope used at Lick Observatory expeditions since 1893 and to employ instead the two lenses, each of 5-inch aperture and 15-foot focal length, used at the 1922 eclipse for the relativity shift. On account of the thin clouds experienced throughout totality, the decision evidently was a wise one. Certainly the definition obtained by Wright in 1932 is quite up to the standard set by Marriott at the 1930 eclipse.

Manifestly, it is much easier to erect at the eclipse camp, sometimes an isolated spot on the globe, a polar axis brought as a unit from home with which to mount a camera (or cameras) of 15-foot focus than to build a tower telescope of four times this length, usually from materials found near the eclipse camp and frequently with native workmen. Only occasionally is it possible, as it was in 1930, to transport from home to a distant foreign country the many thousands of feet of lumber needed for the tower telescope. To build from native materials requires much experience and resourcefulness.

We may indeed well ask ourselves why it is that huge focal lengths have been used to photograph the corona. Is it necessary to have the diameter of the moon seven inches in the original photograph? Why not secure negatives with one-quarter the focal length and then enlarge four times if large scale is desired? Little of detail in the original photographs is lost in the enlarging processes while on the contrary faint coronal details may be strengthened and accentuated. The striving after great focal lengths reflects the trend in all as-

[1] *Publ. A. S. P.*, *44*, 341, 1932.

tronomical observations to use telescopes of greater and still greater sizes. For observational work on stars, large apertures and great focal lengths are necessary in order to secure the most refined results. With the eclipsed sun there is no need for great apertures. Is there need for great scale in the original coronal photographs?

The answer to this question depends on many considerations, but primarily on the scientific uses to which coronal photographs are to be put. The relative costs of the two procedures should be taken into consideration. At the 1932 eclipse it was a comparatively simple matter to put all of the eclipse equipment on a motor truck at Mount Hamilton, then transport to a steamer in San Francisco and then from New York to Fryeburg. It was even simpler for the Mount Wilson expedition to put all of its equipment on a motor truck in Pasadena and transport it directly across the whole continent, a distance of 6000 kms, to the eclipse site at Lancaster. But in the coming generation, American astronomers must travel an average of one-third the way around the globe for each total eclipse. In future eclipses the relative costs of transporting a heavy polar axis from home and then erecting in the field or transporting from home the lighter equipment of the tower telescope and then building the double tower from local materials will differ little. The cost should not be a deciding factor — but in most observatories it must be considered. After all, the problem is one primarily for American observatories (and very few are involved) for the reason that the heavy polar axes to carry wide-angle lenses to measure the gravitational displacement of stars during totality and tower telescopes are the contributions of America to eclipse history.

Each observatory interested will have to provide its own answer regarding the best equipment for it to use for direct photography of the corona. Unquestionably, there is much to be said in favor of securing photographs of great scale by tower telescopes.

The present writer would like to raise another question, namely, what scientific use is being made (other than that of providing important historical documents) of the long series of large-scale coronal photographs accumulated by Lick,

Greenwich and other observatories? At the same time we might ask, what scientific uses are being made of photographs to the hundreds of thousands, of solar details like sun spots and prominences accumulated through long series of years by many observatories? The answers to these questions and to many others of similar nature dealing with solar problems can furnish topics of valuable discussions for many future sessions of the Solar Physics Commission of the International Astronomical Union.

Fortunately, photographs of all scales, except the small, are of great value for determining the shape of the corona by the method devised by Ludendorff and described on page 401. Recently two articles on this subject have appeared, one [1] by Ludendorff, the other [2] by Bergstrand. On account of a slight error in Ludendorff's published formula, the values given on page 402 must be altered in two cases. For the 1922 eclipse, the corrected value for the quantity a is $+ 0.01$, and for b, $+ 0.22$; and for the 1930 eclipse the values should read $+ 0.03$ and $+ 0.22$, respectively. For the 1932 eclipse, Miss Williams derived values for the ellipticity from measures of photographs, and found a was $+ 0.05$ and b, $+ 0.22$, the value for the spot numbers being 6. For the same eclipse Ludendorff, from fewer photographs, derived values for a and b of $+ 0.06$ and $+ 0.14$, respectively. Measures of 1932 coronal photographs, however, showed quite conclusively that values of a and b from photographs by one camera were quite consistent among themselves but differed systematically from values from other photographs. Details cannot be given here but will appear in the forthcoming volume of *Handbuch der Astrophysik*.

For the eclipse of February 14, 1934, the shape of the corona was determined from an excellent photograph by Satome reproduced facing page 500 and from isophotes furnished in advance of publication by Cohn from photographs polarized in directions at right angles to each other. The ellipticity of the corona in one of Cohn's photographs differs radically from the ellipticity of the companion photograph polarized in a

[1] *Sitz. Preuss. Akad., Phys. Math. Kl.*, 16, 1934.
[2] *Arkiv. Mat. Astr. och Physik.*, 25A, No. 4, 1934.

plane perpendicular, thus showing the same effect displayed by the pair of photographs taken at the 1908 eclipse, reproduced facing page 388. The shape of the 1934 corona was derived by following the Ludendorff method and determining separately the ellipticities of the photographs polarized in the two directions, and then taking the mean, the results being confirmed by measures from the Satome photograph. For this eclipse, a was found to be $+ 0.08$ and b, $+ 0.14$, making $a + b$ equal to $+ 0.22$. The value of " spot numbers " (page 402) is 8, the spots being on the increase. The eclipse took place 0.2 years after the sun-spot minimum.

As shown on page 402, values of a differ little from eclipse to eclipse while there is quite a range in the values of b. Ludendorff has adopted $a + b$, or the ellipticity of the corona at a distance of one radius from the edge of the sun, or two radii from the sun's center, as the best measure for ascertaining the shape of the corona. To derive the relative activity of the sun in the 11-year cycle there is a wealth of published material on both spots and prominences. For the sun spots, it has seemed wisest always to derive a mean from values spread over a whole synodic revolution of twenty-seven days, that is, a mean of thirteen days both before and after and also the day in question. The values for " spot numbers " on page 402, derived from information published at Zurich, correspond to average values referred to the day of the eclipse. In a similar way, Ludendorff derives values for the spot numbers for one, two and three months, respectively, before the eclipse, for the average annual value, and also for the phase in the sun-spot cycle depending on the times of sun-spot maxima and minima. In a similar way (*loc. cit.*) corresponding values are found from prominences. With these values in hand, Ludendorff plots as abscissae various sun-spot (or prominence) numbers, and as ordinates the values $(a + b)$ representing the shapes of the various coronas. An attempt to pass a smooth curve through any one of the individual graphs shows that the 1918 eclipse falls far distant from any curve. With a total of eighteen eclipses starting in 1893, measured and plotted in a uniform manner (leaving out of consideration the discordant eclipse of 1918) it is difficult to say which of the dozen curves makes the best fit or shows the closest correlation between the shape of the corona and solar activity.

Bergstrand uses Ludendorff's measures of a and b and derives for each eclipse a coronal number p and an ellipticity E_1. These are plotted against the different numbers found from spots and prominences, and smooth curves are again drawn. It is true that these plotted points lie closer to the curves (the 1918 eclipse is still discordant) than do the similar points plotted against $(a + b)$ by Ludendorff. However, the numbers p and E_1 used by Bergstrand are not very sensitive to changes in the shape of the corona as is shown by the fact that throughout the period from one year after spot maximum until one year after spot minimum there is very little variation in values of either of the two Bergstrand quantities.

For the eclipse of 1900, the value of $(a + b)$ which is equal $+ 0.30$ is the greatest of any of the eighteen eclipses considered, or the eclipse in the last year of the 19th century showed the most pronounced "minimum-type" corona with long equatorial streamers and pronounced solar brushes. The average value of the probable error of $a + b$ is $\pm .03$. Those familiar with the theory of least squares realize that individual values may readily have discordances of two or three times the probable error. In addition to the accidental errors (represented by the probable error) there may be systematic errors as was shown by the 1932 eclipse. If there is an active sun spot or prominence at the sun's edge there may be a long streamer in the corona, as was shown (page 404) by the 1930 eclipse, which might cause changes in the measured shape of the corona.

Taken all in all, it is too much to expect that the rough method of deriving the shape of the corona could possibly give points that would lie close to a smooth curve. With a total of only eighteen points, one of them (1918) very discordant, it is not worth while to waste time to find which of the many curves gives the closest correlation between coronal shape and sun-spot or prominence data.

The accompanying curve gives as ordinates the values of $100\ (a + b)$ and as abscissae the values of the phases in the sun-spot cycle, O in the middle of the diagram representing spot minimum. The — signs signify that the spots are decreasing from maximum, at the left of the diagram, towards minimum, and the + sign that the spots are increasing in numbers after minimum has passed. The curve was drawn to fit as well as possible all of the observations including the corona of 1918. It is unfortunate, however, that the six points on the left side of the diagram representing the shapes of the corona in the years following maximum of sun spots are on the whole quite discordant while the other twelve eclipses to the right are more consistent and hug the curve much more closely. Two eclipses of the six, those of 1896 and 1898, tell quite different stories in that the corona of the earlier year was more elliptical in shape than that of two years later. The

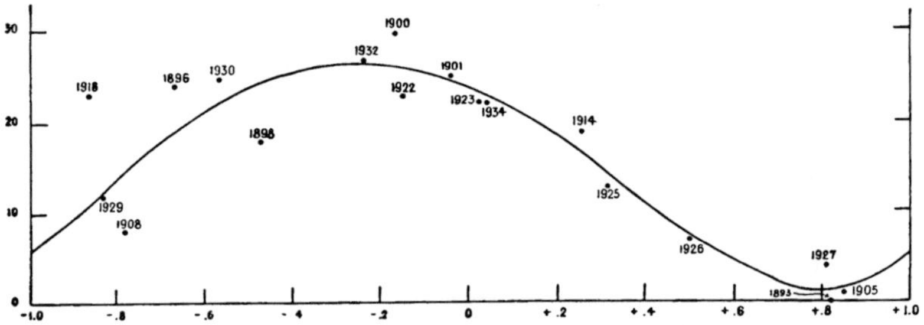

only method of determining a more exact connection between the shape of the corona and solar activity will be to acquire more observations, particularly on the left side of the curve, which may be done only by observing future eclipses in the nineteen-forties.

Various other graphs were made connecting coronal shapes with spot or prominence numbers, but all of the curves without exception, no matter what spot or prominence numbers are plotted, tell the same story exhibited in the curve plotted to sun-spot phase. The most pronounced minimum type of corona does not take place exactly at sun-spot minimum. From the curve it is seen that two and a half years before spot minimum the corona is quite as elongated as it is at the time of minimum of spots, or phase O in the curve. The most elongated corona is found one and a quarter years before minimum of spots, and likewise the corona closest in shape to a

The Corona of 1934
Photographed by Satome of the Tokyo Expedition.

Cohn and Japanese Assistants (above) and Johnson (below)
at the 1934 Eclipse

circle takes place one and a quarter years before spot maximum.

The recent minimum occurred at the year 1933.9 and hence the 1932 eclipse took place 1.2 years before spot minimum, almost at the highest part of the curve. The 1934 eclipse took place very close (0.2 years after) to spot minimum. With $a + b$ equal $+ 0.22$ it is evident that the corona had lost its minimum type characteristics. The 1936 corona taking place 2.4 years after spot minimum should be intermediate between minimum and maximum types. It and the 1937 eclipse will fill gaps in the curves connecting the shape of the corona and sun-spot numbers.

When arranged according to phase in the sun-spot cycle, the eighteen eclipses fall naturally into six groups of three eclipses each as given in the table. The mean values for the phases are given below. The three eclipses, 1900, 1922 and

THE SHAPE OF THE CORONA AND SUN-SPOT PHASE

1908	− .78	1896	− .67	1900	− .17	1901	− .04	1914	+ .25	1893	+ .82
1918	− .87	1898	− .47	1922	− .15	1923	+ .02	1925	+ .31	1905	+ .85
1929	− .83	1930	− .57	1932	− .24	1934	+ .04	1926	+ .50	1927	+ .81
Mean	− .83		− .57		− .19		+ .01		+ .35		+ .83

1932, each of which showed so beautifully the so-called "minimum type" of corona had a mean phase of — .19, while the "maximum type" of corona of the three eclipses 1893, 1905 and 1927 had + .83 for the average mean phase. From the table it is evident that eclipses should be divided into maximum and minimum types and into two types (rather than one type) intermediate between maximum and minimum and between minimum and maximum.

The astronomers of the world have been greatly thrilled by the skillful work of Lyot in photographing the corona without an eclipse. A later publication [1] than those alluded to on page 419 gives the details of his valuable work. The great triumph of Lyot resulted from his ingenious arrangement of his telescope to reduce to a minimum the amount of instrumental scattered light. A mountain observatory on the Pic du Midi (9100 feet) permitted photographs at times when the sky was

[1] *Zs. f. Astroph.*, 5, 73, 1932.

exceptionally clear, or when the intensity of scattered sky light was less than five-millionths that of the sun. By combining the coronagraph with a spectrograph it was possible to determine the wave-lengths of both the green and red coronal lines. A concave grating gave a dispersion of 1.2 A per mm and a prism arranged Littrow type gave a dispersion in the green of 5 A per mm. This large scale of wave-lengths permitted Lyot to measure the width of green line with the surprising result that the total width is 1.2 angstroms.

As a consequence of having 36 spectra of large dispersion, it was naturally assumed that the wave-length of the green line had an accuracy greater than had resulted from eclipse spectra. Similar conclusions were drawn for the accuracy of the red line. At the 1930 eclipse, Mitchell had eight spectra, on all but two of which measures were made of the wave-length of the green line on both E and W limbs of the sun. The wave-length is therefore the result of fourteen independent measures with a dispersion of 10.9 A per mm. For the red line there were eight separate measures on six spectra. The wave-lengths and probable errors are given, together with similar values by Lyot.

Lyot	*Mitchell*
5302.85 ± .03	5302.91 ± .02
6374.75 ± .15	6374.28 ± .03

It is evident that Lyot's claim of greater accuracy and a similar claim in *Monthly Notices, 93*, 278, 1933, of " an accuracy far greater than that obtained hitherto " cannot be substantiated as is shown by the sizes of the probable errors.

By means of a Lyot telescope constructed [1] at Mount Wilson, Pettit and Slocum secured direct photographs of prominences.

No coronal wave-lengths have yet (1935) been published for the 1932 and 1934 eclipses. After many unsuccessful attempts at previous eclipses (page 385), a photograph of the green line has finally been secured by the interferometer. At the 1932 eclipse this method was tried by the members of three different expeditions. Farthest west, Carroll had a

[1] *Publ. A. S. P., 45*, 187, 1933.

beautifully designed instrument, but again he had bad luck with the weather, and his record now is four cloudy eclipses in four attempts. At Fryeburg there were two interferometers, at the University of Michigan and Lick expeditions. By means of a double slit on the Lick[1] interferometer, an attempt was made to photograph fringes both E and W of the sun. At one limb there was no trace of the green line but on the other it was strongly photographed. Although fringes were very distinct in the comparison lines, no evidence of fringes were visible in the coronal line and measures with a microphotometer revealed a mere suggestion of the correct periodicity. Lyot's photographs explain the absence of fringes, namely, that the surprising width of the green coronal line of one angstrom would permit distinct fringes only with a very small separation of the interferometer plates, which unfortunately would entail a very low accuracy in the wave-lengths. Hence in the moment of triumph of securing the first photographs of the coronal line with the interferometer there is the realization that further attempts must be abandoned. This must be a keen disappointment to those astronomers who have spent so much time and energy and money in perfecting the interferometer to secure accurate wave-lengths of the green line. The disappointment, however, is part of the game of being a scientist.

At the 1930 eclipse, Mitchell's photographs (page 377) showed both the green and red coronal lines in the short exposures given for the flash spectrum. It is therefore not surprising to find that with the Mt. Wilson tower telescope Babcock has obtained the green coronal line on photographs of the chromosphere without an eclipse. On the original photographs (which the writer has seen) the 5303 line is so wide and hazy that an accurate wave-length is impossible in spite of the large scale of the photographs. Adams and Joy[2] at Mt. Wilson make the surprising observation that five coronal lines, 3987, 4086, 4231, 5303 and 6374, were present at one stage of the outburst of the Nova-like RS Ophiuchi.

The best wave-length for the green line is 5302.90 ± .02

[1] *Publ. A. S. P.*, *44*, 358, 1932.
[2] *Ibid.*, *45*, 301, 1933.

which is derived by combining, with equal weights given to each, the Lick-Mt. Wilson value of 5302.93 (page 374) and the Lyot and Mitchell values. For the red line, perhaps the best method will be to combine the Lyot and Mitchell values with weights depending on the probable errors. The derived value is 6374.30 ± .03, in close agreement with the laboratory value of neutral oxygen at 6374.29 A.

The great difference in the wave-lengths of the two best values for the red line amounting to nearly half an angstrom shows the need of new determinations. Until more accurate values are obtained for the whole coronal spectrum, it seems a waste of time to devote much energy to deciphering the riddle of coronium. Since the date of the attempts recorded on page 382, other individuals have made widely published claims. The consensus of well-informed opinion [1] seems to be that although it is quite probable that neutral oxygen may eventually be found to be one of the chief constituents of coronium, at present that fact is far from being proved.

In the first burst of enthusiasm over Lyot's remarkable work, it was thought by many that another eclipse problem, that of coronal wave-lengths, could be attacked when there was no eclipse. Success on an isolated mountain peak like Pic du Midi requires remarkable transparency of sky and conditions of seeing which come but seldom. In cost such an expedition is commensurate with that of an eclipse expedition. Through perseverance, Mt. Wilson may finally succeed in deriving accurate wave-lengths for the green and red coronal lines. Of equal importance, however, are better wave-lengths for the balance of the coronal lines. At present there seems to be no method of securing this information so badly needed for finding the origin of coronium, than by the use of well-designed spectrographs during the total phase of an eclipse. Photographs which give accurate wave-lengths of the fainter lines of the corona will of necessity give a greater accuracy for the brighter green and red lines so that eventually we may be able to ascertain something about the rotation of the corona, a subject concerning which we know practically nothing.

At the 1932 eclipse, photo-electric cells were used at many

[1] R. Frerichs, *Die Naturwiss.*, *48*, 849, 1933.

expeditions in attempts to measure the light of the corona. At Magog, we had two separate set-ups of Weston Illumination Meters, and we were fortunate in that E. F. Weston was there in person. An automatic record was made of the sky illumination near the North Pole starting before daylight and continuing throughout the whole day until after dark. Another set of sixteen cells was directed towards the sun and the field of illumination was confined to a circle of 8° in diameter with the eclipsed sun in the center. Although clouds prevailed during totality, the records were sent to Dr. M. Minnaert who in part reports, " It is easy to show that during the partial phases the illumination of the sky surrounding the sun was chiefly due to direct solar radiation, but during totality the illumination was due mainly to reflection of the light from points outside the path of totality." Although the clouds at Magog were denser than they were (for instance) at Fryeburg, one seems compelled to raise the question whether the Fraunhofer lines shown in the spectra of the corona taken by Moore at this eclipse are the result of coronal radiation or whether they are not in part caused by sunlight reflected by the thin clouds which were present throughout totality. On the photographs, the effects would be greatest at the beginning and the ending of totality.

In this connection it is well to again raise the question, already raised on page 397, as to what information we have from previous eclipses regarding the total light of the corona. Regarding the measures of the illumination of corona *plus* sky by means of illuminometers, one competent photometric authority states,[1] " There can be no objection to anyone measuring the normal illumination on a completely exposed test plate at the time of totality if he is convinced of the utility of so doing. Trouble, however, may ensue when the results of such measurement are seriously put forward as indicative of the amount of light which the corona is giving."

The light of the corona which has been seen to extend to the enormous distance of ten million miles from the sun's surface has three separate spectra (page 382): (1) the bright line spectrum of " coronium," (2) a continuous spectrum in the

[1] *Jour. Opt. Socy. America*, *23*, 234, 1933.

brighter inner corona, and (3) a Fraunhofer spectrum in the middle and outer corona. Recently very substantial progress has been made in investigating the continuous spectrum and also the dark line spectrum, and thereby valuable information concerning the physical origin of the corona has been given.

At the eclipse of 1929, with three different spectrographs, Grotrian [1] secured spectra of the corona which were compared with the solar spectrum taken with the same instruments but diminished in intensity by known amounts. Measurements made with the microphotometer give the following interesting results. (1) The position of the energy maximum in the corona is independent of the height in the corona and agrees exactly with the energy maximum of the solar spectrum. (2) Within errors of measurement, the distribution of energy in the coronal spectrum is independent of height in the corona and is identical with the intensity of radiation in the solar spectrum. These measures confirm the conclusions of Ludendorff [1] from the 1923 eclipse. After many long years of uncertainty, at last we seem to know for certain that the color of the corona is identical with that of the sun, or, in other words, the corona appears to shine from scattered sunlight.

In the remarkably clear sky at the 1922 eclipse (page 383), Moore seemed to prove conclusively that the Fraunhofer lines in the middle and outer corona have their origin in the corona itself and are not caused by sunlight scattered by the earth's atmosphere. By means of a microphotometer, Grotrian [3] measured the widths of the stronger Fraunhofer lines in the coronal spectrum taken with one prism at the 1923 eclipse. These widths were compared with those of the solar spectrum taken by the same spectrograph and weakened by known amounts. In every case the widths of lines in the coronal spectrum were compared with the widths in both stronger and weaker solar spectra. All observers heretofore had thought that the lines in the coronal spectrum were both wider and less distinct than similar lines in the sun. On the contrary, Grotrian's measures show that although the dark lines in the co-

[1] *Zs. für Astrophysik*, *3*, 199, 1931.
[2] *Sitz. d. preuss. Akad. d. Wiss.*, *5*, 83, 1925.
[3] *Zs. für Astrophysik*, *8*, 124, 1934.

ronal spectrum certainly appear less distinct than in the solar spectrum, and moreover that the dark lines are much more prominent in the outer corona than in the inner, nevertheless, the widths of the coronal lines are the same at all distances out from the sun's surface, and within errors of measurement, are identical in width with the corresponding lines in the solar spectrum. From similar measures carried out on the 1922 and 1932 eclipse spectra, Moore [1] confirms Grotrian's measures of the 1923 eclipse spectra.

Leaving out of consideration for the time being the bright line spectrum of coronium, which remains a perplexing eclipse problem, it is evident that the coronal spectrum consists of two spectra overlapping each other, the continuous and the Fraunhofer. The 1923 spectrum showed Fraunhofer lines superposed on the dark moon. On the assumption that these were caused by scattered coronal light, either in the earth's atmosphere or in the spectrograph, Grotrian was able to apply corrections to the measured intensities of the dark lines in the coronal spectrum. These corrected intensities show that near the sun's edge the continuous spectrum is considerably stronger than the Fraunhofer spectrum so that no dark lines are visible, due to lack of contrast, though no doubt they are there. At increasing distances from the sun, the intensity of the continuous spectrum falls off more rapidly than the dark line spectrum. At a distance of about one solar radius, the intensities of the two spectra are equal and the Fraunhofer lines then become the more readily visible in the outer corona. From what has been said on page 505, it is quite unsafe to assume, especially under the conditions experienced at the 1923 eclipse, that the dark lines of the spectrum superposed on the face of the moon were actually caused by scattered coronal light rather than by sunlight scattered by the earth's atmosphere especially at times near the beginning and ending of totality. This, however, is a mere detail that affects the relative intensities of the continuous and Fraunhofer spectra at different distances from the sun's edge, but it in no way detracts from the remarkably fine work of Grotrian.

[1] *Publ. A. S. P.*, *46*, 298, 1934.

According to Rosseland,[1] the corona has "stimulated speculation to the breaking point, it being even suggested that there we witness our recognized physical laws set at naught by nature herself." As a result of many theories of the corona (page 415), it is now generally recognized that the electron must play an important rôle in explaining the radiation of the corona. If the continuous spectrum is caused by free electrons which scatter the solar radiation, Rosseland finds that at a temperature of 4000° K, the average thermal velocity of the electron would cause a Doppler effect of the order of ten angstroms. As a result, all of the dark Fraunhofer lines will be wiped out, with the exception of the broad wings of the H and K lines.

Electron scattering therefore, furnishes an adequate explanation of the absence of dark lines in the spectrum of the inner corona and also is in conformity with the fact that there is no difference in color of the corona from the sun. But how explain the dark lines in the outer parts of the corona? If the electron is to be the agent, then it is necessary to find some process whereby the random velocities of the electrons may be slowed down to such an extent that the Doppler effects will not obliterate the Fraunhofer lines. It is scarcely possible to assume a temperature in the middle corona less than 2000° K which would cause a Doppler effect of about seven angstroms. We might, if we please, assume some unknown (as yet) influence from positrons or that the sun's magnetic field or some other unknown source might gradually slow up the electron velocities at distances of 10′ from the sun's surface. If, indeed, there were a gradual slowing up of the electrons, this would cause the Fraunhofer lines to be broader in the middle corona and sharper in the outer corona — which is not in accord with the facts.

Under the assumption that the light of the corona is caused by the superposition of a continuous spectrum over a dark line spectrum, it is necessary to conclude that electrons are present everywhere throughout the corona. As atoms and ions are present in the inner corona causing the bright line spectrum of coronium, it has been thought by many that

[1] *University Observatory, Oslo, Publ.*, No. 5, 1933.

ionized atoms must be the cause of the dark lines of the coronal spectrum. We do not see how it would be possible (page 417) for atoms to exist in the outer corona in appreciable numbers and moreover they would cause a Rayleigh scattering and a distribution of intensity in the corona different from that in the sun — which again is not in accord with the facts.

After discussing many possibilities, Grotrian is forced to the conclusion that the sun's light is scattered by solid particles with diameters greater than three times the wave-length of light which would cause a scattering of sunlight without change in color. Close to the sun (page 411), solid particles would be immediately vaporized but at greater distances there would be an increasing number of solid particles; in fact, in the zodiacal light we have independent evidence of the existence of such material in interplanetary space. By making plausible assumptions, Grotrian shows that the surface brightness of the outer corona bears to the brightness of the zodiacal light a ratio of about the right order of magnitude. This marks a partial return to the theory of Arrhenius (page 410). Within one radius of the sun's surface we must regard the electron as the scattering mechanism but farther out the interplanetary dust does most of the scattering.

The eclipses of 1932 and 1934 have cleared up many questions regarding the polarization of light surrounding the sun at the time of totality. Polarization measures must be the result of a combination of the partly polarized light from the corona and the unpolarized light in the earth's atmosphere. The measures of Young (page 388) showed that the percentage of polarization in photographic light was about three times greater than in visual light, which was explained by the assumption that the coronal radiation is Rayleigh scattering from atoms and ions. If however, the corona is due to scattering by the electron, then the amount of polarization must be the same in photographic as in visual light. Dufay and Grouiller[1] show that at the eclipse of 1932 the amount of polarization is independent of wave-length and is a maximum of 26 per cent at 10′ from the sun's edge. At

[1] *C. R.*, *196*, 1574, 1933.

the eclipse of 1934,[1] from visual measurements, Johnson finds a maximum of 28 per cent polarization at 8′.5 from the sun's limb, a result in good agreement with the 1932 eclipse.

Fortunately many of the difficulties that retarded progress in our understanding of the corona have been cleared away. The assumption that the corona consists essentially of free electrons permits us to apply the theory of Minnaert.[2] When knowledge progresses still farther and we learn more about the physical laws that govern matter under solar conditions, we may be able to find an explanation for the dark lines in the coronal spectrum that is simpler than the assumption that they are caused by the scattering of the sun's light by interplanetary dust.

For future progress we need more and better spectra of the dark lines in the coronal spectrum taken under clear skies devoid of water vapor in order to be certain that the Fraunhofer lines are coronal in origin and do not come from sunlight scattered in the earth's atmosphere. We need more and better spectra of large dispersion in order to get more accurate wave-lengths of the bright line spectrum by which to find the origin of coronium and the rotation of the corona. In spite of the splendid achievement of Lyot, new information regarding the corona will come entirely from eclipse expeditions, and observations will be accumulated at the average rate of one minute of time per year.

Eclipse expeditions are still needed in the future to secure additional information about the chromosphere. In pages 294–98, the author gave recommendations for future work. He would like to clarify his meaning by urgently recommending the use of both fixed and moving plates for securing photographs of the flash spectrum. In addition he would like to quote the opinions of the Greenwich eclipse observers.[3] "The great need for the future appears to be accurate photometric intensities of the lines, especially of multiplets, and these can only be determined when a comparison spectrum can be secured with the same instrument. The comparison

1 *Publ. A. S. P.*, *46*, 226, 1934.
2 *Zs. für Astrophysik*, *1*, 209, 1930.
3 *Monthly Notices*, *93*, 1, 1932.

spectrum must be obtained from the uneclipsed sun or an artificial source, and for this a collimator and slit are necessary. But it is not necessary to use a narrow slit in taking the eclipse spectra. In 1927, Minnaert used a wide slit, and there appears to be no reason why the slit should not be widened still further, perhaps even to take in the whole of the sun." In 1932, it might be added, Minnaert employed a modest little instrument without coelostat mirror or clock mechanism. At future eclipses, the aluminizing of mirrors should permit spectra to be photographed to shorter wavelengths in the violet.

INDEX

Abbe, 150, 152, 370
Abbot, 105, 111, 116, 120, 123, 166, 174, 182, 391, 396, 406, 411, 419
aberration, 427
Abetti, 279, 328
Abney, 373, 396
absorption in sun's atmosphere, 116, 147
absorption of water vapor, 88
absorption spectra, 88
accelerations of moon's motion, 69
accuracy needed in gratings, 88
accuracy of eclipse predictions, 67
Adams, C. E., 224, 376
Adams, E. D., 199, 202
Adams, W. S., 73, 283, 289, 292, 293, 298, 321, 326, 328, 337, 345, 374, 458, 465, 503
advantages of gratings, 104
age of earth, xiii
age of pyramids, 23
age of sun, xiii
airplane photography, 205, 491
Airy, 31, 103, 133, 134
Albrecht, 374
Aldrich, 397
alkali metals, 309
Almagest, 10, 18, 68
Ames, 165, 193
Anderson, J. A., 213, 216, 280, 296
Anderson, W., 415
Andrews, 394
Angstrom, 94, 100, 101
anomalous dispersion, 458
appearance of sun's surface, 115
Aristotle, 38
Arnold, 216, 221
Arrhenius, 409, 410
astigmatism, 105
Aston, 236
astrology, 17, 18
astronomical instruments, early, 22, 25
astronomical unit, 108
astronomy, early, 17; Chaldean, 17; Chinese, 17; Egyptian, 17, 19; Greek, 17, 37; Roman, 18; Arabian, 18
atmosphere of moon, 65
atmosphere of sun, 116, 147, 149
atom, Bohr-Sommerfeld, 239, 242, 276, 292, 299; structure of, 226; transformation of, 230; Rutherford model, 234; positive-ray analysis, 236; mass confined to nucleus, 236; chemical and physical properties, 237; Lewis-Langmuir, 237; explanation of 4686 A, 245
atomic numbers, 235
atomic weights, 235
aurora borealis, 123, 336, 409
authentic dates, 19, 20
Auwers, 123

Baade, 219
Babcock, 463, 503
Babylonian calendar, 18
Babylonian eclipses, 9
Babylonian lens, 17
Bacon, 396
Baily's beads, 129, 267
Baker, 372
Balmer series, 240, 243, 244, 245, 309
Barnard, 159, 165, 174, 182
Bartoli, 409
Bassett, 480
Baume Pluvinel, 487
Becker, 391
Becquerel, 227, 228
beginnings of astronomy, 17
Belanovsky, 394
Bergmann, 240
Bergstrand, 391, 393, 395, 401, 402, 406, 497, 499
Bernheimer, 403
Bessel, 409
Bigelow, 408
black body, 97
black drop, 110, 130
Block, 374
Bohr, 312
Bohr-Sommerfeld atom, 239, 242, 276, 292, 299
Bosler, 374
Bourget, 385
Bowen, 381, 416
Brackett, 244
Bradley, 427
Bredichin, 409
Briggs, 396
bright line spectrum, 76, 95, 97
Brown, 40, 68, 70, 71, 111
Buisson, 385, 457, 459

Burns, 280, 348, 352, 374
Burton, 175, 181
Burwell, 289
Butler, 60, 130, 200

calcium spectrum, 144
calculation of an eclipse, 62
calendar, of Chinese, 5; of Babylon, 18; of Egypt, 21, 24; revision of, 21
Campbell, 163, 168, 189, 198, 209, 221, 251, 260, 262, 273, 279, 368, 370, 374, 375, 376, 377, 417, 469, 484
Carrasco, 374
Carrington, 320
Carroll, 503
Cavendish, 465
celestial latitude, 48
celestial longitude, 48
changes in prominences, 261
changes in sun-spots, 84, 109
Chant, 467
chemical elements, 234, 306
Chester, 183
Chevalier, 117
Chinese eclipses, 1
Chinese science, 5
chromosphere, 108, 135, 147
chronology, 2, 5, 9, 19, 20, 27
chronology, change of Biblical necessary, 19, 27
circulation of solar gases, 342, 464
Clerke, 16, 82, 119, 129, 370
clock, Egyptian, 22
Coblentz, 213, 215, 396
Cohn, 493, 497
collimator, 254
color of sky during totality, 201
comet discovered during eclipse, 153
comets' tails, 409
composition of chromosphere, 353
composition of sun, 72, 98, 100
composition of sun's atmosphere, 345
concave gratings, 104, 248
Confucius, 6
constellations, astronomical, 15, 23
constitution of sun, 72, 98, 100
contacts, 65, 259
continuous spectrum, 97
Copernicus, 423
corona, xv, 23, 108, 131, 139, 140, 143, 147, 163, 360
corona as Egyptian symbol, 24
corona compared with intensity of full moon, 395
corona, intensity laws, 390
corona photographed, 8, 9, 12, 13, 148, 164, 197, 209, 212, 213, 273, 349, 364, 365, 400, 425, 448, 449
corona, polarization of, 387, 497
corona, shape of, 364, 399, 417, 497
corona, spectrum, 140, 148, 152, 154, 156, 161, 373, 382, 502, 506
corona, total light, 395, 505
corona without an eclipse, 155, 418, 501
coronal structure seen in different wave-lengths, 371, 373, 380
coronal theories, meteoric, 407, 509; mechanical, 407; electrical, 408; magnetic, 408; electro-magnetic, 408; radiation-pressure, 409; electron, 414, 508
coronium, 140, 149, 152, 374, 376, 381, 417, 504
coronium lines in Nova, 503
Cortie, 289
Cottingham, 447
Courvoisier, 471
Cowell, 9, 30, 31, 38
creation, 14
crescent lines, 221
Crew, 165
Crommelin, 446
Crookes, 231
Curie, xiv, 227
Curtis, 210, 213, 215, 280, 368, 374, 385
curvature of space, 466
cylindrical lens, 80

darkening at limb of sun, 116
dark lines in solar spectrum, 75, 79, 91, 95, 97, 141
date of Easter, 37
dates, earliest authentic, 18, 20
Davidson, 249, 250, 278, 280, 374, 375, 376, 378, 380, 446, 486
Debierne, 230
de Bruin, 382
decimal system of Egypt, 21
delicacy of spectroscopic test, 82
De Lury, 321, 322
density gradients in chromosphere, 284, 356
density of sun, 112, 123
de Sitter, 478, 479
Deslandres, 105, 139, 161, 373, 418
development of photographs, 363, 495
development of spot, 84, 109
diameter of sun, 48, 111, 123
diameters of eclipse tracks, 44
diffraction of light, 84, 157
diffraction spectrum, 84, 104
Dingle, 382
Dinwiddie, 178
dissociation theory, 146, 156, 162
distance of moon, 39
distance of sun, 108
distances of stars, 73, 77

disturbance in corona, 209
D-lines, 82, 93, 96, 137, 307
Doppler effect, 340, 343, 384, 456, 458, 460
Dorsey, 388
Dufay, 488, 509
Dunér, 320, 321
duodecimal system, 16
duration of partial phases of eclipses, 46
duration of totality, 45
Dyson, 165, 218, 374, 376, 380

early Chinese records, 5
earth a sphere, 38
Ebert, 408
eclipse of 1652, 360; 1706, 131; 1715, 128; 1778, 128, 132; 1780, 129; 1836, 91, 129; 1842, 130, 135; 1851, 134, 135; 1860, 134, 387; 1868, 137, 142; 1869, 139, 142, 148, 370; 1870, 141, 148, 151, 247; 1871, 142, 147, 148, 151; 1872, 142; 1874, 142; 1878, 151, 388, 451; 1882, 153, 373; 1883, 155, 373; 1886, 158, 373, 396; 1887, 159; 1889, 159, 396; 1893, 160, 364, 370, 373, 396, 402; 1896, 161, 370, 376, 402; 1898, 162, 260, 370, 373, 376, 396, 402; 1900, 164, 167, 260, 364, 373, 374, 376, 387, 402; 1901, 168, 370, 373, 374, 376, 402; 1905, 183, 251, 252, 260, 374, 376, 396, 402; 1908, 251, 260, 373, 374, 376, 396, 402; 1912 (partial), 125, 287; 1914, 125, 364, 374, 402; 1916, 364; 1918, 191, 364, 370, 374, 376, 396, 402; 1919, 421, 444, 446, 454, 461; 1921 (partial), 288; 1922, 364, 373, 374, 376, 383, 396, 402, 449, 467; 1923, 126, 203, 370, 402; 1925, 69, 126, 211, 252, 364, 374, 376, 396, 402; 1926, 215, 249, 250, 364, 374, 376, 396, 402; 1927, 216, 250, 364, 402; 1929, 220, 364, 374, 376, 402, 474; 1930, 127, 222, 251, 364, 371, 374, 376, 402; 1932, 483 *et seq.*; 1934, 492 *et seq.*
eclipses, Chinese, 1; Babylonian, 9; Egyptian, 23; Biblical, 27; classical, 29; Thales, 30; Xerxes, 32; Julius Caesar, 35; eclipse of B.C., 2137, 1; 1063, 9; 791, 27; 776, 8; 771, 27; 770, 27; 763, 10, 19, 27, 54; 720, 8; 669, 19; 661, 19; 603, 32; 592, 7; 585, 31; 552, 7; 549, 7; 495, 7; 431, 32; 424, 33; 413, 34; 364, 34; 310, 34
eclipses of past and future, 55
eclipses that did not occur, B.C., 645, 592, 552, 549, 7; of Xerxes, 32; of Julius Caesar, 35
eclipses visible in a year, 58
eclipses visible in the United States, 56; in the British Isles, 58; in Spain, 58
ecliptic, inclination of moon's orbit to, 39, 48
ecliptic limits, 49
Eddington, 346, 411, 416, 420, 435, 441, 447, 466, 479
Egyptian astronomy, 19
Egyptian chronology, 20
Eichelberger, 174
Einstein, 205, 421, 443, 479
electrons, 102, 108, 232
elements, 234, 306
elevations, differences caused by, 268
enhanced, 163, 246, 311
Epynom canon, 18
equivalence, 440
ether, 425
Evans, 245
events, 433
Evershed, 162, 165, 297, 321, 326, 340, 343, 349, 458, 460, 464, 467
evolution, 72, 113, 318
excitation potential, 313, 339
Exner and Haschek, 105, 303
expenses of expeditions, 139

Fabry, 385, 396, 412, 457, 459
faculae, 118, 123
Faye, 320, 322
first photograph, 135
Fisher, 492
Fitzgerald, 427
flash spectrum at partial eclipses, 287
flash spectrum, discovered, 141; photography necessary, 147; improvements in observing, 157; photographed in 1896, 161; photographed in 1898, 162; enhanced lines, 163; gratings used in 1900, 168; advantages of various spectrographs, 248; method of timing photographs, 258; moving plate vs. fixed plate, 261; jumping film, 489; intensities compared with Fraunhofer spectrum, 268, 309; determination of heights, 271; effects of seeing, 274; discussions of, 277; self-absorption in, 283, 286; density gradients, 284, 356; at a partial eclipse, 287; general conclusions, 293; recommendations for future eclipses, 294, 510; without an eclipse, 289
flocculi, 108, 300

focus of eclipse spectra, 248
Forbes, 91, 130
Fotheringham, 3, 9, 30, 31
Foucault, 93, 100
four-dimensional space, 422, 466
Fowle, 123
Fowler, 105, 160, 241, 245, 275, 276, 287, 314, 343, 373, 374, 376, 378
Fox, 92, 134, 209
Fraunhofer, 76, 101, 123
Fraunhofer lines, 75, 79, 91, 95, 136, 143, 152, 268, 269, 275, 277, 281, 293, 303, 350, 382, 383, 406, 407, 506
Freeman, 381
Frerichs, 382, 504
Fresnel, 425
Freundlich, 216, 220, 475, 477
Frost, 165, 374
furnace lines, 315
future eclipses, 55, 510

Gale, 458
Galileo, 423
Gallo, 210
Gare, 396
Geiger, 231, 234
geodesic, 435
Gill, 110
Graff, 391, 396, 401
granulation, 116
gratings, 84, 104, 157, 168
gratings and prisms compared, 104, 248
grating spectrum, 85, 104
gravitation, 421, 443
Grotrian, 312, 374, 375, 376, 378, 379, 506, 509
Grouiller, 488, 509

Hale, 105, 139, 289, 328, 331, 418
Hall, 141, 455
Halley, 128, 131, 134, 451
Halm, 321, 322, 323, 325, 458
Hammond, 197
Hansen, 31, 68, 71
Hansky, 117, 367, 418
Harkness, 140, 141, 151
Hartmann, 79
Harwood, 393
Hastings, 156, 481
Hayn, 125, 126
heat of corona, 174
heights of solar vapors, 165, 271, 319
heliometer, 77, 110, 123
helium, xiv, 137, 239, 270, 309, 359
Herodotus, reliability as historian, 32
high-level lines in sun, 304
high-temperature lines, 308
Hills, 161, 162, 374, 376
Hind, 31, 63
Hipparchus, 17, 31, 38
historian, 2, 32
Holden, 396
Hooke, 425
Hopfield, 382
Horn d'Arturo, 216, 369
Hubble, 414
Hubrecht, 321
Huff, 168
Huggins, 103, 138, 155, 158, 239, 408, 418
Hull, 409
Humphreys, 105, 165, 457, 481
Hussey, 368
hydrogen, 136, 149, 152, 239, 270, 310, 355
Hylleraas, 382

intensities in flash spectrum, 268, 304
intensities of spectra of electric arc and sunlight, 93
intensity of coronal light, 395, 506
intra-Mercurial planets, 153, 174
invisible sun-spots, 332
ionization, 299, 311
ionized atoms, 233, 246, 311
irradiation, 124, 126, 127
isotopes, 235

Jackson, 475, 486
Janssen, 68, 103, 116, 137, 141, 152, 156
Jeffreys, 455
Jewell, 105, 174, 178, 254, 456, 457
Johnson, 221, 493, 510
Jones, 467
Joy, 503
Julius, 458

Kaufmann, 228
Kayser, 105, 240, 253, 275, 282, 425
Kelvin, 421, 425
Kepler, 18, 73, 131, 423, 451
Kimura, 120
King, A. S., 105, 314, 345
King, E. S., 213, 393, 397
King, L. W., 9, 12
Kirchhoff, 73, 92, 94, 99, 135, 144
Kirchhoff laws, 97
Knopf, 396
knowledge, how derived, 434
Kohlschütter, 210
Krakatoa, 170, 411
Kunz, 395, 396, 419

Lalande, 451
Langley, 105, 115, 118, 141, 151, 168
Larmor, 120, 427, 466
law of coronal intensity, 390

law of diffraction spectra, 85
law of gravitation, 421, 443
laws of spectrum analysis, 97
Lebedew, 409
length of month, 40
Lemaître, 479
length of seasons, 39
length of shadow cast by earth, 42
length of shadow cast by moon, 43
length of year, 40
Lescarbault, 451
Leuschner, 396
level of sun-spots, 119
levels at which spectral lines originate, 319, 342, 350
levels in chromosphere, 165, 277, 319, 338
Leverrier, 451
Lewis, 374, 376, 380, 390
Lewis-Langmuir, 237
Lick Observatory, 18, 33, 160, 163, 167, 189, 198, 209, 221, 260, 261, 264, 265, 266, 278, 361, 367, 372, 383, 384, 410, 467, 489, 503
life on earth, xiii
light-pressure, 409, 445
light-waves, 101
Lindemann, 418
Lockyer, 90, 103, 135, 139, 142, 143, 144, 152, 161, 162, 247, 261, 317, 374, 376, 378
Lockyer, W. J. S., 163, 165, 402
Lohse, 217
long and short lines, 145
Lorentz, 427, 437
low-pressure lines, 308
Ludendorff, 210, 401, 402, 415, 497, 499, 506
luminosity of corona, 157, 395
lunar ecliptic limits, 49
lunar tables, 68
Lyle, 105
Lyman, 244
Lyot, 419, 501, 504

magnetic field of sun, 330, 335
magnetic storms, 123, 336
Marriott, 127, 224, 225, 362, 371, 380, 401, 495
Marsden, 234
mass of sun, 112
Maunder, 23, 163
Maxwell, 409, 425, 445
Mees, 193
Meggers, 105, 276, 314
Melotte, 467
Menzel, 260, 261, 267, 275, 279, 280, 284, 347, 353, 355, 356, 490
Mercury, 424, 437, 443, 451
Merfield, 105
meteoric hypothesis, 150
meteors, 114, 149
meter, 109, 432
method of viewing sun, 115
Metonic cycle, 37
metric system, 108
Michelson, 105, 120, 207, 425
Miller, D. C., 426
Miller, J. A., 10, 210, 212, 220, 361, 369, 370, 410, 484, 485, 495
Miller, W. H., 92, 103
Millikan, 243
Milne, 300, 353, 358, 414, 416, 463
Minnaert, 219, 225, 250, 251, 273, 278, 281, 282, 283, 353, 356, 484, 505, 510, 511
Minkowski, 434, 437
Mitchell, W. M., 329
Mohler, 457
month, Babylonian, 18; Egyptian, 21; length of different, 40
moon away from predicted place, 67
moon, distance of, 39
moon interferes with astronomical observations, 44
moon, motion of, 39, 68
moon's shadow, 284, 485, 491
Moore, C. E., 283, 314, 345
Moore, J. H., 370, 372, 374, 375, 376, 377, 401, 506
Morley, 426
Moseley, 234
motion in chromosphere, 342
motion in corona, 365, 384
motion in line of sight, 73, 342, 464
motion of the moon, 39, 68
motions of solar gases, 342, 464
moving plate spectrograph, 260
moving plate vs. fixed plate, 261, 271
multiplet groups, 314, 344, 348
music of the spheres, 37

Naegamwala, 374
nebulæ, 413
nebulium, 416
neutral atom, 307
Newall, 162, 166, 182, 217, 288, 322, 325, 374, 376, 390
Newcomb, 31, 36, 41, 54, 62, 68, 123, 141, 152, 405, 424, 452
Newton, 73, 101, 407, 421, 423, 446
Nichols, 409
Nicholson, J. W., 381, 393
Nicholson, S. B., 111, 120, 213, 337, 396
Nineveh, eclipse, 8, 11, 18, 26, 54, 70
nodes, 39

normal spectrum, 85, 104
northern lights, 123, 336, 409
number of eclipses each year, 59
number of eclipses in Saros, 52

objective prisms, 82
Oppolzer, 1, 29, 36, 51, 60, 62, 63

Pannekoek, 219, 250, 251, 273, 278, 281, 282, 283, 353, 356, 381
parallax of the moon, 48
parallax, solar, 48, 110
parallax, stellar, 73, 77, 124
Parkhurst, 213, 396
partial eclipse, 46
Paschen, 244, 245
past eclipses, 55, 494
Payne, 347
percentage of ionization, 301
Perepelkin, 394
periodic table of elements, 234, 306
period of sun-spots, 119
Perrine, 173, 181, 367, 368, 370, 383, 388, 396
personal errors, 124
Pettit, 213, 393, 396, 502
photographing the sun, 115, 125
photography first applied, 134
photosphere, 98, 113
Pickering, 140, 141, 244, 396
Pitman, 127
Planck, 238, 242
plane gratings, 104
Plaskett, H. H., 321, 353
Plaskett, J. S., 209, 321
polarity of spots, 331, 333
polarized light, 148, 387, 509
pole-effect, 459
Poor, 453
potassium in sun, 308
prediction of eclipses, 5; Thales, 31; from ecliptic limits, 50; by the Saros, 31, 50; from Oppolzer maps, 51
pressure in photosphere, 114
pressures in sun's atmosphere, 307, 346
principles of spectrum analysis, 97
prism, 73
prismatic camera, 82, 153
prismatic spectrum, 85, 104
prisms compared with gratings, 104, 248
problems of the corona, 360, 506
Proctor, 23
prominences, 25, 123, 128, 136, 149, 158, 165, 196, 261, 272, 320, 321, 369, 370, 401
protons, 232
Ptolemy, 10, 19
pyramid not an observatory, 23

quantum, 102, 238

Radau, 68
radiation of corona, 166, 395
radiation-pressure, 409, 445
radium, xiv, 228, 316
Ramsay, 230
Ranyard, 399
Rawlinson, 11, 18
red flames, 131, 136
relative numbers of lunar and solar eclipses, 61
relative numbers of total and annual eclipses, 44
relativity, 205, 215, 221, 280, 349, 421, 443
reversal of polarities, 333
reversed spectrum, 97, 98
reversing layer, 98, 141, 143, 147, 163, 247
Riccò, 418
Rigge, 64
Röntgen, 227
Rosenthal, 382
Ross, 397, 471
Rosseland, 217, 358, 508
rotation of corona, 161, 164, 182, 384, 504
rotation of sun, 319
rotation of sun's magnetic axis, 336
Rowland, H. A., 88, 103, 113, 162, 168, 205, 282, 456
Rowland, J. P., 289
Rowland gratings, 104, 390
Royds, 230, 321, 460
rubidium in sun, 308
Runge, 105, 240, 253
Russell, 105, 126, 276, 283, 300, 308, 314, 345, 349, 354, 381, 449
Russell, Dugan and Stewart, 241
Rutherford, 228, 230, 231, 242
Rydberg, 105, 240, 241
Rydberg constant, 240, 241, 243

Saha, 268, 270, 276, 292, 294, 299, 315, 413
St. John, 105, 111, 321, 322, 326, 340, 341, 344, 348, 352, 374, 463
Saros, 17, 31, 37, 41, 50, 53, 62
Satome, 493, 497
Saunders, 105, 252, 314
scale of intensities, 354
scattering of light, 387, 412
Schaeberle, 159, 370, 407, 410
Schlesinger, 321
Schorr, 210, 218
Schuster, 120, 154, 373
Schwabe, 119, 133
Schwarzschild, 302, 391, 396, 415, 416

seasons, length of, 39
Secchi, 103, 135, 387
seeing, 274
sensitiveness of spectroscopic test, 83, 232
series, in spectra, 240
Shackleton, 161, 376
shadow bands, 480
shadow cast by earth, length of, 42
shadow cast by moon, length of, 43
shadow of moon approaching, 284
shadow, width of, 44
shape of corona, 123, 364, 399, 417, 497
shape of sun, 124
shift to red in spectrum, 457
sierra, 134
size of atom, 237
size of electron, 226
size of molecule, 233
size of universe, 226
Skinner, 174
Slipher, 374
slitless spectroscope, 82, 149, 153, 250
Slocum, 76, 80, 138, 139, 449, 502
Soddy, 230
sodium lines conspicuous, 82
solar ecliptic limits, 49
solar parallax, 48, 110
source of knowledge, 434
space curved, 466
space of four dimensions, 422, 466
spectral series, 240
spectroscope, 72
spectroscopic binaries, 73
spectroscopic tests, sensitiveness of, 83, 232
spectrum analysis, principles, 97
spectrum lines on dark moon, 154
spectrum of corona, 140, 148, 152, 154, 156, 161, 373, 382, 504
spectrum of Venus, 81
spherical sun, 124
spot-vortices, 329, 340
Stark-effect, 285, 336, 459
Stebbins, 209, 213, 368, 395, 396, 419
stellar parallax, 73, 77, 123
stellar spectrum, 80, 317
Stetson, 213, 215, 217, 221, 394, 396, 397
Stevens, 491
Stewart, 94, 100
Stokes, 93, 100
Storey and Wilson, 321
Störmer, 217
Stratton, 225, 249, 250, 278, 280, 374, 375, 376, 378, 380, 484
stripped atoms, 246
sun-spot photographs, 109, 117, 320, 333, 340
sun-spots, 112; appearance, 117; level of, 119; period of, 119; distribution of, 121, 334; related to other phenomena, 121; spectrum, 308; vortices, 329; polarity, 331, 333; invisible spots, 332
surface of sun, 115
Swift, 153

Tacchini, 139, 158
telluric lines, 90
temperature, change of spectrum with, 144, 145
temperature change during eclipse, 492
temperature of sun, 114, 302, 308, 346
temperature of sun-spots, 308
temperatures of stars, 72, 145
tenth-meter, 101
terrestrial magnetism, 122
Thales, 17, 30, 70
theory of eclipses, 42
Thomson, 232, 236
Thorpe, 396
time available for observations, xvii
times of contacts, 66, 259
titanium oxide, bands of, 343
Todd, 161
transit of Venus, 110, 130
Trumpler, 468, 475
Turner, 158, 162, 166, 391, 396
Tut-ankh-Amen, 24
Tycho Brahe, 128, 423

Ueta, 493
ultimate lines, 315
unexplained facts about corona, 406
Unsöld, 286, 353

valence, 237
variable star, 120
Vegard, 219
velocity of light, 111
velocity of moon's shadow over earth, 45
von Brunn, 476, 477
von Klüber, 369, 394, 401, 402, 475, 477
von Soldner, 465
vortices, sun-spot, 329, 340

Ware, 321, 326, 341, 374
Watson, 141, 153
wave-lengths, 79, 87, 97, 101, 105, 461
weather and sun-spots, 121
week of seven days, 18; of ten days, 21
Wesley, 166, 365
Weston, 505
wet-plate photography, 115

Weyl, 442
white prominences, 158
Williams, E. T. R., 355, 356, 401
Williams, J., 9, 134
Wilson, A., 119
Wilson, C. T. R., 233
Wilson, H. C., 209, 279
Wollaston, 76
Wood, 244, 387, 389, 418
Wooley, 353
Wright, A. W., 388
Wright, W. R., 385, 477, 478, 495

year, Babylonian, 18; Egyptian, 21; tropical, 21; Julian, 21; Gregorian, 21; length of various, 41
Young, C. A., 97, 103, 113, 139, 147, 149, 152, 158, 168, 247, 260, 329
Young, R. K., 388, 391, 467

Zanstra, 416
Zech, 31
Zeeman, 105, 329, 459
zodiac, 15, 26
zodiacal light, 409, 454

COLUMBIA UNIVERSITY PRESS
Columbia University
New York

FOREIGN AGENT
OXFORD UNIVERSITY PRESS
Humphrey Milford
Amen House, London, E. C. 4

Bei Fragen zur Produktsicherheit wenden Sie sich bitte an:
If you have any questions regarding product safety,
please contact:

Walter de Gruyter GmbH
Genthiner Straße 13
10785 Berlin
productsafety@degruyterbrill.com